Environmental Chemistry for a Sustainable World

Volume 64

Series Editors
Eric Lichtfouse ⓘ, Aix-Marseille University, CNRS, IRD, INRAE, Coll France, CEREGE, Aix-en-Provence, France
Jan Schwarzbauer, RWTH Aachen University, Aachen, Germany
Didier Robert, CNRS, European Laboratory for Catalysis and Surface Sciences, Saint-Avold, France

Other Publications by the Editors

Books
Environmental Chemistry
http://www.springer.com/978-3-540-22860-8

Organic Contaminants in Riverine and Groundwater Systems
http://www.springer.com/978-3-540-31169-0

Sustainable Agriculture
Volume 1: http://www.springer.com/978-90-481-2665-1
Volume 2: http://www.springer.com/978-94-007-0393-3

Book series
Environmental Chemistry for a Sustainable World
http://www.springer.com/series/11480

Sustainable Agriculture Reviews
http://www.springer.com/series/8380

Journals
Environmental Chemistry Letters
http://www.springer.com/10311

More information about this series at http://www.springer.com/series/11480

Saravanan Rajendran • Mu. Naushad
Lorena Cornejo Ponce • Eric Lichtfouse
Editors

Metal, Metal-Oxides and Metal-Organic Frameworks for Environmental Remediation

Editors
Saravanan Rajendran
Department of Mechanical Engineering
University of Tarapacá
Arica, Chile

Mu. Naushad
Department of Chemistry
King Saud University
Riyadh, Saudi Arabia

Lorena Cornejo Ponce
Department of Mechanical Engineering,
Faculty of Engineering
University of Tarapacá
Arica, Chile

Eric Lichtfouse
Aix-Marseille University, CNRS, IRD, INRAE,
Coll France, CEREGE
Aix-en-Provence, France

ISSN 2213-7114   ISSN 2213-7122 (electronic)
Environmental Chemistry for a Sustainable World
ISBN 978-3-030-68978-0   ISBN 978-3-030-68976-6 (eBook)
https://doi.org/10.1007/978-3-030-68976-6

© The Editor(s) (if applicable) and The Author(s), under exclusive license to Springer Nature Switzerland AG 2021
This work is subject to copyright. All rights are solely and exclusively licensed by the Publisher, whether the whole or part of the material is concerned, specifically the rights of translation, reprinting, reuse of illustrations, recitation, broadcasting, reproduction on microfilms or in any other physical way, and transmission or information storage and retrieval, electronic adaptation, computer software, or by similar or dissimilar methodology now known or hereafter developed.
The use of general descriptive names, registered names, trademarks, service marks, etc. in this publication does not imply, even in the absence of a specific statement, that such names are exempt from the relevant protective laws and regulations and therefore free for general use.
The publisher, the authors, and the editors are safe to assume that the advice and information in this book are believed to be true and accurate at the date of publication. Neither the publisher nor the authors or the editors give a warranty, expressed or implied, with respect to the material contained herein or for any errors or omissions that may have been made. The publisher remains neutral with regard to jurisdictional claims in published maps and institutional affiliations.

This Springer imprint is published by the registered company Springer Nature Switzerland AG
The registered company address is: Gewerbestrasse 11, 6330 Cham, Switzerland

Acknowledgments

First and foremost, we would like to thank Almighty **God** for giving us this opportunity and good strength to complete this book successfully.

Our honest thanks to series editors and the advisory board for accepting our book, entitled ***Metal, Metal-Oxides and Metal-Organic Frameworks for Environmental Remediation***, as part of the series **Environmental Chemistry for a Sustainable World** and their continuous support to complete this hard task successfully. We convey our deepest appreciations to authors and reviewers. We convey our truthful gratitude to Springer. We have great pleasure in acknowledging various publishers and authors for permitting us to use their figures and tables. We would still like to offer our deep apologies to any copyright holder if unknowingly their right is being infringed.

Dr. R. Saravanan thanks SERC–Chile (CONICYT/FONDAP/15110019) for the financial support. Finally, he expresses his sincere thanks to the Faculty of Engineering in the Department of Mechanical Engineering at the University of Tarapacá, Arica, Chile.

Dr. Mu. Naushad extends his appreciation to the Deanship of Scientific Research at King Saud University for the support.

Lorena Cornejo Ponce expresses her sincere thanks to the Faculty of Engineering in the Department of Mechanical Engineering at the University of Tarapacá, Arica. She extends her thanks for the financial support from Ayllu Solar and SERC–CHILE (CONICYT/FONDAP/15110019).

Universidad de Tarapacá, Arica, Chile	Saravanan Rajendran
King Saud University, Riyadh, Saudi Arabia	Mu. Naushad
Universidad de Tarapacá, Arica, Chile	Lorena Cornejo Ponce
Aix-Marseille University, Aix-en-Provence, France	Eric Lichtfouse

Contents

1. **Metal Oxides as Decontaminants of Water and Wastewater** 1
   Kingshuk Dutta

2. **Photo-Assisted Antimicrobial Activity of Transition Metal Oxides** 29
   Rajini P. Antony, L. K. Preethi, and Tom Mathews

3. **Metal and Metal Oxide Nanomaterials for Wastewater Decontamination** 63
   Mohd. Tauqeer, Mohammad Ehtisham Khan, Radhe Shyam Ji, Prafful Bansal, and Akbar Mohammad

4. **MoS_2 Based Nanocomposites for Treatment of Industrial Effluents** 97
   Manjot Kaur, Unni Krishnan, and Akshay Kumar

5. **Removal of Priority Water Pollutants Using Adsorption and Oxidation Process Combined with Sustainable Energy Production** 117
   Sheen Mers Sathianesan Vimala, Omar Francisco González-Vázquez, Ma. del Rosario Moreno-Virgen, Sathish-Kumar Kamaraj, Sheem Mers Sathianesan Vimala, Virginia Hernández-Montoya, and Rigoberto Tovar-Gómez

6. **Metal Oxides for Removal of Arsenic Contaminants from Water** 147
   Tamil Selvan Sakthivel, Ananthakumar Soosaimanickam, Samuel Paul David, Anandhi Sivaramalingam, and Balaji Sambandham

7	**Earth Abundant Materials for Environmental Remediation and Commercialization**...................................... J. Nimita Jebaranjitham, Adhimoorthy Prasannan, K. Sankarasubramanian, K. S. Prakash, and Baskaran Ganesh Kumar	195
8	**Arsenic Contamination: Sources, Chemistry and Remediation Strategies**............................... Pankaj K. Parhi, Snehasish Mishra, Ranjan K. Mohapatra, Puneet K. Singh, Suresh K. Verma, Prasun Kumar, and Tapan K. Adhya	219
9	**Mycoremediation: An Elimination of Metal and Non-metal Inclusions from Polluted Soil**................................ Jegadeesh Raman, Jang Kab-Yeul, Hariprasath Lakshmanan, Kong Won-Sik, and Babu Gajendran	239
10	**Photocatalytic Degradation of Dyes in Wastewater Using Metal Organic Frameworks**......................... Thabiso C. Maponya, Mpitloane J. Hato, Edwin Makhado, Katlego Makgopa, Manika Khanuja, and Kwena D. Modibane	261

Contributors

Tapan K. Adhya School of Biotechnology, KIIT (Deemed University), Bhubaneswar, Odisha, India

Rajini P. Antony Water and Steam Chemistry Division, BARC – F, Kalpakkam, Tamil Nadu, India

Prafful Bansal Department of Chemistry, Aligarh Muslim University, Aligarh, Uttar Pradesh, India

Samuel Paul David HiLASE Centre, Institute of Physics of the Czech Academy of Science, Dolni Brezany, Czech Republic
Department of Physics, Kalasalingam Academy of Research and Education, Krishnankoil, Tamil Nadu, India

Kingshuk Dutta Advanced Polymer Design and Development Research Laboratory (APDDRL), School for Advanced Research in Polymers (SARP), Central Institute of Petrochemicals Engineering and Technology (CIPET), Bengaluru, Karnataka, India

Babu Gajendran State Key Laboratory of Functions and Applications of Medicinal Plants and Chinese Academy of Sciences, Guizhou Medical University, Guiyang, China

Rigoberto Tovar-Gómez Instituto Tecnológico de Aguascalientes (ITA), Tecnológico Nacional de México (TecNM), Aguascalientes, Mexico

Omar Francisco González-Vázquez Instituto Tecnológico de Aguascalientes (ITA), Tecnológico Nacional de México (TecNM), Aguascalientes, Mexico

Mpitloane J. Hato Nanotechnology Research Lab, Department of Chemistry, School of Physical and Mineral Sciences, University of Limpopo (Turfloop), Polokwane, South Africa

Radhe Shyam Ji Discipline of Chemistry, Indian Institute of Technology Indore, Indore, Madhya Pradesh, India

Jang Kab-Yeul Mushroom Research Division, National Institute of Horticultural and Herbal Science, Rural Development Administration, Eumsung, Republic of Korea

Sathish-Kumar Kamaraj Instituto Tecnológico El Llano Aguascalientes (ITEL), Tecnológico Nacional de México (TecNM), Aguascalientes, Mexico

Manjot Kaur Advanced Functional Materials Laboratory, Department of Nanotechnology, Sri Guru Granth Sahib World University, Fatehgarh Sahib, Punjab, India

Mohammad Ehtisham Khan Department of Chemical Engineering Technology, College of Applied Industrial Technology (CAIT), Jazan University, Jazan, Kingdom of Saudi Arabia

Manika Khanuja Centre for Nanoscience and Nano Technology, Jamia Millia Islamia (A Central University), New Delhi, India

Unni Krishnan Advanced Functional Materials Laboratory, Department of Nanotechnology, Sri Guru Granth Sahib World University, Fatehgarh Sahib, Punjab, India

Akshay Kumar Advanced Functional Materials Laboratory, Department of Nanotechnology, Sri Guru Granth Sahib World University, Fatehgarh Sahib, Punjab, India

Baskaran Ganesh Kumar Department of Chemistry, P.S.R. Arts and College (Affiliated to Madurai Kamaraj University, Madurai), Sivakasi, Tamil Nadu, India Department of Science and Humanities, P.S.R. Engineering College (Affiliated to Anna University, Chennai), Sivakasi, Tamil Nadu, India

Prasun Kumar Department of Chemical Engineering, Chungbuk National University, Cheongju, Chungbuk, Republic of Korea

Hariprasath Lakshmanan Department of Biochemistry, Karpagam Academy of Higher Education, Coimbatore, India

Katlego Makgopa Department of Chemistry, Faculty of Science, Tshwane University of Technology (Acardia Campus), Pretoria, South Africa

Edwin Makhado Nanotechnology Research Lab, Department of Chemistry, School of Physical and Mineral Sciences, University of Limpopo (Turfloop), Polokwane, South Africa

Thabiso C. Maponya Nanotechnology Research Lab, Department of Chemistry, School of Physical and Mineral Sciences, University of Limpopo (Turfloop), Polokwane, South Africa

Tom Mathews Thin Films and Coatings Section, Surface and Nanoscience Division, Materials Science Group, Indira Gandhi Centre for Atomic Research, Kalpakkam, Tamil Nadu, India

Snehasish Mishra Bioenergy Lab and BDTC, School of Biotechnology, KIIT (Deemed University), Bhubaneswar, Odisha, India

Kwena D. Modibane Nanotechnology Research Lab, Department of Chemistry, School of Physical and Mineral Sciences, University of Limpopo (Turfloop), Polokwane, South Africa

Akbar Mohammad School of Chemical Engineering, Yeungnam University, Gyeongsan-si, Gyeongbuk, South Korea

Ranjan K. Mohapatra School of Biotechnology, KIIT (Deemed University), Bhubaneswar, Odisha, India

Virginia Hernández-Montoya Instituto Tecnológico de Aguascalientes (ITA), Tecnológico Nacional de México (TecNM), Aguascalientes, Mexico

Pankaj K. Parhi Department of Chemistry, Fakir Mohan University, Balasore, Odisha, India

K. S. Prakash Department of Chemistry, Bharathidasan Government College for Women (Autonomous) (Affiliated to Pondicherry University, Pondicherry), Muthialpet, Puducherry U.T, India

Adhimoorthy Prasannan Graduate Institute of Applied Science and Technology, National Taiwan University of Science and Technology, Taipei, Taiwan

L. K. Preethi Centre for Nanoscience and Nanotechnology, Sathyabama Institute of Science and Technology, Chennai, Tamil Nadu, India

Jegadeesh Raman Department of Biochemical and Polymer Engineering, Chosun University, Gwangju, Republic of Korea
Mushroom Research Division, National Institute of Horticultural and Herbal Science, Rural Development Administration, Eumsung, Republic of Korea

J. Nimita Jebaranjitham P.G. Department of Chemistry, Women's Christian College (An Autonomous Institution Affiliated to University of Madras), Chennai, Tamil Nadu, India

Tamil Selvan Sakthivel Department of Materials Science and Engineering (MSE), Advanced Materials Processing and Analysis Center (AMPAC), University of Central Florida, Orlando, FL, USA

Balaji Sambandham Department of Materials Science and Engineering, Chonnam National University, Gwangju, South Korea

K. Sankarasubramanian School of Physics, Madurai Kamaraj University, Madurai, Tamil Nadu, India

Puneet K. Singh Bioenergy Lab and BDTC, School of Biotechnology, KIIT (Deemed University), Bhubaneswar, Odisha, India

Anandhi Sivaramalingam Department of Physics, Sathyabama Institute of Science and Technology, Chennai, Tamil Nadu, India

Ananthakumar Soosaimanickam Institute of Materials (ICMUV), University of Valencia, Valencia, Spain

Mohd. Tauqeer Department of Chemistry, Aligarh Muslim University, Aligarh, Uttar Pradesh, India

Suresh K. Verma Division of Molecular Toxicology, Institute of Environmental Medicine, Karolinska Institute, Stockholm, Sweden

Sheem Mers Sathianesan Vimala Department of Basic Engineering, Government Polytechnic College, Nagercoil, Tamil Nadu, India

Sheen Mers Sathianesan Vimala Department of Chemical Engineering, Indian Institute of Technology, Chennai, Tamil Nadu, India

Ma. del Rosario Moreno-Virgen Instituto Tecnológico de Aguascalientes (ITA), Tecnológico Nacional de México (TecNM), Aguascalientes, Mexico

Kong Won-Sik Mushroom Research Division, National Institute of Horticultural and Herbal Science, Rural Development Administration, Eumsung, Republic of Korea

About the Editors

Saravanan Rajendran has received his Ph.D. in physics-material science in 2013 from the Department of Nuclear Physics, University of Madras, Chennai, India. He was awarded the University Research Fellowship (URF) during the years 2009–2011 by the University of Madras. After working as an assistant professor at Dhanalakshmi College of Engineering, Chennai, India, during the year 2013–2014, he was awarded SERC and CONICYT-FONDECYT postdoctoral fellowship, University of Chile, Santiago, in the years 2014–2017. Dr. Rajendran has worked (2017–2018) in the research group of Professor John Irvine, School of Chemistry, University of St Andrews, UK, as a postdoctoral research fellow within the framework of a EPSRC-Global Challenges Research Fund for the removal of blue-green algae and their toxins. He is currently working as an assistant professor in the Faculty of Engineering, Department of Mechanical Engineering, University of Tarapacá, Arica, Chile. Additionally this, he is also working as a research associate at SERC, Santiago, Chile. He is the associate editor for the *International Journal of Environmental Science and Technology* (Springer). His research interests focus in the area of nanostructured functional materials, photophysics, surface chemistry, and nanocatalysts for renewable energy and waste water purification. Dr. Rajendran has published in several international peer-reviewed journals and authored eight book chapters and seven books by renowned international publishers.

Mu. Naushad is presently working as an associate professor in the Department of Chemistry, College of Science, King Saud University (KSU), Riyadh, Kingdom of Saudi Arabia. He obtained his M.Sc. and Ph.D. degrees in analytical chemistry from Aligarh Muslim University, Aligarh, India, in 2002 and 2007, respectively. He has a vast research experience in the multidisciplinary fields of analytical chemistry, materials chemistry, and environmental science. Dr. Naushad holds several US patents, over 290 publications in international journals of repute, 20 book chapters, and several books published by renowned international publishers. He has >11000 citations with a Google Scholar H-Index of >60. He has been included in the list of Highly Cited Researchers 2019. Dr. Naushad has successfully run several research projects funded by National Plan for Science and Technology (NPST) and King Abdulaziz City for Science and Technology (KACST), Kingdom of Saudi Arabia. He is the editor/editorial member of several reputed journals like *Scientific Report* (Nature); *Process Safety & Environmental Protection* (Elsevier); *Journal of Water Process Engineering* (Elsevier); and *International Journal of Environmental Research & Public Health* (MDPI). Dr. Naushad is also the associate editor for *Environmental Chemistry Letters* (Springer) and *Desalination & Water Treatment* (Taylor & Francis). He has been presented with the Scientist of the Year Award–2015 by the National Environmental Science Academy, Delhi, India, the Almarai Award–2017 by Saudi Arabia, and Best Research Quality Award–2019 by King Saud University, Saudi Arabia.

Lorena Cornejo Ponce is currently working as a professor in the Faculty of Engineering, Department of Mechanical Engineering, at University of Tarapacá, Arica, Chile. She obtained her master's degree in chemistry (analytical chemistry) and Ph.D. (science – chemistry) from State University of Campinas, Campinas, Brazil, between 1988 and 1995. Further, she continued her postdoctorate at the same university in the year 1998. Dr. Ponje is interested in the field of spectroscopic analysis, environmental issues, and novel catalysts, developing new technology in the area of decontamination, water disinfection, and solar water treatment. She has published several articles in ISI publications, book chapters, and conference proceedings. Dr. Ponje handled successfully several national and international research projects financed by different government and private funding agents such as FONDECYT, FONDAP, FIC, CONICYT, OAS, and CYTED. She obtained international recognition awards – two MERCOSUR (science and technology – integration category) in the year 2006 and 2011. Also, she is coordinator in the Laboratory of Environmental Research of Arid Zones, LIMZA. In addition, she holds the position of principal investigator and coordinator of Line No. 1: Solar Water Treatment of SERC-CHILE. Dr. Ponje also participates in the AYLLU SOLAR Project, Solar Energy: Sustainable Development for Arica, and Parinacota as a member of the steering committee and as the coordinator of the "Solar Water Treatment" area.

Eric Lichtfouse is geochemist and professor of scientific writing at Aix-Marseille University, France, and visiting professor at Xi'an Jiaotong University, China. He has discovered temporal pools of molecular substances in soils, invented carbon-13 dating, and published the book *Scientific Writing for Impact Factor Journals*. He is chief editor and founder of the journal *Environmental Chemistry Letters* and the book series Sustainable Agriculture Reviews and Environmental Chemistry for a Sustainable World. He has awards in analytical chemistry and scientific editing. He is World XTerra Vice-Champion.

Chapter 1
Metal Oxides as Decontaminants of Water and Wastewater

Kingshuk Dutta

Abstract One of the most emerging threat to the present and future of the surface and aquatic lives is the contamination of water and wastewater. The aquatic and surface lives are severely affected through the contamination by detrimental chemicals in recent years; and, as a consequence, the decontamination techniques are attracting significant attention worldwide. In this regard, metal oxide nanoparticles have been witnessing rapid emergent as a promising decontaminant of water and wastewater. Decontamination by metal oxide nanoparticles is consisted of mainly photocatalysis, adsorption and disinfection techniques. More precisely, the metal oxide nanoparticles are enabled to function as nonophotocatalysts, nanosorbents and bioactive nanoparticles, through the application of which various problems, involving water quality, can be resolved or significantly ameliorated.

Herein, systematic emphasis has been provided on the progress in utilization of various metal oxide nanoparticles for decontamination of water and wastewater. Since, metal oxides have been used to decontaminate a wide variety and nature of pollutants from water and wastewater; therefore, examples from each category have be presented, along with the mechanistic view of the decontamination process wherever necessary. It is interesting to note that almost all the metal oxides reviewed herein follows the Langmuir adsorption isotherm model and pseudo second order kinetics, irrespective of the nature of contaminants adsorbed/removed. In terms of contaminant removal efficiency, metal oxides have been consistently able to achieve values of >95; and have, in some cases, reached 100% removal efficiencies.

Keywords Metal oxides · Water and wastewater decontamination · Photocatalysis · Adsorption · Disinfection · Dye and heavy metal · Salt and oil · Fluoride and phosphate · Phenol and pathogen · Single and multicomponent removal

---

K. Dutta (✉)
Advanced Polymer Design and Development Research Laboratory (APDDRL), School for Advanced Research in Polymers (SARP), Central Institute of Petrochemicals Engineering and Technology (CIPET), Bengaluru, Karnataka, India

© The Author(s), under exclusive license to Springer Nature Switzerland AG 2021
S. Rajendran et al. (eds.), *Metal, Metal-Oxides and Metal-Organic Frameworks for Environmental Remediation*, Environmental Chemistry for a Sustainable World 64, https://doi.org/10.1007/978-3-030-68976-6_1

1.1 Introduction

Since, water is the most essential compound for survival of majority of lives on earth; therefore, its contamination in any form is going to cause an irreversible, irreparable and detrimental effect on lifeforms on our planet. As humans, we use water on a daily basis for various purposes – the most important of which is its consumption. Therefore, owing to the ever-increasing levels of pollution of water bodies around the world at an alarming rate, we are forced to consume harmful and toxic substances that are present in water. Apart from consumption, owing to the various daily activities like washing, cleaning, etc., we regularly come in contact with water along with the contaminants that are present in it. The pollutant particles present in water are harmful, most often toxic and in certain cases carcinogenic and/or mutagenic as well (Sciacca and Conti 2009; Saxena et al. 2016; Naushad et al. 2018; Lu and Astruc 2018). Regular intake and contact with these substances can lead to severe health issues, even leading to death in the extreme cases (Landrigan et al. 2018; Saha et al. 2017). With already high level of pollution, followed by continuous addition of pollutants each passing moment, the concentration of the different categories of polluting materials are often present in excess of the prescribed safety limit.

It is a general perception that underground water is probably one of the least contaminated sources of water due to the fact that it has the least possibility of getting exposed to pollutants, as it is considered to be protected within the natural underground reservoirs and also getting naturally filtered out of pollutants while passing through different layers of rocks present underground. However, various human activities have led to slow leaching and percolation of chemicals, in the forms of fertilizers, pesticides, arsenic, etc., to these underground water reservoirs and, in the process, contaminating the underground water. Moreover, the underground water reserves are in grave danger and facing near total exhaustion in most parts of the world, including India and parts of west and southeast Asia as well as parts of Africa, Central, North and South America, Europe and Australia. The future prediction on the underground fresh water scarcity issue is threatening enough to immediately start taking remedial steps with extremely high intensity and rapid pace. Now, alternatives to underground water are water available in waterbodies, such as oceans, seas, rivers, lakes, ponds, etc. However, these are the most contaminated source of water, and is most often not fit for consumption or other daily usages. These water bodies keep on getting increased level of pollutants mainly from industrial, municipal, domestic, agricultural and animal wastewater effluents. On the other hand, the largest water reservoirs in the world are the oceans and seas. However, we have still not been able to utilize, to a satisfactory extent, the water contained in these bodies mainly because of their inherently high salt content, apart from the presence of pollutants. Therefore, the present fresh water crisis scenario can be summarized as arising due to three following reasons: (a) rapid depletion and near extinction of underground water, (b) huge level of polluting contamination of water bodies and (c) inherent salinity of water present in seas and oceans.

Based on the above discussion and realization, the need-of-the-hour is to develop water purification technologies and increase their water purification capacities at a pace matching the requirement of catering pure and fresh water to the entire world's rapidly increasingly population. In this regard, there are two major aspects: (a) to develop new and efficient technologies other than those already established in the market and (b) to enhance the capacity of already existing technologies by means of development of new materials, designs, operating parameters, etc. Now, there are already several technologies available in the commercial market that have proven their worth in effecting water purification. These include membrane filtration, adsorption, photocatalysis, desalination, chemical precipitation, electrochemical treatment, disinfection and solvent extraction to name a few (Dutta and De 2017a, b; Dutta and Rana 2019). However, we need innovative and highly efficient materials in each category that can perform their respective function in a commercial scale. In this respect, metal oxides have stood out in terms of their performance efficiency and their applicability within the domains of a number of the abovementioned technologies; for example, adsorption, photocatalysis, electro-chemical treatment, disinfection, desalination, etc. (Ali 2012; Herrmann et al. 1993; Kasprzyk-Hordern 2004; Lee and Park 2013; Qu et al. 2013). In addition, metal oxides have found employment in treating a number of varied polluting materials, like heavy metals, salts, dyes, phosphates, fluorides, phenols, microbes and oil (Lee et al. 2016; Mano et al. 2015; Upadhyay et al. 2014; Xu et al. 2012). Some high-performance metals have also been realized to treat two or more categories of pollutants simultaneously or otherwise.

Herein, the subject of utilization of metal oxides in decontamination of water and wastewater has been dealt with in a case-by-case manner, for the purpose of analyzing the efficacy of the materials used under various categories of polluting contaminants. However, the intention here is not to provide an exhaustive literature review on the subject. Accordingly, under each separate category, some representative published reports have been presented in order to bring out the overall as well as comparative picture for readers to get an overview of this field of research.

1.2 Dye

Dyes are a class of compounds that have very high usage around the world in industries, such as textiles, leather, furniture, paint, etc. They are produced in huge quantities owing to their very high demand. This invariably leads to it eventually getting drained out as part of effluents; in the process, contaminating water bodies. Dyes are often toxic, as they are prone to environmental oxidative degradation producing toxic products like carbon monoxide, nitrogen oxides, etc. Therefore, it should be ensured that they are separated from the effluent stream before getting deposited in water bodies. Keeping this in mind, various researchers have devised methods and materials to ensure complete removal of dye molecules from wastewater comprising industrial effluents. Among the materials utilized, metal oxides

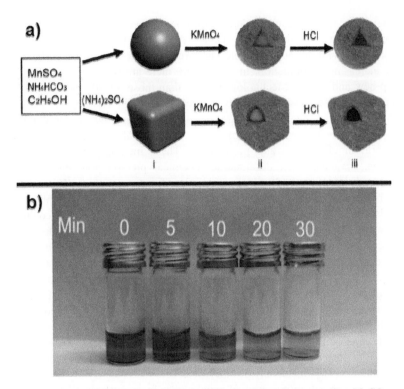

Fig. 1.1 (a) Preparation of MnO_2 hierarchical hollow particles: (i) intermediate $MnCO_3$ crystal templates with different morphologies, (ii) MnO_2 shell structures with $MnCO_3$ cores, and (iii) as-prepared MnO_2 hierarchical hollow nanostructures. (b) Absorption of Congo red dye by the synthesized MnO_2 microspherical hollow hierarchical particles at different time intervals. (Reprinted from Fei et al. (2008), with permission from Wiley)

constitute a very important part owing to their semiconducting nature, easy availability, easy synthesis, low toxicity and high dye degradation and removal efficiencies (Chan et al. 2011). This section will deal with some of the important studies, and will present the results obtained.

Fei et al. (2008) synthesized MnO_2 microcubes and microspheres, having hierarchical hollow morphology, via intermediate $MnCO_3$ crystal templating. The synthetic process has been shown on Fig. 1.1a. These uniquely prepared templated MnO_2 structures demonstrated promising capability of adsorptive removal of Congo red dye from aqueous solution (Fig. 1.1b). Moreover, the adsorption rate demonstrated by these synthesized MnO_2 particles were higher than that obtained upon using commercial MnO_2 and commercial γ-Fe_2O_3 particles. In addition, the synthesized MnO_2 particles could be renewed and reused at least three times with negligible dip in the adsorption efficiency. Also, the removal of dye-adsorbed MnO_2 particles were found to be easy owing to their size in micrometers. In another use of manganese oxide as a material of choice, Chen and He (2008) showed that manganese oxide nanostructures can more efficiently adsorb methylene blue dye in terms

Table 1.1 The Freundlich and Langmuir isotherm parameters for the adsorption process of the neutral red dye by the magnetic and hollow Fe_3O_4 nanospheres

Langmuir isotherm				Freundlich isotherm		
K_L (Lg^{-1})	α_L (Lmg^{-1})	q_{max} (mgg^{-1})	R^2	log K_f	$1/n$	R^2
4.81	0.0451	105	0.992	1.026	0.46	0.994

Reproduced from Iram et al. (2010), with permission from Elsevier

of rate of adsorption (85% of the dye in 30 min at room temperature) as well as adsorption/removal capacity (68.4 mgg^{-1}), compared to the common adsorbents red mud and MCM-22. The authors also made two important observations: (a) the adsorbent can be reused for up to 3 times without significant reduction of its capacity and (b) the adsorption capacity was independent of the specific surface area of the adsorbent.

Oxides of iron are an important class of material primarily owing to their magnetic attribute. This unique feature has been utilized for complete removal of the adsorbent, with adsorbed pollutant materials, after completion of the adsorption process by application of a magnetic field. In addition, iron oxides are also very good adsorbents. Therefore, a combination of both of these features result in a superior adsorbent material. In a typical study, Iram et al. (2010) synthesized hollow Fe_3O_4 nanospheres for magnetically driven adsorption of neutral red dye from aqueous solution. The authors achieved a 90% dye removal with a monolayer adsorption capacity of 105 mgg^{-1} at a pH value of 6 and a temperature of 25 °C. The result was found to fit the Langmuir isotherm model. Table 1.1 presents the Freundlich and Langmuir isotherm parameters for the adsorption process of the dye by Fe_3O_4 nanospheres. The hollow morphology, nanosized particles and high specific surface area were found to be responsible for the observed high adsorption behavior of the magnetic nanospheres.

In an interesting work, hierarchical three-dimensional nanostructures of iron oxide were prepared by Fei et al. (2011) via iron hydroxide precursor. The authors obtained an adsorption efficiency of 66.7 mgg^{-1} of Congo red dye at a pH value of 7.6 and at room temperature. The adsorption was attributed to the electrostatic attraction taking place between the adsorbent and the dye molecules. Wang et al. (2010) synthesized magnetic Fe_3O_4 nanoparticles that possessed high aqueous dispersibility owing to their small particle size of ~16.5 nm and high surface area of ~82.5 m^{2}g^{-1}. A removal efficiency of 90% of Rhodamine B dye by the synthesized magnetic nanoparticles were observed in presence of hydrogen peroxide, the latter getting activated by the magnetic adsorbent at 40 °C and 5.4 pH and contributing towards the dye degradation process. The mechanisms of the peroxide activation and dye degradation processes have been presented in Fig. 1.2. A unique ordered core-shell nanomorphology of magnetic iron oxide-manganese oxide was developed by Zhai et al. (2009) and demonstrated 42 mgg^{-1} adsorption capacity towards Congo red dye, with facile removal efficiency by application of a magnetic field. Reusing capability of the adsorbent was also realized upon recovery by combustion at 300 °C.

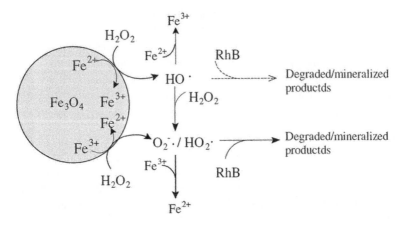

Fig. 1.2 The mechanisms involved in the activation of hydrogen peroxide by the magnetic Fe_3O_4 nanoparticles, followed by degradation of Rhodamine B dye. (Reprinted from Wang et al. (2010), with permission from Elsevier)

Apart form manganese oxide and magnetic iron oxides, other oxides have also found use in dye treatment and removal application. For example, hollow urchin-like nanostructure of tungsten oxide has been used to adsorb methylene blue dye from aqueous solution (Jeon and Yong 2010). This nanoadsorbent showed adsorption capacity of 138.88 mgg^{-1} towards methylene blue, which was found to be higher than that demonstrated by commercial tungsten oxide (12.37 mgg^{-1}) and most other commercial adsorbents like zeolite MCM-22 (67.3 mgg^{-1}), red mud (2.5 mgg^{-1}), fly ash (70.37 mgg^{-1}), etc. towards the same dye. Khan et al. (2012) studied the photocatalytic efficiency of two different oxides, namely ZnO and Al_2O_3, in combination with CdS and graphene oxide. The CdS/Al_2O_3/graphene oxide photocatalyst displayed ~90% photodegradation efficiency towards methyl orange dye while CdS/ZnO/graphene oxide showed ~99% efficiency, both within 60 min. These high efficiencies were found to be due to the presence of graphene oxide (1 wt %), with sheet-like structure, that showed high separation of charge carriers formed upon photo-irradiation, leading to reduced recombination, as well as high surface area. The involved mechanism of separation and transfer of charge, along with the degradation of methyl orange, under photo-irradiation using the CdS/ZnO/graphene oxide photocatalyst has been presented in Fig. 1.3.

In another work, TiO_2/reduced graphene oxide and SnO_2/reduced graphene oxide nanoparticles were synthesized and used to photo-degrade Rhodamine B dye under irradiated visible light (Zhang et al. 2011). The reduced graphene oxide nanosheet support was found to be effectively disperse the deposited metal oxide photocatalytic nanoparticles and increase the photocatalytic efficiency of the photocatalysts. The photo-degradation efficiency was found to be higher for the SnO_2/reduced graphene oxide system, compared to the TiO_2/reduced graphene oxide and a commercial TiO_2 photocatalyst systems. The better result observed for the former composite was attributed to higher charge separation and electrical conductivity. In a different

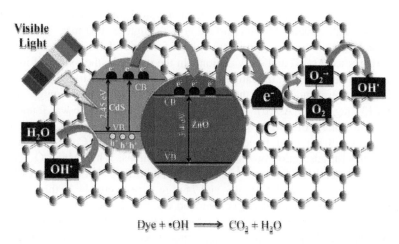

Fig. 1.3 The involved mechanism of separation and transfer of charge, along with the degradation of methyl orange, under photo-irradiation using the CdS/ZnO/graphene oxide photocatalyst. (Reprinted from Khan et al. (2012), with permission from The Royal Society of Chemistry)

approach, Khairy and Zakaria (2014) performed doping of TiO_2 by 2 wt% of Cu and Zn and showed that the type of dopant metal influenced the photodegradation efficiency of TiO_2 towards methyl orange dye. Performed under both visible and UV light irradiation, the photocatalytic activity was found to be in the order: Cu-doped TiO_2 > Zn-doped TiO_2 > undoped TiO_2. It was inferred that the Cu-doped TiO_2 was the most efficient in preventing recombination of electron-hole, which led to its best performance.

1.3 Heavy Metal

Heavy metals are in general highly toxic contaminants, often leading to serious health issues if their concentrations are more than safety limits. Therefore, we need to ensure that we restrict any increase in their concentration in water and remove the already present metals from water and wastewater to the maximum possible extent. Metal oxides nanomaterials have been playing a major role in this process of treating and removing heavy metal ions (Hua et al. 2012; Trivedi and Axe 2000).

Manganese oxide has been used in combination with reduced graphene oxide to scavenge Hg(II) ions (Sreeprasad et al. 2011). The adsorption process was found to follow the pseudo-first-order equation, and demonstrated a 100% removal efficiency towards Hg(II) ions from ground water. In another study, hydrous manganese and iron oxides were made use of in order to adsorb Pb(II), Cd(II), Tl^+ and Zn(II) (Gadde and Laitinen 1974). The adsorption of the polluting metals was found to be in the order Pb(II) > Zn(II) > Cd(II) > Tl^+. Manganese oxides have been used in several occasions in combination with iron oxides either as a simple mixture or in a core-shell structure or as a bimetallic oxide. In a typical study, Ociński et al. (2016) residuals of water treatment with surface area as high as 120 m^{2}g^{-1}, which get generated during processing of infiltration water for removal of manganese and iron, as adsorbents for As(III) and As(V) anions. In the residuals, the oxides of iron and manganese were present in the ratio of 5:1. The adsorption was found to fit into the Langmuir isotherm model, and the capacities were found to be 77 mgg^{-1} for As(V) and 132 mgg^{-1} for As(III). The presence of manganese oxide was found to be critical for oxidation of As(III), which led to its high adsorption. In a similar study, Zhang et al. (2010) used bimetallic oxides MnFe$_2$O$_4$ and CoFe$_2$O$_4$ magnetic nanoparticles for adsorption of As(III) and As(V) anions. While MnFe$_2$O$_4$ showed adsorption capacities of 94 mgg^{-1} for As(III) and 90 mgg^{-1} for As(V), CoFe$_2$O$_4$ exhibited higher values of 100 mgg^{-1} and 74 mgg^{-1}, respectively. In this study too, the presence of the second metal, i.e. Mn and Co, was found to be critical behind exhibition of higher adsorption capacities, because they enabled formation of surface hydroxyl moieties. This performance-enhancing effect of the second metal is evident from the fact that Fe$_3$O$_4$ could produce only 50 mgg^{-1} for As(III) and 44 mgg^{-1} for As(V). Moreover, the bimetallic nanoadsorbent demonstrated good desorption of higher than 90% for As(V) and 80% for As(III).

In another study, a graphene oxide/bimetallic MnFe$_2$O$_4$ nanoparticle hybrid composite was utilized for removing As and Pb contaminants from water (Kumar et al. 2014). The presence of graphene oxide layers ensured availability for higher surface area for better dispersion of the magnetic bimetallic particles. This nanohybrid composite exhibited adsorption capacities of 673 mgg^{-1} for Pb(II), 207 mgg^{-1} for As(V) and 146 mgg^{-1} for As(III), which was much higher than that obtained for only the bimetallic nanoparticles (without the presence of graphene oxide) with the respective values of 488 mgg^{-1}, 136 mgg^{-1} and 97 mgg^{-1}. Desorption of the adsorbed contaminants was realized to be 99%, 93% and 99% for As(V), As(III) and Pb(II), respectively. Most importantly, the nanohybrid could be used for five cycles of adsorption, without noticeable change in the efficiency. On the other hand, Kim et al. (2013) designed magnetic Fe$_3$O$_4$ nanoparticles coated with amorphous MnO$_2$, which exhibited a hierarchical core-shell nanocomposite three-dimension flower-like structure (Fig. 1.4). This nanocomposite showed enhanced adsorption efficiency towards Cu(II), Zn(II), Pb(II) and Cd(II) ions over bare Fe$_3$O$_4$ nanoparticles. From Langmuir isotherm model, the adsorption capacity of the nanocomposite towards Cd(II) was found to be 53.2 mgg^{-1}, with a recycling capacity of up to 5 cycles.

Oxides of iron have been the material of choice in heavy metal treatment application. For example, mesoporous nanocomposites of iron/iron oxide have been used for removal of chromate anions from solution in water (Kim et al. 2012). From the Langmuir model, the maximum value of adsorption capacity was found to be 34.1 mgg^{-1}. Nalbandian et al. (2016) used nanofibers of Fe$_2$O$_3$ for adsorptive removal of chromate from aqueous solution. With a specific surface area of 59.2 m^{2}g^{-1}, the nanofibers with 23 nm average diameter demonstrated an adsorption capacity of 90.9 mgg^{-1}. This result was better than that obtained for the commercial Fe$_2$O$_3$ nanoparticles (49.3 mgg^{-1}). Both the specific surface area

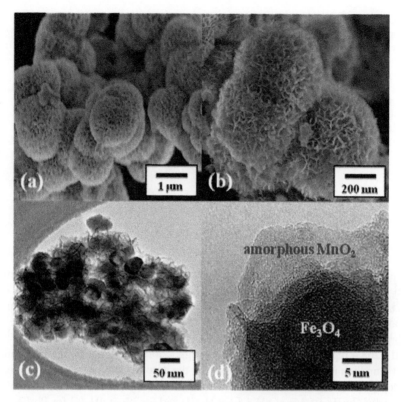

Fig. 1.4 Scanning electron (**a** and **b**) and transmission electron (**c** and **d**) micrographs of the synthesized amorphous MnO_2-coated magnetic Fe_3O_4 nanoparticles, exhibiting hierarchical core-shell nanocomposite three-dimension flower-like structure. (Reprinted with permission from Kim et al. (2013). © 2013 American Chemical Society)

and the adsorption capacity of the nanofibers were found to increase with reduction in the fiber diameter. Palanisamy et al. (2013) reported the use of flaxseed and olive oil stabilized nanoparticles of iron oxide for removal of Cr, Ni and Cu ions from aqueous solutions. Mesoporous Fe_3O_4 nanoparticles functionalized by amine showed very high adsorption of Cu(II), Cd(II) and Pb(II) ions (Xin et al. 2012). From Langmuir isotherm model, the maximum adsorption capacity was found to be 523.6 mgg^{-1}, 446.4 mgg^{-1} and 369.0 mgg^{-1}, respectively for Cu, Cd and Pb ions. The adsorbent also showed a removal efficiency of >98% for all the polluting ions.

Magnetite was composited with reduced graphene oxide by Chandra et al. (2010) to remove As(V) and As(III) anions from aqueous solution. Use of reduced graphene oxide support resulted in minimizing the magnetite aggregates, leading to high dispersion of nanoparticles of magnetite. This, in turn, increased the number of adsorption sites in the nanocomposites, leading to enhanced capacity of binding with the contaminating metal ions. As a result, almost 100% removal of arsenic by the magnetite/reduced graphene oxide nanocomposite was observed. Also, the magnetic

Fig. 1.5 Easy separation of contaminant-loaded magnetic adsorbent from aqueous medium upon employment of an external magnetic field. (Reprinted with permission from Chandra et al. (2010). © 2010 American Chemical Society)

nature of the composite enabled easy separation of the contaminant-loaded adsorbent from aqueous medium upon employment of an external magnetic field (Fig. 1.5). Fe_3O_4/graphene oxide magnetic nanocomposite was functionalized by EDTA for the purpose of adsorptive removal of Cu(II), Hg(II) and Pb(II) ions (Cui et al. 2015). The adsorption capacity was found to be high, with obtained values of 301.2 mgg^{-1} for Cu(II), 268.4 mgg^{-1} for Hg(II) and 508.4 mgg^{-1} for Pb(II) after fitting into Temkin and Freundlich isotherm models. Moreover, the adsorption process was found to be pH dependent, with demonstration of the maximum removal efficiencies of 96.5% at pH 5.1 for Cu(II), 95.1% at pH 4.1 for Hg(II) and 96.2% at pH 4.2 for Pb(II).

Wu et al. (2012) demonstrated a unique material composed of magnetic Fe_2O_3 encapsulated in mesoporous carbon matrix for adsorptive capture and removal of arsenic. This novel material produced 29.4 mgg^{-1} of adsorption capacity with fast adsorption pseudo-second-order rate. Moreover, it showed easy removal owing to its magnetic nature, as well as potential to be reused. The processes of iron oxide encapsulation and arsenic adsorption have been presented in Fig. 1.6. Other usages of iron oxides have been as sand coated with iron oxide as well as zerovalent iron for removal of arsenate and chromate (Mak et al. 2011); as a three-dimensional nanostructure comprising of carbon nanotube, graphene and iron oxide for adsorptive removal of arsenic (Vadahanambi et al. 2013); as sewage sludge from municipality, coated with iron oxide, for removal of Pb(II), Ni(II), Cd(II) and Cu(II) ions (Phuengprasop et al. 2011); and polymer/iron oxide hybrid nanocomposites for removal of chromate, arsenate, Cu(II), Cd(II) and Pb(II) ions from aqueous solutions (Peter et al. 2017; Pan et al. 2010).

Several other metal oxides have been tested as potential decontaminating agents. Among them, TiO_2 has gained some popularity owing to its performance efficiency. For instance, Lee and Yang (2012) fabricated a hybrid composite constituting TiO_2 and graphene oxide. Upon using exfoliated graphene oxide nanosheets for supporting TiO_2 nanoparticles, the formed hybrid presented a flower-like morphology. This nanohybrid adsorbent were used to adsorb Zn(II), Cd(II) and Pb(II) ions

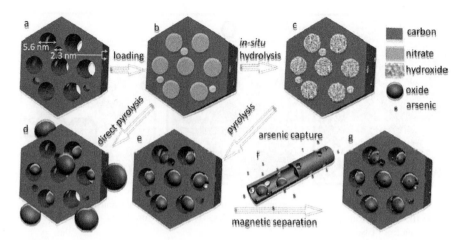

Fig. 1.6 Synthesis and arsenic capture processes for the ordered mesoporous Fe_2O_3@C encapsulates: (**a**) the bimodal mesoporous carbon, (**b**) carbon loaded with hydrated iron nitrate precursor, (**c**) carbon loaded with iron hydroxide obtained by *in situ* hydrolysis under ammonia atmosphere, (**d**) iron oxide@carbon composites obtained by direct pyrolysis, (**e**) the Fe_2O_3@C encapsulates obtained by pyrolysis following the pre-hydrolysis, (**f**) arsenic capture and (**g**) arsenic-enriched encapsulates. (Reprinted from Wu et al. (2012), with permission from Wiley)

from aqueous solution. At a pH value of 5.6 after 12 h treatment, the observed adsorption capacities were ~ 65.6 mgg^{-1} for Pb(II), ~72.8 mgg^{-1} for Cd (II) and ~ 88.9 mgg^{-1} for Zn(II). These values were much higher than that observed for colloidal graphene oxide. In another study, synthesized TiO_2 hierarchical spheres demonstrated higher As(V) removal capacity compared to commercial TiO_2, commercial CeO_2, synthesized CeO_2 hierarchical nanostructures and commercial α-Fe_2O_3 (Hu et al. 2008). On the other hand, the synthesized CeO_2 produced the best result in terms of Cr(VI) adsorption capacity among the oxides studied. Apart from the ones discussed above, other metal oxides used with some success for the purpose of heavy metal decontamination from water are CeO_2 (Zhong et al. 2007), Al_2O_3 (Yamani et al. 2012; Kuan et al. 1998), SiO_2 (Sheet et al. 2014; Karnib et al. 2014), MgO (Cao et al. 2012), ZrO_2 (Hristovski et al. 2007), and NiO (Hristovski et al. 2007).

1.4 Salt

We have vast resources of water in the forms of seas and oceans. However, we cannot use water from these resources for our consumption and daily household needs because they contain large concentration of dissolved salts. Nevertheless, owing to the severe scarcity of water underground and fresh water bodies, we have in recent times turned our attention to device methods and techniques for utilizing sea and ocean waters for our daily activities. There are a number of different

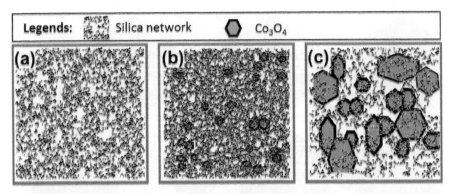

Fig. 1.7 Silica network microstructure and porous texture of (**a**) pure silica, and cobalt oxide silica of the (**b**) microporous and (**c**) mesoporous samples. (Reprinted from Elma et al. (2015), with permission from Elsevier)

technologies that have been developed over the years for this purpose under the common phenomenon named "desalination". In terms of materials used, metal oxides have found some potential use in more than one desalination technologies. This last aspect will be the subject matter of this section.

Membrane filtration is one of the highly used technologies involved in desalination. Lin et al. (2012) had fabricated cobalt oxide silica filtration membranes for removal of salt from aqueous solution. The authors used three initial concentration of salt solutions, representing three different contaminated sources – brine solution, seawater and brackish water. It was observed that with increasing salt concentration, the water flux decreased. Most importantly, higher than 99% salt rejection was realized by using these membranes. In a similar approach, Elma et al. (2015) showed that mesoporous membranes formed from cobalt oxide silica can reach even more than 99.7% of salt rejection with significant water flux. This efficient result was attributed to the combination of mesopores and structural integrity of the membranes (Fig. 1.7).

Another popular technology of desalination that makes use of metal oxides is capacitive desalination. In a typical work, Myint et al. (2014) utilized a number of micro and nanostructured materials of ZnO, like microspheres, microsheets, nanorods and nanoparticles, as grafts on activated carbon cloth to form electrodes for capacitive desalination of brackish water. The use of microsheets of ZnO produced a desalination efficiency of 22%, a desalination capacity of 8.5 mgg^{-1} and a regeneration/salt removal efficiency of 19%. On the other hand, use of nanorods of Zno demonstrated respective values of 22%, 8.5 mgg^{-1} and 21%. In another similar study, a composite electrode made of Zno and activated carbon produced a charge efficiency of 80.5% and a desalination capacity of 9.4 mgg^{-1} (Liu et al. 2015). It was further shown that TiO_2 supported on graphene aerogel, with a loading TiO_2 of 60.4 wt%, can also be used as a highly efficient three-dimensional electrode material for capacitive desalination of aqueous saline solution

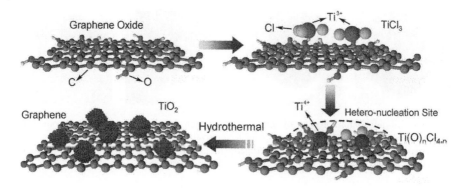

Fig. 1.8 Fabrication of graphene aerogel/TiO_2 hybrid electrode material. (Reprinted from Yin et al. (2013), with permission from Wiley)

(Yin et al. 2013). The fabrication of this hybrid electrode material has been presented in Fig. 1.8.

1.5 Oil

Oil contamination of water is a very serious issue. Large volume of oil enters the water bodies from industries, such as petroleum refining and oil extraction, and from oil spill incidents. Efficient removal of oil from water via oil/water separation is, therefore, required to ensure proper treatment of oily wastewater. This section is dedicated to this issue, containing discussions on selected major reported research works in this area involving the use of metal oxides. For detailed literature review on this subject, the readers may refer to Ma et al. (2016).

Adopting a typical membrane filtration method, Lu et al. (2016) used a number of metal oxides as deposits on ultrafiltration membranes made of ceramics. The metal oxides used were CeO_2, CuO, MnO_2, Fe_2O_3 and TiO_2, all having an average size of 10 nm. It was inferred that the more the hydrophilic nature of the metal oxide, the more efficient it is as a material for filtration layer of a ceramic membrane. As a consequence, the most hydrophilic oxide Fe_2O_3 (among the studied oxides)-deposited membrane showed the least tendency towards irreversible fouling, high chemical oxygen demand rejection percentage and the highest normalized initial permeate flux for up to 7 cycles of filtration. Zhu et al. (2010) fabricated an innovation polysiloxane-coated magnetic Fe_2O_3@C core-shell nanoparticulate adsorbent to remove oil from oil/water mixture. These corrosion-resistant nanoadsorbent showed oil uptake up to ~4 times of their weight, in addition to their floatable nature that is important considering that oil (light oils) in general floats on water, and so, interaction between the contaminant oil and the nanoadsorbent will be more feasible. A photograph of the oil adsorption and removal processes have been shown in Fig. 1.9. On the other hand, Wang et al. (2013) designed and

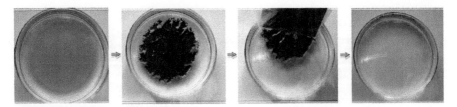

Fig. 1.9 Oil adsorption process by the polysiloxane-coated magnetic Fe_2O_3@C core-shell nanoparticulate adsorbent, followed by complete removal of the oil-adsorbed nanoparticles with the help of an external magnetic field. (Reprinted with permission from Zhu et al. (2010). © 2010 American Chemical Society)

fabricated a number of sponges and fabrics formed upon growing of nanocrystals of transition metals and metal oxides, with further modification by thiol. The metal oxides used were that of copper, nickel, cobalt and iron. The fabricated fibers possessed superhydrophibic/superoleophilic surface wettability, and were found to be effective to remove both light and heavy oils. On the other hand, the fabricated sponges possessed very high adsorption capacities, selectivity and recyclability.

1.6 Fluoride

Below the prescribed concentration, intake of fluoride is beneficial to our bodies. However, if taken above the safe limit, fluorides can seriously affect our bones and teeth in the form of fluorosis, cause damage to our brain and kidney, as well as affect the metabolism of elements in our body and cause neurological damage. Fluoride gets added to the surface and groundwater via effluents from industries, including fertilizer, semiconductor, cosmetics, drugs and ceramic manufacturing plants and power plants running on coal. Several metal oxides have been successfully used for the purpose of decontaminating fluoride from water (Velazquez-Jimenez et al. 2015). This section is dedicated to this aspect of water treatment.

The mostly used metal oxide for defluoridation of water is aluminum oxide, either as the sole decontaminant or in conjunction with other metal oxides. The main advantage of using Al_2O_3 is the fact that it possesses a very high internal surface area; thus, a high number of adsorption sites are available for trapping contaminant moieties. For example, Kumar et al. (2011) made us of nano-Al_2O_3 for adsorption of fluoride from aqueous solution and achieved a maximum adsorption capacity of 14 mgg^{-1} at a pH value of 6.15 and at a temperature of 25 °C. It was further noticed that the fluoride adsorption process was influenced by the solution pH, temperature and by the presence of other contaminating ions like phosphate, sulfate and carbonate. Table 1.2 presents the Freundlich and Langmuir isotherms of the adsorption process at two different temperatures. It can be realized from the table that the Langmuir isotherm model better explains the adsorption process. In another related study with nano-AlOOH, maximum adsorption (fitting the Langmuir isotherm model) was observed to happen at ~pH 7 and desorption at pH 13, revealing the

Table 1.2 Freundlich and Langmuir isotherms of the defluoridation by nano-Al_2O_3 at two different temperatures

Temperature (°C)	Langmuir constants				Freundlich constants		
	q_m (mgg^{-1})	b (Lmol^{-1})	R_L	R^2	$1/n$	K_F (mgg^{-1}) (Lmg^{-1})$^{1/n}$	R^2
10	14.10	2.36 × 10^3	0.31	0.9980	0.94	0.81	0.9823
25	15.43	3.24 × 10^3	0.27	0.9912	0.98	1.01	0.9626

Reproduced from Kumar et al. (2011), with permission from Elsevier

reusing potential of the nanoadsorbent (Wang et al. 2009a). Thakre et al. (2010) reported the use of chitosan-templated mesoporous beads of a binary metal oxide of Ti and Al and reported a maximum fluoride adsorption capacity of 2.22 mgg^{-1} and a working pH range of less than or equal to 9. Maliyekkal et al. (2006) showed that a manganese oxide-coated Al_2O_3 adsorbent can potentially adsorb fluoride, following a pseudo-second-order kinetics, with a maximum capacity of 2.85 mgg^{-1} between pH 4 to pH 7; while, activated Al_2O_3 could only produce a value of 1.08 mgg^{-1}. In another study, nano-MgO has also been shown to produce good fluoride adsorption and adsorbent reusing capability (Devi et al. 2014).

1.7 Phosphate

High concentration of phosphate in water bodies lead to excessive and localized increase of nutrients, causing high density growth of plants within that localized area. This phenomenon, called eutrophication, leads to deterioration of the quality of affected water. In addition, keeping in mind that phosphates are essential constituents of fertilizer, this eutrophication leads to wastage of this important class of compound. Therefore, we must ensure their removal and recovery in order to use them in desired applications. For this decontamination purpose, metal oxides have been utilized to a certain extent. The various reports on the use of metal oxides for removal of phosphates from water and wastewater have been discussed below.

Amorphous nanoparticles of ZrO_2 was able to adsorb phosphate at a maximum capacity of 99.01 mgg^{-1} at a pH value of 6.2, following the Langmuir isotherm model; however, the process was found to be pH independent between pH values of 2 and 6 (Su et al. 2013). Acelas et al. (2015) carried out comparative analysis of three hydrated metal oxides, namely hydrated copper oxide, hydrated zirconium oxide and hydrated ferric oxide, towards their adsorption potential of phosphates. They reported that the hydrated ferric oxide produced the best result with a maximum adsorption capacity of 111.1 mgg^{-1}, followed by hydrated zirconium oxide (with 91.74 mgg^{-1}) and hydrated copper oxide (with 74.07 mgg^{-1}). The interaction between the phosphate anion and the hydrated metal oxides, leading to adsorption of the former on the latter, has been depicted in Fig. 1.10. Hydrated ferric oxide nanomaterials have also been utilized with exhibition of promising results by Pan et al. (2009). In another comparative study, Delaney et al. (2011) could achieve

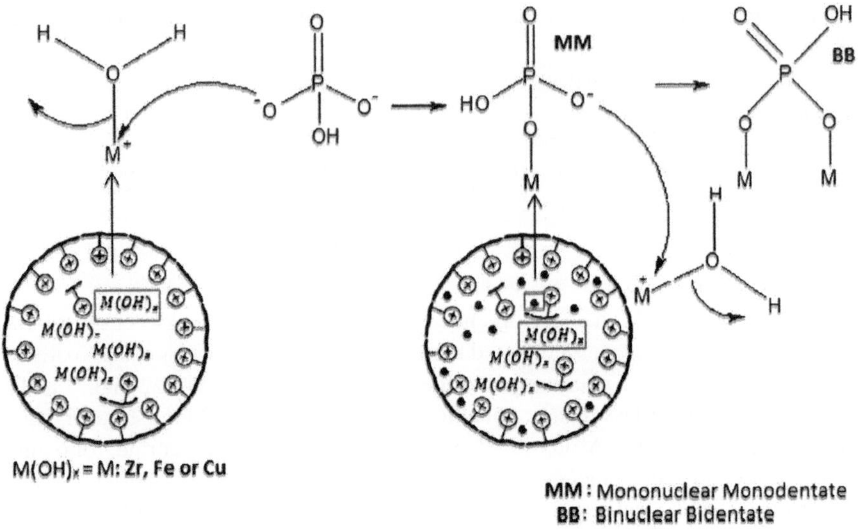

Fig. 1.10 Interaction between the phosphate anion and the hydrated metal oxides, leading to adsorption of the former on the latter. (Reprinted from Acelas et al. (2015); with permission from Elsevier)

100% removal of phosphate contaminant from water, with a 4.5 mgg^{-1} maximum adsorption capacity, upon use of a number of metal oxides adsorbents, namely zirconium oxide, iron oxide, aluminum oxide and titanium dioxide, doped with mesoporous silica, having surface areas in the range of 600 m^{2}g^{-1} to 700 m^{2}g^{-1}.

In order to achieve higher adsorption pf phosphate with better rate, binary and ternary metal oxides have been fabricated and used. Li et al. (2014) and Zhang et al. (2009a) used binary oxides, having iron as one component. While, the Fe-Mn binary oxide showed a maximum phosphate adsorption capacity of 36 mgg^{-1} at a pH value of 5.6, following a pseudo-second-order kinetics and fitting in the Freundlich isotherm model; the Fe-Cu binary oxide exhibited a maximum adsorption capacity of 35.2 mgg^{-1} at a pH value of 7, following a pseudo-second-order kinetic model and fitting in the Langmuir isotherm model. Lǔ et al. (2013) tried to analyze the adsorption efficacy of a ternary metal oxide composed of iron oxide, aluminum oxide and manganese oxide, present at a 3:3:1 molar ration of metals Fe, Al and Mn. With a maximum phosphate adsorption capacity of 48.3 mgg^{-1} at 25 °C (fitted best to the Freundlich isotherm model), this ternary oxide adsorbent showed an inverse dependency on pH (between 4 and 10.5) while executing its adsorption function.

1.8 Phenol

Phenol is an organic pollutant that can serious health consequences. It produced chlorophenols, a carcinogen, upon reacting with chlorine during normal treatment of water. Industries like pharmaceuticals, pulp, plastics, paper, dyes, textiles, pesticides, detergents, coke plants and oil refineries release phenolic compounds to the water bodies. Therefore, thorough treatment of water to get rid of these extremely toxic phenolic compounds is an essential requirement. Among other techniques, metal oxide-based electrocatalytic oxidative degradation of phenol is a very successful one in terms of phenol decontamination efficiency. Some representative reports have been discussed below.

Upon using activated carbon impregnated with CuO as the catalyst, Liou and Chen (2009) achieved >98% removal of phenol and > 90% removal of chemical oxygen demand upon stepwise addition of H_2O_2. In a similar study, Shukla et al. (2010) found that Co_2O_3 supported on activated carbon could produce a decomposition of 100% and a total organic carbon removal of 80% in presence of sulfate radicals within 60 min. Again, it was found that phenol can be treated by using catalytic nanoparticles of Fe_3O_4 that were superparamagnetic in nature, with a total organic carbon removal of 42.79% in presence of H_2O_2 (Zhang et al. 2009b). Using a mixed metal oxide of Ru as the anode catalyst, Yavuz and Koparal (2006) reported a removal efficiency of phenol as high as 99.7% from synthetic wastewater. However, when the authors used real wastewater from petroleum refinery, they could achieve a 94.5% phenol removal and 70.1% chemical oxygen demand removal efficiencies. Similarly, Yang et al. (2009) designed a mixed metal oxide-coated Ti electrode, for oxidative removal of phenol from water. A 78.6% removal was observed at a pH of 7 and a temperature of 20 °C, which got increased to 97.2% upon addition of chloride. In this study, the mixed metal oxide was comprised of PbO_2, Nb_2O_5, Sb_2O_3 and SnO_2. Similar studies have been performed by others, using a number of metal oxides, and have reported good performance (Feng and Li 2003; Wang et al. 2009b).

1.9 Other Contaminants

Other than the contaminants dealt with above, there are certain other harmful contaminants present in water and wastewater that need to be treated. Among them, the most important are nitrates and pathogens. Nitrates general enter the waterbodies through agricultural wastewater, owing to the use of excessing fertilizers. Consumption of nitrate-containing water at above safety limit can lead to diseases like methemoglobinemia. Therefore, decontaminating water from nitrates is essential. In this respect, metal oxides have played a role (Mook et al. 2012). For example, zero valent iron nanoparticles and its composite with TiO_2 have been used effectively for this purpose (Huang et al. 2013; Pan et al. 2012).

Pathogens are microorganisms, mainly viruses and bacteria, that can cause diseases upon entering our bodies through contaminated water. Therefore, their removal from drinking water is of utmost importance before intake. In this regard, metal oxide nanoparticles such as TiO_2 doped with silver has been used to inactivate virus Bacteriophage MS2 (ATCC 15597-B1) (Liga et al. 2011). Increase formation of hydroxyl radical owing to the presence of silver dopant was found to be the chief cause behind the virus deactivation process. Similarly, virus MS2 coliphage was successfully treated via a photo-Fenton process using semiconductor iron oxides photocatalyst, like magnetite, maghemite and wüstite, in presence of H_2O_2 (Giannakis et al. 2017). Biosand filters amended with iron oxide have also served as an effective virus decontaminant (Bradley et al. 2011). On a similar note, sand coated with iron oxide can effectively adsorb and lead to photoinactivation (Pecson et al. 2012). Other important works in this area have been reported in Brown and Sobsey (2009), Yang et al. (2013) and Ahammed and Davra (2011).

1.10 Dual- and Multi-contaminants

Some authors have reported use of metal oxide nanoparticles for removal of more than one contaminant. This fact shows the real potency of metal oxides in water and wastewater treatment. For instance, nanosheets of graphene oxide-supported nanoparticles of TiO_2 was successfully used to degrade azo dyes by photocatalytic oxidation as well as convert Cr(VI) to Cr(III) by photocatalytic reduction (Jiang et al. 2011). Similarly, hollow spheres of α-FeOOH has been used as an adsorbent to remove Congo red dye (maximum adsorption capacity: 275 mgg^{-1}) and heavy metal ions As (V) (maximum adsorption capacity: 58 mgg^{-1}) and Pb(II) (maximum adsorption capacity: 80 mgg^{-1}) (Wang et al. 2012). On the other hand, manganese oxide- and iron hydroxide-coated sand filter was used to remove bacteria (removal capacity: 99%) and zinc (removal capacity: 96%) simultaneously (Ahammed and Meera 2010).

On the other hand, Singh et al. (2011) demonstrated the use of Fe_3O_4 magnetic nanoadsorbent as a decontaminant of bacteria *Escherichia coli* as well as heavy metals As(III), Pb(II), Cd(II), Cu(II), Ni(II), Co(II) and Cr(III). The inverse spinel magnetic nanostructures were functionalized with thiol, amine and carboxyl functionalizations. The mechanism for metal ion adsorption by the functionalized magnetic nanoparticles has been presented in Fig. 1.11. Other important results relating to dual-contaminant removal can be found in Gollavelli et al. (2013), Mishra and Ramaprabhu (2010), Xiong et al. (2011), Yang et al. (2012) and Zhong et al. (2006). Targeting a multi-contaminant removal approach, Ma et al. (2012) prepared nanocomposites of Bi_2WO_6 modified by reduced graphene oxide. This nanocomposites was found to effectively remove Cr(VI), phenol and Rhodamine B dye from their respective aqueous solutions under visible and UV light irradiations.

1.11 Conclusion

In summary, we have tried to bring out the wide use of metal oxides and their composites in various categories of decontaminant removal from water and wastewater. Table 1.3 presents a comparative summary of the different metal oxides that have been used in water treatment applications. We have seen that metal oxides and their composites have found high use in treatment and removal of polluting contaminants, such as heavy metals, salts, dyes, oils, phenols, phosphates, nitrates,

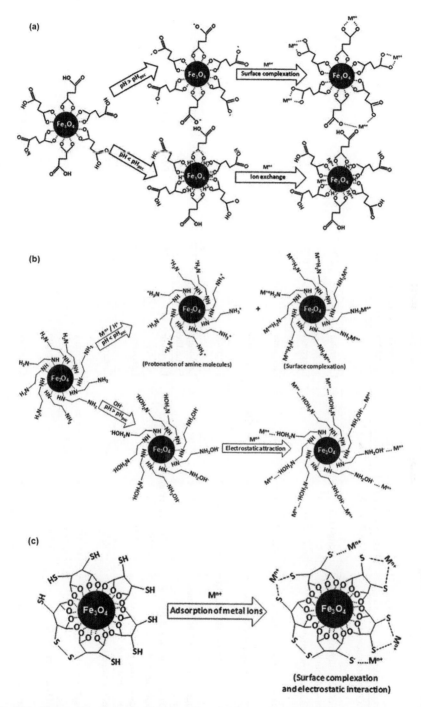

Fig. 1.11 The possible mechanism for adsorption of metal ions by (**a**) carboxyl-, (**b**) amine- and (**c**) thiol-functionalized Fe_3O_4 magnetic nanoparticles. (Reprinted from Singh et al. (2011), with permission from Elsevier)

Table 1.3 A comparative summary of the different metal oxides that have been used in water treatment applications

Metal oxide used	Pollutant removed	Removal efficiency (%)	Adsorption isotherm model followed	Adsorption capacity (mgg^{-1})	Reaction kinetics followed	Reference
MnO_2 nanostructures	Methylene blue dye	85	–	68.4	–	Chen and He (2008)
α-FeOOH hierarchical nanostructures	Congo red	–	Langmuir	239	–	Fei et al. (2011)
Fe_3O_4 hollow nanospheres	Neutral red	90	Langmuir	105	Pseudo second order	Iram et al. (2010)
Tungsten oxide nanostructures	Methylene blue	99.4	Langmuir	139	–	Jeon and Yong (2010)
Iron and manganese oxides	As(III) and As(V)	85 and 100	Langmuir	132 and 77	Pseudo second order	Ociński et al. (2016)
Flower-like MgO nanostructures	Pb(II) and Cd(II)	–	Langmuir	1980 and 1500	–	Cao et al. (2012)
Magnetite/reduced graphene oxide	As(III) and As(V)	99.9	Langmuir and Freundlich	13.10 and 5.83	Pseudo second order	Chandra et al. (2010)
EDTA-functionalized Fe_3O_4/graphene oxide	Pb(II), Hg(II) and Cu(II)	96.2, 95.1 and 96.5	Langmuir	508.4, 268.4 and 301.2	Pseudo second order	Cui et al. (2015)
Mesoporous iron/iron oxide nanocomposites	Cr(VI)	–	Langmuir	34.1	Pseudo first order	Kim et al. (2012)
MnO_2-coated Fe_3O_4/MnO_2	Cd(II)	~97	Langmuir	53.2	Pseudo second order	Kim et al. (2013)
Graphene oxide/$MnFe_2O_4$	Pb(II), As(III) and As(V)	100, 96 and 99.5	Langmuir	673, 146 and 207	Pseudo second order	Kumar et al. (2014)
Fe_2O_3 nanofibers	Cr(VI)	–	Langmuir	90.9	Pseudo second order	Nalbandian et al. (2016)
Iron oxide-coated sewage sludge	Cu(II), Cd(II), Ni(II) and Pb(II)	41.8, 25.4, 14.2 and 59.7	Langmuir	17.3, 14.7, 7.8 and 42.4	Pseudo second order	Phuengprasop et al. (2011)

1 Metal Oxides as Decontaminants of Water and Wastewater

Adsorbent	Contaminant	Removal %	Isotherm	Capacity	Kinetics	Reference
Mesoporous Fe_2O_3@C	As(III)	98	Langmuir	~29.4	Pseudo second order	Wu et al. (2012)
Amine-functionalized mesoporous Fe_3O_4	Cu(II), Cd(II) and Pb(II)	>98 for all	Langmuir	523.6, 446.4 and 369	Pseudo second order	Xin et al. (2012)
Nano-MgO	Fluoride	90	Langmuir	14	Pseudo second order	Devi et al. (2014)
Nano-Al_2O_3	Fluoride	>85	Langmuir	14	Pseudo second order	Kumar et al. (2011)
MnO-coated Al_2O_3	Fluoride	~91	Langmuir	2.85	Pseudo second order	Maliyekkal et al. (2006)
Chitosan-based mesoporous Ti-Al binary oxide beads	Fluoride	~94	Langmuir	2.22	Pseudo second order	Thakre et al. (2010)
Nano-AlOOH	Fluoride	96.7	Langmuir	3.26	Pseudo second order	Wang et al. (2009a)
Nanostructured Fe(III)-Cu(II) binary oxide	Phosphate	–	Langmuir	35.2	Pseudo second order	Li et al. (2014)
HFeO	Phosphate	70	Langmuir and Freundlich	111.1	Pseudo second order	Acelas et al. (2015)
HZrO	Phosphate	83	Langmuir and Freundlich	91.74	Pseudo second order	Acelas et al. (2015)
HCuO	Phosphate	44	Langmuir and Freundlich	74.07	Pseudo second order	Acelas et al. (2015)
Nanostructured Fe-Al-Mn trimetal oxide	Phosphate	>90	Freundlich	48.3	Pseudo second order	Lü et al. (2013)
Amorphous nano-ZrO_2	Phosphate	>99	Langmuir	99	Pseudo second order	Su et al. (2013)
Fe-Mn binary oxide	Phosphate	–	Freundlich	36	Pseudo second order	Zhang et al. (2009a)

Note: Unavailable or unclear values have been presented as '–'

pathogens and fluorides, from water and wastewater with high efficiency in most of the cases. Within metal oxides, iron oxides have found the most utilization owing to their magnetic nature that invariably leads to complete separation of the pollutant-loaded adsorbent/catalyst from the aqueous medium with the help of an external magnetic field. A huge progress has taken place in this field over the years; therefore, the most important objective of the future is to see facile scaling up and widescale commercialization and utilization of these metal oxide-based decontaminants in real life water and wastewater treatment. In this respect, the most important materials are likely to be based on iron oxides and titanium dioxide, with support from graphene and carbon-based nanomaterials (including conjugated polymers).

References

Acelas NY, Martin BD, López D, Jefferson B (2015) Selective removal of phosphate from wastewater using hydrated metal oxides dispersed within anionic exchange media. Chemosphere 119:1353–1360. https://doi.org/10.1016/j.chemosphere.2014.02.024

Ahammed MM, Davra K (2011) Performance evaluation of biosand filter modified with iron oxide-coated sand for household treatment of drinking water. Desalination 276:287–293. https://doi.org/10.1016/j.desal.2011.03.065

Ahammed MM, Meera V (2010) Metal oxide/hydroxide-coated dual-media filter for simultaneous removal of bacteria and heavy metals from natural waters. J Hazard Mater 181:788–793. https://doi.org/10.1016/j.jhazmat.2010.05.082

Ali I (2012) New generation adsorbents for water treatment. Chem Rev 112:5073–5091. https://doi.org/10.1021/cr300133d

Bradley I, Straub A, Maraccini P, Markazi S, Nguyen TH (2011) Iron oxide amended biosand filters for virus removal. Water Res 45:4501–4510. https://doi.org/10.1016/j.watres.2011.05.045

Brown J, Sobsey MD (2009) Ceramic media amended with metal oxide for the capture of viruses in drinking water. Environ Technol 30:379–391. https://doi.org/10.1080/09593330902753461

Cao C-Y, Qu J, Wei F, Liu H, Song W-G (2012) Superb adsorption capacity and mechanism of flowerlike magnesium oxide nanostructures for lead and cadmium ions. ACS Appl Mater Interfaces 4:4283–4287. https://doi.org/10.1021/am300972z

Chan SHS, Wu TY, Juan JC, Teh CY (2011) Recent developments of metal oxide semiconductors as photocatalysts in advanced oxidation processes (AOPs) for treatment of dye waste-water. J Chem Technol Biotechnol 86:1130–1158. https://doi.org/10.1002/jctb.2636

Chandra V, Park J, Chun Y, Lee JW, Hwang I-C, Kim KS (2010) Water-dispersible magnetite-reduced graphene oxide composites for arsenic removal. ACS Nano 4:3979–3986. https://doi.org/10.1021/nn1008897

Chen H, He J (2008) Facile synthesis of monodisperse manganese oxide nanostructures and their application in water treatment. J Phys Chem C 112:17540–17545. https://doi.org/10.1021/jp806160g

Cui L, Wang Y, Gao L, Hu L, Yan L, Wei Q, Du B (2015) EDTA functionalized magnetic graphene oxide for removal of Pb(II), Hg(II) and Cu(II) in water treatment: adsorption mechanism and separation property. Chem Eng J 281:1–10. https://doi.org/10.1016/j.cej.2015.06.043

Delaney P, McManamon C, Hanrahan JP, Copley MP, Holmes JD, Morris MA (2011) Development of chemically engineered porous metal oxides for phosphate removal. J Hazard Mater 185:382–391. https://doi.org/10.1016/j.jhazmat.2010.08.128

Devi RR, Umlong IM, Raul PK, Das B, Banerjee S, Singh L (2014) Defluoridation of water using nano-magnesium oxide. J Exp Nanosci 9:512–524. https://doi.org/10.1080/17458080.2012.675522

Dutta K, De S (2017a) Aromatic conjugated polymers for removal of heavy metal ions from wastewater: a short review. Environ Sci Water Res Technol 3:793–805. https://doi.org/10.1039/c7ew00154a

Dutta K, De S (2017b) Smart responsive materials for water purification: an overview. J Mater Chem A 5:22095–22112. https://doi.org/10.1039/c7ta07054c

Dutta K, Rana D (2019) Polythiophenes: an emerging class of promising water purifying materials. Eur Polym J 116:370–385. https://doi.org/10.1016/j.eurpolymj.2019.04.033

Elma M, Wang DK, Yacou C, Motuzas J, da Costa JCD (2015) High performance interlayer-free mesoporous cobalt oxide silica membranes for desalination applications. Desalination 365:308–315. https://doi.org/10.1016/j.desal.2015.02.034

Fei J, Cui Y, Yan X, Qi W, Yang Y, Wang K, He Q, Li J (2008) Controlled preparation of MnO_2 hierarchical hollow nanostructures and their application in water treatment. Adv Mater 20:452–456. https://doi.org/10.1002/adma.200701231

Fei J, Cui Y, Zhao J, Gao L, Yang Y, Li J (2011) Large-scale preparation of 3D self-assembled iron hydroxide and oxide hierarchical nanostructures and their applications for water treatment. J Mater Chem 21:11742–11746. https://doi.org/10.1039/c1jm11950h

Feng YJ, Li XY (2003) Electro-catalytic oxidation of phenol on several metal-oxide electrodes in aqueous solution. Water Res 37:2399–2407. https://doi.org/10.1016/S0043-1354(03)00026-5

Gadde RR, Laitinen HA (1974) Studies of heavy metal adsorption by hydrous iron and manganese oxides. Anal Chem 46:2022–2026. https://doi.org/10.1021/ac60349a004

Giannakis S, Liu S, Carratalà A, Rtimi S, Amiri MT, Bensimon M, Pulgarin C (2017) Iron oxide-mediated semiconductor photocatalysis vs. heterogeneous photo-Fenton treatment of viruses in wastewater. Impact of the oxide particle size. J Hazard Mater 339:223–231. https://doi.org/10.1016/j.jhazmat.2017.06.037

Gollavelli G, Chang C-C, Ling Y-C (2013) Facile synthesis of smart magnetic graphene for safe drinking water: heavy metal removal and disinfection control. ACS Sustain Chem Eng 1:462–472. https://doi.org/10.1021/sc300112z

Herrmann J-M, Guillard C, Pichat P (1993) Heterogeneous photocatalysis: an emerging technology for water treatment. Catal Today 17:7–20. https://doi.org/10.1016/0920-5861(93)80003-J

Hristovski K, Baumgardner A, Westerhoff P (2007) Selecting metal oxide nanomaterials for arsenic removal in fixed bed columns: from nanopowders to aggregated nanoparticle media. J Hazard Mater 147:265–274. https://doi.org/10.1016/j.jhazmat.2007.01.017

Hu J-S, Zhong L-S, Song W-G, Wan L-J (2008) Synthesis of hierarchically structured metal oxides and their application in heavy metal ion removal. Adv Mater 20:2977–2982. https://doi.org/10.1002/adma.200800623

Hua M, Zhang S, Pan B, Zhang W, Lv L, Zhang Q (2012) Heavy metal removal from water/wastewater by nanosized metal oxides: a review. J Hazard Mater 211-212:317–331. https://doi.org/10.1016/j.jhazmat.2011.10.016

Huang YH, Peddi PK, Tang C, Zeng H, Teng X (2013) Hybrid zero-valent iron process for removing heavy metals and nitrate from flue-gas-desulfurization wastewater. Sep Purif Technol 118:690–698. https://doi.org/10.1016/j.seppur.2013.07.009

Iram M, Guo C, Guan Y, Ishfaq A, Liu H (2010) Adsorption and magnetic removal of neutral red dye from aqueous solution using Fe_3O_4 hollow nanospheres. J Hazard Mater 181:1039–1050. https://doi.org/10.1016/j.jhazmat.2010.05.119

Jeon S, Yong K (2010) Morphology-controlled synthesis of highly adsorptive tungsten oxide nanostructures and their application to water treatment. J Mater Chem 20:10146–10151. https://doi.org/10.1039/c0jm01644f

Jiang G, Lin Z, Chen C, Zhu L, Chang Q, Wang N, Wei W, Tang H (2011) TiO_2 nanoparticles assembled on graphene oxide nanosheets with high photocatalytic activity for removal of pollutants. Carbon 49:2693–2701. https://doi.org/10.1016/j.carbon.2011.02.059

Karnib M, Kabbani A, Holail H, Olama Z (2014) Heavy metals removal using activated carbon, silica and silica activated carbon composite. Energy Proc 50:113–120. https://doi.org/10.1016/j.egypro.2014.06.014

Kasprzyk-Hordern B (2004) Chemistry of alumina, reactions in aqueous solution and its application in water treatment. Adv Colloid Interf Sci 110:19–48. https://doi.org/10.1016/j.cis.2004.02.002

Khairy M, Zakaria W (2014) Effect of metal-doping of TiO_2 nanoparticles on their photocatalytic activities toward removal of organic dyes. Egypt J Pet 23:419–426. https://doi.org/10.1016/j.ejpe.2014.09.010

Khan Z, Chetia TR, Vardhaman AK, Barpuzary D, Sastri CV, Qureshi M (2012) Visible light assisted photocatalytic hydrogen generation and organic dye degradation by CdS-metal oxide hybrids in presence of graphene oxide. RSC Adv 2:12122–12128. https://doi.org/10.1039/c2ra21596a

Kim J-H, Kim J-H, Bokare V, Kim E-J, Chang Y-Y, Chang Y-S (2012) Enhanced removal of chromate from aqueous solution by sequential adsorption-reduction on mesoporous iron-iron oxide nanocomposites. J Nanopart Res 14:1010. https://doi.org/10.1007/s11051-012-1010-6

Kim E-J, Lee C-S, Chang Y-Y, Chang Y-S (2013) Hierarchically structured manganese oxide-coated magnetic nanocomposites for the efficient removal of heavy metal ions from aqueous systems. ACS Appl Mater Interfaces 5:9628–9634. https://doi.org/10.1021/am402615m

Kuan W-H, Lo S-L, Wang MK, Lin C-F (1998) Removal of Se(IV) and Se(VI) from water by aluminum-oxide-coated sand. Water Res 32:915–923. https://doi.org/10.1016/S0043-1354(97)00228-5

Kumar E, Bhatnagar A, Kumar U, Sillanpää M (2011) Defluoridation from aqueous solutions by nano-alumina: characterization and sorption studies. J Hazard Mater 186:1042–1049. https://doi.org/10.1016/j.jhazmat.2010.11.102

Kumar S, Nair RR, Pillai PB, Gupta SN, Iyengar MAR, Sood AK (2014) Graphene oxide-$MnFe_2O_4$ magnetic nanohybrids for efficient removal of lead and arsenic from water. ACS Appl Mater Interfaces 6:17426–17436. https://doi.org/10.1021/am504826q

Landrigan PJ, Fuller R, Acosta NJR, Adeyi O, Arnold R, Basu N, Baldé AB, Bertollini R, Bose-O'Reilly S, Boufford JI, Breysse PN, Chiles T, Mahidol C, Coll-Seck AM, Cropper ML, Fobil J, Fuster V, Greenstone M, Haines A, Hanrahan D, Hunter D, Khare M, Krupnick A, Lanphear B, Lohani B, Martin K, Mathiasen KV, McTeer MA, Murray CJL, Ndahimananjara JD, Perera F, Potočnik J, Preker AS, Ramesh J, Rockström J, Salinas C, Samson LD, Sandilya K, Sly PD, Smith KR, Steiner A, Stewart RB, Suk WA, van Schayck OCP, Yadama GN, Yumkella K, Zhong M (2018) The *Lancet* commission on pollution and health. Lancet 391:462–512. https://doi.org/10.1016/S0140-6736(17)32345-0

Lee S-Y, Park S-J (2013) TiO_2 photocatalyst for water treatment applications. J Ind Eng Chem 19:1761–1769. https://doi.org/10.1016/j.jiec.2013.07.012

Lee Y-C, Yang J-W (2012) Self-assembled flower-like TiO_2 on exfoliated graphite oxide for heavy metal removal. J Ind Eng Chem 18:1178–1185. https://doi.org/10.1016/j.jiec.2012.01.005

Lee KM, Lai CW, Ngai KS, Juan JC (2016) Recent developments of zinc oxide based photocatalyst in water treatment technology: a review. Water Res 88:428–448. https://doi.org/10.1016/j.watres.2015.09.045

Li G, Gao S, Zhang G, Zhang X (2014) Enhanced adsorption of phosphate from aqueous solution by nanostructured iron(III)-copper(II) binary oxides. Chem Eng J 235:124–131. https://doi.org/10.1016/j.cej.2013.09.021

Liga MV, Bryant EL, Colvin VL, Li Q (2011) Virus inactivation by silver doped titanium dioxide nanoparticles for drinking water treatment. Water Res 45:535–544. https://doi.org/10.1016/j.watres.2010.09.012

Lin CXC, Ding LP, Smart S, da Costa JCD (2012) Cobalt oxide silica membranes for desalination. J Colloid Interface Sci 368:70–76. https://doi.org/10.1016/j.jcis.2011.10.041

Liou R-M, Chen S-H (2009) CuO impregnated activated carbon for catalytic wet peroxide oxidation of phenol. J Hazard Mater 172:498–506. https://doi.org/10.1016/j.jhazmat.2009.07.012

Liu J, Lu M, Yang J, Cheng J, Cai W (2015) Capacitive desalination of ZnO/activated carbon asymmetric capacitor and mechanism analysis. Electrochim Acta 151:312–318. https://doi.org/10.1016/j.electacta.2014.11.023

Lu F, Astruc D (2018) Nanomaterials for removal of toxic elements from water. Coord Chem Rev 356:147–164. https://doi.org/10.1016/j.ccr.2017.11.003

Lü J, Liu H, Liu R, Zhao X, Sun L, Qu J (2013) Adsorptive removal of phosphate by a nanostructured Fe-Al-Mn trimetal oxide adsorbent. Powder Technol 233:146–154. https://doi.org/10.1016/j.powtec.2012.08.024

Lu D, Zhang T, Gutierrez L, Ma J, Croue J-P (2016) Influence of surface properties of filtration-layer metal oxide on ceramic membrane fouling during ultrafiltration of oil/water emulsion. Environ Sci Technol 50:4668–4674. https://doi.org/10.1021/acs.est.5b04151

Ma H, Shen J, Shi M, Lu X, Li Z, Long Y, Li N, Ye M (2012) Significant enhanced performance for Rhodamine B, phenol and Cr(VI) removal by Bi_2WO_6 nanocomposites via reduced graphene oxide modification. Appl Catal B Environ 121-122:198–205. https://doi.org/10.1016/j.apcatb.2012.03.023

Ma Q, Cheng H, Fane AG, Wang R, Zhang H (2016) Recent development of advanced materials with special wettability for selective oil/water separation. Small 12:2186–2202. https://doi.org/10.1002/smll.201503685

Mak MSH, Rao P, Lo IMC (2011) Zero-valent iron and iron oxide-coated sand as a combination for removal of co-present chromate and arsenate from groundwater with humic acid. Environ Pollut 159:377–382. https://doi.org/10.1016/j.envpol.2010.11.006

Maliyekkal SM, Sharma AK, Philip L (2006) Manganese-oxide-coated alumina: a promising sorbent for defluoridation of water. Water Res 40:3497–3506. https://doi.org/10.1016/j.watres.2006.08.007

Mano T, Nishimoto S, Kameshima Y, Miyake M (2015) Water treatment efficacy of various metal oxide semiconductors for photocatalytic ozonation under UV and visible light irradiation. Chem Eng J 264:221–229. https://doi.org/10.1016/j.cej.2014.11.088

Mishra AK, Ramaprabhu S (2010) Magnetite decorated multiwalled carbon nanotube based supercapacitor for arsenic removal and desalination of seawater. J Phys Chem C 114:2583–2590. https://doi.org/10.1021/jp911631w

Mook WT, Chakrabarti MH, Aroua MK, Khan GMA, Ali BS, Islam MS, Hassan MAA (2012) Removal of total ammonia nitrogen (TAN), nitrate and total organic carbon (TOC) from aquaculture wastewater using electrochemical technology: a review. Desalination 285:1–13. https://doi.org/10.1016/j.desal.2011.09.029

Myint MTZ, Al-Harthi SH, Dutta J (2014) Brackish water desalination by capacitive deionization using zinc oxide micro/nanostructures grafted on activated carbon cloth electrodes. Desalination 344:236–242. https://doi.org/10.1016/j.desal.2014.03.037

Nalbandian MJ, Zhang M, Sanchez J, Choa Y-H, Nam J, Cwiertny DM, Myung NV (2016) Synthesis and optimization of Fe_2O_3 nanofibers for chromate adsorption from contaminated water sources. Chemosphere 144:975–981. https://doi.org/10.1016/j.chemosphere.2015.08.056

Naushad M, Sharma G, Kumar A, Sharma S, Ghfar AA, Bhatnagar A, Stadler FJ, Khan MR (2018) Efficient removal of toxic phosphate anions from aqueous environment using pectin based quaternary amino anion exchanger. Int J Biol Macromol 106:1–10. https://doi.org/10.1016/j.ijbiomac.2017.07.169

Ociński D, Jacukowicz-Sobala I, Mazur P, Raczyk J, Kociołek-Balawejder E (2016) Water treatment residuals containing iron and manganese oxides for arsenic removal from water – characterization of physicochemical properties and adsorption studies. Chem Eng J 294:210–221. https://doi.org/10.1016/j.cej.2016.02.111

Palanisamy KL, Devabharathi V, Sundaram NM (2013) The utility of magnetic iron oxide nanoparticles stabilized by carrier oils in removal of heavy metals from waste water. Int J Res Appl Nat Social Sci 1:15–22

Pan B, Wu J, Pan B, Lv L, Zhang W, Xiao L, Wang X, Tao X, Zheng S (2009) Development of polymer-based nanosized hydrated ferric oxides (HFOs) for enhanced phosphate removal from waste effluents. Water Res 43:4421–4429. https://doi.org/10.1016/j.watres.2009.06.055

Pan B, Qiu H, Pan B, Nie G, Xiao L, Lv L, Zhang W, Zhang Q, Zheng S (2010) Highly efficient removal of heavy metals by polymer-supported nanosized hydrated Fe(III) oxides: behavior and XPS study. Water Res 44:815–824. https://doi.org/10.1016/j.watres.2009.10.027

Pan JR, Huang C, Hsieh W-P, Wu B-J (2012) Reductive catalysis of novel TiO_2/Fe^0 composite under UV irradiation for nitrate removal from aqueous solution. Sep Purif Technol 84:52–55. https://doi.org/10.1016/j.seppur.2011.06.024

Pecson BM, Decrey L, Kohn T (2012) Photoinactivation of virus on iron-oxide coated sand: enhancing inactivation in sunlit waters. Water Res 46:1763–1770. https://doi.org/10.1016/j.watres.2011.12.059

Peter KT, Johns AJ, Myung NV, Cwiertny DM (2017) Functionalized polymer-iron oxide hybrid nanofibers: electrospun filtration devices for metal oxyanion removal. Water Res 117:207–217. https://doi.org/10.1016/j.watres.2017.04.007

Phuengprasop T, Sittiwong J, Unob F (2011) Removal of heavy metal ions by iron oxide coated sewage sludge. J Hazard Mater 186:502–507. https://doi.org/10.1016/j.jhazmat.2010.11.065

Qu X, Alvarez PJJ, Li Q (2013) Applications of nanotechnology in water and wastewater treatment. Water Res 47:3931–3946. https://doi.org/10.1016/j.watres.2012.09.058

Saha N, Rahman MS, Ahmed MB, Zhou JL, Ngo HH, Guo W (2017) Industrial metal pollution in water and probabilistic assessment of human health risk. J Environ Manag 185:70–78. https://doi.org/10.1016/j.jenvman.2016.10.023

Saxena G, Chandra R, Bharagava RN (2016) Environmental pollution, toxicity profile and treatment approaches for tannery wastewater and its chemical pollutants. In: de Voogt P (ed) Reviews of environmental contamination and toxicology volume 240, Reviews of Environmental Contamination and Toxicology (Continuation of Residue Reviews), vol 240. Springer, Cham. https://doi.org/10.1007/398_2015_5009

Sciacca S, Conti GO (2009) Mutagens and carcinogens in drinking water. Mediterr J Nutr Metab 2:157–162. https://doi.org/10.1007/s12349-009-0052-5

Sheet I, Kabbani A, Holail H (2014) Removal of heavy metals using nanostructured graphite oxide, silica nanoparticles and silica/graphite oxide composite. Energy Proc 50:130–138. https://doi.org/10.1016/j.egypro.2014.06.016

Shukla PR, Wang S, Sun H, Ang HM, Tadé M (2010) Activated carbon supported cobalt catalysts for advanced oxidation of organic contaminants in aqueous solution. Appl Catal B Environ 100:529–534. https://doi.org/10.1016/j.apcatb.2010.09.006

Singh S, Barick KC, Bahadur D (2011) Surface engineered magnetic nanoparticles for removal of toxic metal ions and bacterial pathogens. J Hazard Mater 192:1539–1547. https://doi.org/10.1016/j.jhazmat.2011.06.074

Sreeprasad TS, Maliyekkal SM, Lisha KP, Pradeep T (2011) Reduced graphene oxide-metal/metal oxide composites: facile synthesis and application in water purification. J Hazard Mater 186:921–931. https://doi.org/10.1016/j.jhazmat.2010.11.100

Su Y, Cui H, Li Q, Gao S, Shang JK (2013) Strong adsorption of phosphate by amorphous zirconium oxide nanoparticles. Water Res 47:5018–5026. https://doi.org/10.1016/j.watres.2013.05.044

Thakre D, Jagtap S, Sakhare N, Labhsetwar N, Meshram S, Rayalu S (2010) Chitosan based mesoporous Ti-Al binary metal oxide supported beads for defluoridation of water. Chem Eng J 158:315–324. https://doi.org/10.1016/j.cej.2010.01.008

Trivedi P, Axe L (2000) Modeling Cd and Zn sorption to hydrous metal oxides. Environ Sci Technol 34:2215–2223. https://doi.org/10.1021/es991110c

Upadhyay RK, Soin N, Roy SS (2014) Role of graphene/metal oxide composites as photocatalysts, adsorbents and disinfectants in water treatment: a review. RSC Adv 4:3823–3851. https://doi.org/10.1039/c3ra45013a

Vadahanambi S, Lee S-H, Kim W-J, Oh I-K (2013) Arsenic removal from contaminated water using three-dimensional graphene-carbon nanotube-iron oxide nanostructures. Environ Sci Technol 47:10510–10517. https://doi.org/10.1021/es401389g

Velazquez-Jimenez LH, Vences-Alvarez E, Flores-Arciniega JL, Flores-Zuñiga H, Rangel-Mendez JR (2015) Water defluoridation with special emphasis on adsorbents-containing metal oxides and/or hydroxides: a review. Sep Purif Technol 150:292–307. https://doi.org/10.1016/j.seppur.2015.07.006

Wang S-G, Ma Y, Shi Y-J, Gong W-X (2009a) Defluoridation performance and mechanism of nano-scale aluminum oxide hydroxide in aqueous solution. J Chem Technol Biotechnol 84:1043–1050. https://doi.org/10.1002/jctb.2131

Wang Y-Q, Gu B, Xu WL (2009b) Electro-catalytic degradation of phenol on several metal-oxide anodes. J Hazard Mater 162:1159–1164. https://doi.org/10.1016/j.jhazmat.2008.05.164

Wang N, Zhu L, Wang D, Wang M, Lin Z, Tang H (2010) Sono-assisted preparation of highly-efficient peroxidase-like Fe_3O_4 magnetic nanoparticles for catalytic removal of organic pollutants with H_2O_2. Ultrason Sonochem 17:526–533. https://doi.org/10.1016/j.ultsonch.2009.11.001

Wang B, Wu H, Yu L, Xu R, Lim T-T, Lou XW (2012) Template-free formation of uniform urchin-like α-FeOOH hollow spheres with superior capability for water treatment. Adv Mater 24:1111–1116. https://doi.org/10.1002/adma.201104599

Wang B, Li J, Wang G, Liang W, Zhang Y, Shi L, Guo Z, Liu W (2013) Methodology for robust superhydrophobic fabrics and sponges from in situ growth of transition metal/metal oxide nanocrystals with thiol modification and their applications in oil/water separation. ACS Appl Mater Interfaces 5:1827–1839. https://doi.org/10.1021/am303176a

Wu Z, Li W, Webley PA, Zhao D (2012) General and controllable synthesis of novel mesoporous magnetic iron oxide@carbon encapsulates for efficient arsenic removal. Adv Mater 24:485–491. https://doi.org/10.1002/adma.201103789

Xin X, Wei Q, Yang J, Yan L, Feng R, Chen G, Du B, Li H (2012) Highly efficient removal of heavy metal ions by amine-functionalized mesoporous Fe_3O_4 nanoparticles. Chem Eng J 184:132–140. https://doi.org/10.1016/j.cej.2012.01.016

Xiong Z, Ma J, Ng WJ, Waite TD, Zhao XS (2011) Silver-modified mesoporous TiO_2 photocatalyst for water purification. Water Res 45:2095–2103. https://doi.org/10.1016/j.watres.2010.12.019

Xu P, Zeng GM, Huang DL, Feng CL, Hu S, Zhao MH, Lai C, Wei Z, Huang C, Xie GX, Liu ZF (2012) Use of iron oxide nanomaterials in wastewater treatment: a review. Sci Total Environ 424:1–10. https://doi.org/10.1016/j.scitotenv.2012.02.023

Yamani JS, Miller SM, Spaulding ML, Zimmerman JB (2012) Enhanced arsenic removal using mixed metal oxide impregnated chitosan beads. Water Res 46:4427–4434. https://doi.org/10.1016/j.watres.2012.06.004

Yang X, Zou R, Huo F, Cai D, Xiao D (2009) Preparation and characterization of Ti/SnO_2-Sb_2O_3-Nb_2O_5/PbO_2 thin film as electrode material for the degradation of phenol. J Hazard Mater 164:367–373. https://doi.org/10.1016/j.jhazmat.2008.08.010

Yang X, Chen C, Li J, Zhao G, Ren X, Wang X (2012) Graphene oxide-iron oxide and reduced graphene oxide-iron oxide hybrid materials for the removal of organic and inorganic pollutants. RSC Adv 2:8821–8826. https://doi.org/10.1039/c2ra20885g

Yang Y, Zhang C, Hu Z (2013) Impact of metallic and metal oxide nanoparticles on wastewater treatment and anaerobic digestion. Environ Sci Processes Impacts 15:39–48. https://doi.org/10.1039/c2em30655g

Yavuz Y, Koparal AS (2006) Electrochemical oxidation of phenol in a parallel plate reactor using ruthenium mixed metal oxide electrode. J Hazard Mater 136:296–302. https://doi.org/10.1016/j.jhazmat.2005.12.018

Yin H, Zhao S, Wan J, Tang H, Chang L, He L, Zhao H, Gao Y, Tang Z (2013) Three-dimensional graphene/metal oxide nanoparticle hybrids for high-performance capacitive deionization of saline water. Adv Mater 25:6270–6276. https://doi.org/10.1002/adma.201302223

Zhai Y, Zhai J, Zhou M, Dong S (2009) Ordered magnetic core-manganese oxide shell nanostructures and their application in water treatment. J Mater Chem 19:7030–7035. https://doi.org/10.1039/b912767d

Zhang G, Liu H, Liu R, Qu J (2009a) Removal of phosphate from water by a Fe-Mn binary oxide adsorbent. J Colloid Interface Sci 335:168–174. https://doi.org/10.1016/j.jcis.2009.03.019

Zhang S, Zhao X, Niu H, Shi Y, Cai Y, Jiang G (2009b) Superparamagnetic Fe_3O_4 nanoparticles as catalysts for the catalytic oxidation of phenolic and aniline compounds. J Hazard Mater 167:560–566. https://doi.org/10.1016/j.jhazmat.2009.01.024

Zhang S, Niu H, Cai Y, Zhao X, Shi Y (2010) Arsenite and arsenate adsorption on coprecipitated bimetal oxide magnetic nanomaterials: $MnFe_2O_4$ and $CoFe_2O_4$. Chem Eng J 158:599–607. https://doi.org/10.1016/j.cej.2010.02.013

Zhang J, Xiong Z, Zhao XS (2011) Graphene-metal-oxide composites for the degradation of dyes under visible light irradiation. J Mater Chem 21:3634–3640. https://doi.org/10.1039/c0jm03827j

Zhong L-S, Hu J-S, Liang H-P, Cao A-M, Song W-G, Wan L-J (2006) Self assembled 3D flowerlike iron oxide nanostructures and their application in water treatment. Adv Mater 18:2426–2431. https://doi.org/10.1002/adma.200600504

Zhong L-S, Hu J-S, Cao A-M, Liu Q, Song W-G, Wan L-J (2007) 3D flowerlike ceria micro/nanocomposite structure and its application for water treatment and CO removal. Chem Mater 19:1648–1655. https://doi.org/10.1021/cm062471b

Zhu Q, Tao F, Pan Q (2010) Fast and selective removal of oils from water surface via highly hydrophobic core-shell Fe_2O_3@C nanoparticles under magnetic field. ACS Appl Mater Interfaces 2:3141–3146. https://doi.org/10.1021/am1006194

Chapter 2
Photo-Assisted Antimicrobial Activity of Transition Metal Oxides

Rajini P. Antony, L. K. Preethi, and Tom Mathews

Abstract Photocatalysis is a widely accepted technology which finds enormous applications in the field of fuel production, water remediation, environmental cleaning, self-cleaning coatings, CO_2 sequestration and microbial disinfection. Photo assisted heterogeneous catalysis is the major route employed for all the above applications where transition metal based semiconducting materials are used for the light assisted production of reactive oxygen species. This in turn finds an important application in microbial destruction at different indoor and outdoor level. The increase in environmental pollution urges the need of a simple and cost-effective route of microbial inactivation. Thus, semiconductor-based photo assisted bacterial inactivation is an appropriate strategy which can be upscaled for commercial application. The present chapter tries to provide physical insights on the photocatalytic based microbial inactivation process. The chapter mainly discusses the basic principle of photocatalysis and mechanism behind bacterial disinfection. Thereafter, the chapter focuses on the discussion of selected transition metals oxides such as TiO_2, ZnO and CuO for bacterial disinfection studies and different types of modification methods employed for the improvement of their antibacterial efficiency.

Keywords Bacterial disinfection · Semiconductors · TiO_2 · ZnO · CuO · Photocatalysis

R. P. Antony
Water and Steam Chemistry Division, BARC – F, Kalpakkam, Tamil Nadu, India

L. K. Preethi
Centre for Nanoscience and Nanotechnology, Sathyabama Institute of Science and Technology, Chennai, Tamil Nadu, India

T. Mathews (✉)
Thin Films and Coatings Section, Surface and Nanoscience Division, Materials Science Group, Indira Gandhi Centre for Atomic Research, Kalpakkam, Tamil Nadu, India
e-mail: tom@igcar.gov.in

© The Author(s), under exclusive license to Springer Nature Switzerland AG 2021
S. Rajendran et al. (eds.), *Metal, Metal-Oxides and Metal-Organic Frameworks for Environmental Remediation*, Environmental Chemistry for a Sustainable World 64, https://doi.org/10.1007/978-3-030-68976-6_2

2.1 Introduction

Pathogens and microbes can spread diseases in human beings through air, water and contaminated surfaces. Disinfection processes of these contaminated surfaces are gaining high importance in order to maintain hygienic work practice and day to day life activities. The case is more severe in health care centers where the contaminated surfaces are frequently handled and have a proximity and interaction with patients. There are different disinfections strategies adopted to prevent the contamination such as sterilization, adsorption, antibiotics, biocides etc. Pathogens such as S. aureus, Escherichia Coli and Pseudomonas aeruginosa can contaminate the dry surfaces and can survive for several weeks to months (Kramer et al. 2006). To prevent pathogenic infections from contaminated surfaces, generally antimicrobial agents are coated on the medical devices or incorporated into coating which also includes metals such as silver and copper. There is a need for alternative approaches due to the following reasons

(a) Development of microbial resistance to metal ion coating with time
(b) Lack of enough efficacy in complete disinfection as the metals can kill the bacteria but cannot destruct the endotoxins
(c) High cost of the coatings and related products

Among various alternative approaches, semiconductor based photocatalytic sterilization is found to be an effective approach for disinfection. This is because, the photogenerated holes, the hydroxyl radicals and the superoxide radicals in the semiconductor can exhaustively destruct biological molecules such as proteins, lipids, enzymes and nucleic acids through a series of oxidative chain reactions. The oxidative power of these photogenerated radicals possess oxidation energy of 120 Kcal mol^{-1}, which is sufficient to break the chemical bonds in the above mentioned organic compounds (Dunlop et al. 2010). Moreover, the nonspecific nature of reactive oxygen species attack the cell structures of the outer layer of the pathogens makes it unlikely for the emergence of resistance towards photocatalytic disinfection (Goulhen-Chollet et al. 2009). Further, the utilization of earth abundant semiconducting materials and sunlight, the cost effectiveness, and the viability for commercialization makes the photocatalytic material-based disinfection, an effective strategy.

The most common semiconductors investigated for photocatalytic disinfection are TiO_2, ZnO, CdS, CuO etc. TiO_2 is one of the most widely explored semiconducting materials in the field of photocatalytic disinfection. Appropriate band alignment, low cost and abundant availability makes it a suitable candidate for the same. Commercially available form of TiO_2, known as Degussa P25 is the most explored one in this aspect. ZnO is another wide band gap semiconducting material having wurtzite crystal structure and have similar properties like TiO_2. Hence, it also finds wide applications in the field of photocatalytic water splitting, hydrogen generation, sensors, antibacterial activity and photo assisted organic destruction.

Thus, the present chapter addresses different aspects of semiconductor photocatalysis for microbial destruction. In this regard, the photocatalytic mechanism behind microbial destruction, their various experimental models, the influencing factors determining the efficiency of disinfection and an overview of most widely explored semiconductors are discussed.

2.2 Semiconductor Photocatalysis

Semiconductor based photocatalysis have attracted considerable attention from different researchers as the process can be employed in different fields such as bioremediation, disinfection, energy conversion and energy storage.

The basic mechanism behind photocatalysis is discussed as follows. When a Semiconductor is irradiated with light having energy higher than the band gap of the material, electrons (e^-) and holes (h^+) are generated in the conduction band and valence band, respectively. These photogenerated charge carriers can migrate to the surface of the semiconductor and participate in the redox reactions. The photogenerated holes possess high oxidizing power and thus can lead to the formation of reactive oxygen species (ROS) such as $OH^.$, $O_2^{.-}$, $HO_2^.$, where $OH^.$ is primarily known to be the most responsible species for the disinfection of bacteria and related pathogens (Laxma Reddy et al. 2017). The schematic representation of a Semiconductor based photocatalytic process explaining the mechanism of charge carrier generation is shown in Fig. 2.1. The reactions involved in Semiconductor based photocatalysis are represented in Eqs. 2.1, 2.2, 2.3, 2.4, 2.5, 2.6, 2.7, and 2.8 given below (Gong et al. 2019):

$$TiO_2 + h\nu \rightarrow e^- (CB) + h^+ (VB) \tag{2.1}$$

$$h^+ (VB) + H_2O (ad) \rightarrow TiO_2 + OH.(ad) + H \tag{2.2}$$

$$h^+ (VB) + OH (ad) \rightarrow TiO_2 + OH. \tag{2.3}$$

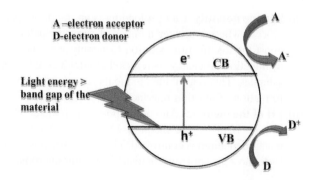

Fig. 2.1 Basic mechanism of semiconductor based photocatalytic disinfection process. A is the electron acceptor and D is the electron donor in the electrolyte. *CB* Conduction band, *VB* valence band

$$e^- (CB) + O_2 \rightarrow TiO_2 + O_2^{-\cdot} \qquad (2.4)$$

$$O_2^{-\cdot} + H^+ \rightarrow HO_2\cdot \qquad (2.5)$$

$$O_2^{-\cdot} + HO_2\cdot \rightarrow OH\cdot + O_2 + H_2O_2 \qquad (2.6)$$

$$2HO_2\cdot \rightarrow O_2 + H_2O_2 \qquad (2.7)$$

$$e^- + H_2O_2 \rightarrow TiO_2 + OH- + OH\cdot \qquad (2.8)$$

2.3 Photocatalytic Antimicrobial Activity

Sun light induced photocatalytic microbial disinfection process is mainly carried out by semiconductors which can absorb visible light. In other words, it should have suitable band gap for visible light absorption ($\lambda > 400$ nm). As discussed previously, the photogenerated holes are strong oxidizing agent and can lead to the formation of reactive oxygen species such as OH·, $O_2^{-\cdot}$ etc., where the former is known to responsible for the bacterial destruction. Based on the structure and complex nature of bacteria, the cellular destruction differs and the cellular component leakage due to cell wall destruction ultimately kills the microbes.

2.3.1 Factors Influencing the Photocatalytic Antimicrobial Activity

The experimental conditions of the photocatalytic disinfection process should meet the demands of commercialization. Hence it is necessary to optimize the experimental parameters such as nature of photocatalysts, type of light used, ambience and pH conditions of the medium to obtain cost effective and efficient microbial destruction. The kinetics of the photocatalytic antimicrobial activity depends on the following parameters:

(a) **Sample quantity**: catalyst loading is an important parameter in the disinfection process where a greater number of semiconductor-microbe-light interface can boost the disinfection activity. Generally with increase in catalyst loading, the disinfection efficiency increases, but after a certain concentration, the efficiency saturates. This is attributed to the increase in turbidity of the solution and lack of formation of reaction interfaces.

(b) **pH of the medium**: Another important parameter which effect the antimicrobial activity is pH of the medium where, the effective charge on the catalyst and reaction medium is controlled. The isoelectric point is the parameter which helps to define a plausible mechanism for the antimicrobial activity. It is defined as the pH value at which the surface charge is zero at the catalyst surface and is also known as point of zero charge.

(c) **Temperature**: In photocatalytic disinfection process, temperature is not considered as an important parameter. However, in some cases, the disinfection process is found to increase with increase in operating temperature. In sunlight based disinfection process, thermal energy is contributed from the light source itself and is presumed to play a minor role in controlling the disinfection process.

(d) **Light wavelength, intensity and irradiation time**: These two important parameters play a crucial role in the photocatalytic disinfection process. Generally, UV assisted bactericidal process is an effective approach where the light energy is quiet higher compared to the band gap of the material. Also, higher irradiation intensity leads to higher reactive oxygen species generation leading to higher efficiency. With increase in irradiation time, the efficiency of the disinfection process increases, but it is desirable to obtain faster photocatalytic disinfection at a commercial level. An important parameter to be considered in this case is the irradiation surface, where the surface area of the semiconducting material is preferred to be high on which the reactive oxygen species has to be generated.

More than these parameters, the structural complexity of the microorganism is an important parameter. This decides the kinetics of the disinfection process, where structural factors such as type of strains, cell wall structure, and type of organ targeted for mineralization etc. play crucial role.

2.4 Different Models Proposed for Photocatalytic Disinfection

The antimicrobial activity can be a cooperative action by all kinds of reactive oxygen species. It was inferred initially that the photo-assisted destruction of coenzyme A in the bacteria is the origin of death of bacteria due to the inhibition of respiratory activity (Matsunaga et al. 1985). Later on, it was observed that the cell wall destruction led to the leaking of potassium ions and other cellular components leading to the destruction of cells (Saito et al. 1992). It was then confirmed by various morphological and structural characterization that, the complete mineralization of the cell walls and components are occurring by photo catalytically generated reactive oxygen species (Bagchi et al. 1993; Jacoby et al. 1998; Kiwi and Nadtochenko 2005; Maness et al. 1999; Sökmen et al. 2001). A peroxidation mechanism of the bacterial cell wall on TiO_2 surface was studied by Kiwi and coworkers and was observed that, the first step of the complete disintegration of the cell wall is the competition between the oxidation process by photo catalytically generated holes and oxidation of lipid polysaccharide layer (Kiwi and Nadtochenko 2005). Nanoparticle interaction with the bacterial cell wall also played a major role in cell damage where electrostatic interaction of the semiconducting surface and cell wall, reactivity of reactive oxygen species on the particle surface and metal ions play a major role in cell lysis by photocatalysis (Matai et al. 2014; Wang et al. 2011). These processes can inhibit DNA replication and gene alteration by metal ions and reactive oxygen species respectively. The kinetics of photocatalytic process is generally explained by models such as Langmuir-Hinselwood model and direct indirect model whose details are available elsewhere (Ganguly et al. 2018).

Based on all these report, it is evident that a kinetic model can help to predict the kinetics of the photocatalytic disinfection process and based on these, several models namely (1) Chick's Model (2) Chick-Watson model (3) Delayed Chick-Watson model and (4) Hom model are proposed by various groups, where the trends in the survival/inactivation rates are portrayed with time. These models are defined based on the following assumptions.

(a) Uniform distribution of the nanoparticles and microorganism
(b) Constant temperature, pH and concentration of the nanoparticles
(c) Uniform mixing of solution to rule out the diffusion limited condition.

2.4.1 Chick's Model

In this model, the bacteria were considered as a molecule and the bacterial inactivation was a chemical reaction. The generalized rate equation for Chick's law is given as Eq. 2.9. The model says that the rate of inactivation is proportional to the number of surviving microbes at a given concentration and the trend is shown in Fig. 2.2 (Chick 1908).

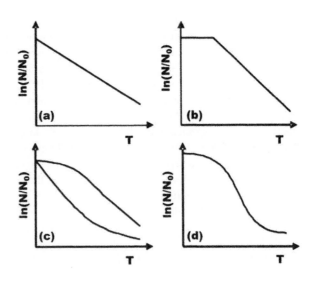

Fig. 2.2 Survival curves observed for different kinetic models in the photocatalytic disinfection process. (Reprinted with permission from Ganguly et al. (2018))

$$\frac{dN}{dt} = -K^*N \qquad (2.9)$$

where, dN/dt is the rate of inactivation, N is the number of survivors at contact time 't' and K is the proportionality constant. The reactions are generally considered as irreversible reactions and follow a pseudo first order reaction. Since concentration of reactive oxygen species also plays a major role in inactivation process, the amount of reactive oxygen species generated during photocatalytic process also plays a role in the inactivation kinetics. Hence the photocatalytic inactivation process does not always follow Chick's law. Therefore, an extended model is proposed which is known as Chick-Watson model.

2.4.2 Chick-Watson Model

Watson in the same year incorporated the time parameter in the rate equation (Eq. 2.10) and the role of the contact time was considered more than the concentration term (Watson 1908)

$$K = C^n T \qquad (2.10)$$

where K is a constant for a particular microbe and its experimental condition, n is a constant whose value is less than 1 which says the importance of contact time than concentration. T is the time required to achieve an activation point. The rate equation is also a pseudo first order reaction and is expressed as Eq. 2.11 and the trend of bacterial disinfection is shown in Fig. 2.3a.

$$\frac{dN}{dt} = -KNC^n \qquad (2.11)$$

Chick-Watson model has its own short comings as it considers the microbes to have a single strain and the inactivation happens on a single hit. Hence the disinfection process deviates from the assumption.

2.4.3 Delayed Chick-Watson Model

An alternate model was proposed in which a time lag (T_{lag}) parameter is introduced to overcome the shoulder phase in the disinfection process (Fig. 2.2b). Using this model, the inactivation of E coli was explained (Cho et al. 2004) where the hydroxyl radical was found to be the dominant inactivating agent. N_0 is the number of survivors at t = 0

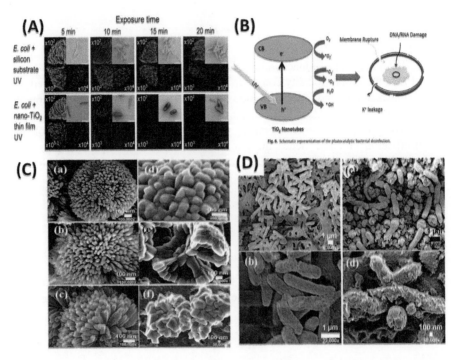

Fig. 2.3 (**A**) Images of surviving luminescent bacteria that were able to yield colonies after exposure to UV-light on Si-monocrystal substrates (upper block of panels) or on nano TiO$_2$ thin films (lower block of panels). Different sectors of agar plates show the colonies of luminescent bacteria in 15 lL of 10^2, 10^3 and 10^4-fold dilutions of UV-exposed bacterial suspension. Inset on each image is an SEM image of the bacteria after respective exposure conditions. SEM images show considerable morphological changes of bacterial cells already after 10-min UV-irradiation on nano-TiO$_2$ thin film. (Joost et al. 2015), (**B**) Schematic representation of the photocatalytic bacterial disinfection. (Podporska-Carroll et al. 2015), (**C & D**) FESEM images of the obtained TiO$_2$ nanorod spheres: (**a**) TiO$_2$–400, (**b**) TiO$_2$–500, (**c**) TiO$_2$–600, (**d**) TiO$_2$–700, (**e**) TiO$_2$–800, and (**f**) TiO$_2$–900 (numbers indicate the calcination temperature [°C]) & FESEM images at different magnifications of (**a, b**) control rodlike E. coli corpses, and (**c, d**) a mixture of rod-like E. coli corpses and TiO$_2$–500 after mixing for 2 h. (Bai et al. 2013)

$$\frac{N}{N_0} = \begin{cases} 1, & T \leq T_{lag} \\ e^{-KC^nT}, & T > T_{lag} \end{cases} \qquad (2.12)$$

2.4.4 Homs Model

Like Chick-Watson model, the Homs model, which was proposed in 1972 (Hom) also considers concentration-time factor and has an additional power factor 'm' as shown in Eq. 2.13 The model accounts for the lag or tailing off process in photocatalytic disinfection process, but cannot be modeled together. When the value of 'm' = 1, the model reduces to Chick-Watson model, when m > 1, it shows a shoulder (lag) and when m < 1, it shows a tailing off behaviour as shown in Fig. 2.2c and d. Further, modified Homs model was derived where initial shoulder, logarithmic decay and tailing off are considered together by a relation given in Eq. 2.14.

$$\ln \frac{N}{N_0} = -KC^n T^m \tag{2.13}$$

$$\ln \frac{N}{N_0} = -K_1[1 - \exp(-K_2 T)]^{K_3} \tag{2.14}$$

2.5 Transition Metal Oxide Based Photocatalyst for Antimicrobial Studies

An ideal photocatalyst should have suitable band gap to absorb visible light, proper band straddling for the desired oxidation reaction, resistance to photo and chemical corrosion and cost efficiency as well. Different types of semiconductor based photocatalysts are studied for antimicrobial disinfection. Among them, TiO_2 and ZnO are the most commonly explored semiconducting materials. The upcoming section presents a brief summary of engineering of these photocatalysts for antimicrobial disinfection application.

2.5.1 Antimicrobial Behavior of Titania

TiO_2 is a wide gap semiconductor with a band gap of 3–3.2 eV having UV light activity. Titania exist mainly in three different phases viz.; anatase, rutile and brookite. The single phase, biphasic and triphasic photocatalytic efficiency of TiO_2 are reported (Preethi et al. 2016, 2017a, b; Preethi and Mathews 2019). Different strategies such as doping (Antony et al. 2012a), composite fabrication (Panthi et al. 2013), heterostructuring (Preethi et al. 2016), metal loading (Liu et al. 2019) are adopted to improve the disinfection mechanism of TiO_2 as it suffers from wide band gap and high electron-hole charge recombination. Hence different types of modifications are adopted to improve its activity.

2.5.1.1 Nanostructured TiO_2

Nanostructuring of TiO_2 based materials can improve the photocatalytic disinfection efficiency by increasing the effective surface and reaction interface. Various researchers adopted different methodologies to improve the surface area of the TiO_2 based structures and studied its role in improving antibacterial efficacy. Joost et al. (2015) prepared high surface area TiO_2 nanoparticles to construct a nanostructured thin film and demonstrated the antimicrobial activity using viable bacterial cells and bacterial plasma membrane fatty acids. They observed the destruction of saturated and unsaturated fatty acids in the bacterial plasma membrane in 10 minutes by fast formation of peroxide through photocatalytic activity. A comparison between silicone substrate and nano TiO_2 thin film in E. coli destruction proved the mechanism and is shown in Fig. 2.3A.

Another way of nanostructuring is construction of TiO_2 nanotube in thin film form by electrochemical anodization and rapid break down anodization (Antony et al. 2011, 2012b). Podporska-Carroll et al. (2015) studied the antibacterial activity of TiO_2 nanotubes prepared by rapid break down anodization process and compared with that of commercial Degussa-P25 TiO_2 powder. The fabricated nanotube powders showed excellent antimicrobial activity towards E. coli and S. aureus under light irradiation for 24 hours. The improved nanostructuring and specific surface area played an important role in increasing the rate of formation of reactive oxygen species in the case of nanotubular structure (Fig. 2.3B). Also, the improved surface area can improve the bacterial adsorption which enhances the bacterial killing. Bai and coworkers fabricated TiO_2 nanorod spheres in large scale by a non-hydrothermal route (Bai et al. 2013) (Fig. 2.3C). Intrinsic antibacterial properties for these structures were observed under dark conditions due to the spikes present on the TiO_2 nanostructure which pierce the bacterial cell wall and kill the bacteria. Under solar light irradiation, bacterial destruction ability was increasing with increase in calcination temperature and 82.12% of bacterial destruction was obtained for sample calcined at 900 °C (Fig. 2.3D). The increase in photocatalytic activity with increase in calcination temperature is attributed to enhanced crystallization leading to redshift thereby narrowing the band gap of TiO_2.

2.5.1.2 Doped TiO_2

The pristine TiO_2 is a wide band gap material and can absorb light in the UV region. In order to make it visible light active, an important strategy adopted in the case of TiO_2 is doping by cations and anions. The present section discusses the bacterial disinfection of doped TiO_2. The photocatalytic property of TiO_2 is utilized for E coli disinfection in wastewater by Majeda and coworkers (Khraisheh et al. 2015). For this activity, Cu doped TiO_2 was prepared by sol gel and wet impregnation method. It was observed that, Cu-TiO_2 prepared by incorporating 10% of $CuCl_2$ as precursor, was showing 100% E. coli destruction. The synergistic effect of oxidative attack of TiO_2 and leaching out of Cu was found to be responsible for the remarkable performance and the photocatalytic disinfection process is shown in Fig. 2.4. Raut *et al* studied the sunlight induced antibacterial effect of $TiO_{2-x-3y}N_{2y}$ thin films deposited on Si(100), quartz and glass substrates by a single step ultrasonic spray pyrolysis route (Raut et al. 2012). The precursor used for N doping was hexamine.

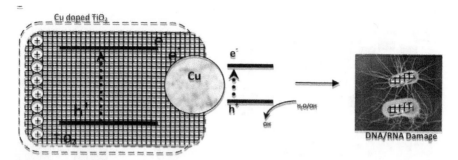

Fig. 2.4 Antibacterial mechanism for Cu doped TiO$_2$. (Khraisheh et al. 2015)

The holes generated in the localized N2p states and oxygen vacancies were attributed to the formation of enhanced redox reactions and sunlight absorption respectively in the aforementioned doped TiO$_2$ thin films. Arenas and coworkers (2013) in 2013 studied the bactericidal effect of doped anodized TiO$_2$, where they constructed a fluoride-anodized TiO$_2$ barrier layers (Ti–6Al–4 V). A significant reduction in the bacterial adhesion was obtained in the case of fluoride doped anodized layer compared to fluoride free anodized layers.

2.5.1.3 Metal Loaded TiO$_2$

Another important approach to improve the bactericidal efficacy is loading of metals which can synergistically improve the activity. Silver is known to be a broad-spectrum antimicrobial agent which can kill antibiotic resistant strains. Ag loaded TiO$_2$ nanorods are fabricated in Ti foil by acid etching, hydrothermal and plasma treatment by Li et al. to investigate the antimicrobial and cytocompatibility of titanium implants. (Li et al. 2014). The negative zeta potential generated on the titania surface has prevented the bacterial adhesion of the implant material through electrostatic repulsion process. The Schottky contact created by the Ag metal on titania surface played a role in microbial destruction. Thus, two defense line mechanisms were proposed where both bacterial adhesion and bacterial growth are prevented under dark conditions. The schematic of the bactericidal mechanism proposed by the group is given as Fig. 2.5a. Recently P/Ag/Ag$_2$O/Ag$_3$PO$_4$/TiO$_2$ photocatalyst was developed by hydrothermal route by Liu et al. to study the E. coli destruction under LED light illumination (Fig. 2.5b). High light intensity (750 Wm^{-2}), ambient temperature, neutral or slightly alkaline and shorter wavelengths are the optimum parameters found in enhancing the photocatalytic efficiency of the composite catalyst. The major reactive species responsible for the bacterial destruction are h$^+$ and ˙O$_2^-$ and no effect due to Ag$^+$ leaking (Liu et al. 2019).

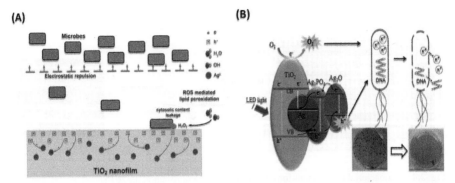

Fig. 2.5 (**a**) Illustration for two-defense-line antimicrobial property with both electrostatic repulsion and lipid peroxidation actions of Ag plasma-modified TiO_2 nanofilm in darkness, mediated by the synergism of negative zeta potential and embedded metallic Ag into TiO_2 nanofilm. (Li et al. 2014) (**b**) Schematic diagram of proposed mechanism of bacterial disinfection by P/Ag/Ag$_2$O/Ag$_3$PO$_4$/TiO$_2$. (Liu et al. 2019)

2.5.1.4 Metal Oxide-Based Composites of TiO_2

In order to enhance the antimicrobial activity of photocatalyst further, their composites with other semiconductor materials are developed. Panthi et al. (2013) constructed Mn_2O_3-TiO_2 nano fiber composite by electrospinning route, where TiO_2 was formed in the rutile phase. Prevention of colonization of gram positive and gram negative bacteria was achieved by these composites, where the activity was proposed to be mediated by Mn_2O_3 phase. He et al. (2017) fabricated CuO-TiO_2 coating on Ti films by magnetron sputtering and annealing process. These films were demonstrated to have improved antibacterial activity, biocompatibility and corrosion resistance (Fig. 2.6A). He et al. (2019) developed reusable magnetic N-TiO_2/Fe_3O_4@SiO_2 photocatalyst for simultaneous degradation of E.Coli and bis phenol A in sewage water under visible light irradiation. In their study, presence of bisphenol A didn't affect the destruction of E. coli but the degradation of former decreased by 10% in presence of bacteria (Fig. 2.6B) due to the same reactive oxygen species ($^{\cdot}O_2^-$ and H_2O_2) to degrade the organic matter and microbes.

2.5.1.5 Carbon Nanostructure Based Composites of TiO_2

Carbon based nanostructures are an important class of materials which play a significant role in photocatalysis by boosting the conductivity and electron transfer in their composites with other semiconductor materials. The carbon-based nanomaterials include carbon nanotubes, fullerene, carbon dots, graphene, graphene oxide etc. Functionalizing TiO_2 with carbon materials is a promising approach for cost effective and efficient antimicrobial activity. Koli et al. (2016) functionalized TiO_2 with multi walled carbon nanotubes using a solution-based method. Multi walled carbon nanotubes can improve/tune the optical band gap of TiO_2 for visible

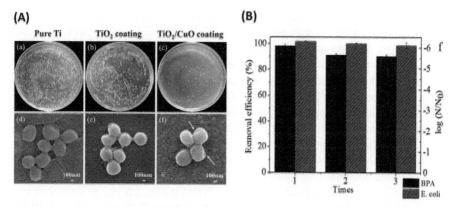

Fig. 2.6 (**A**) The typical photographs of bacterial colonies on (**a**) Ti, (**b**) TiO$_2$ coating and (**c**) TiO$_2$/CuO coating after re-cultivating on agar plates; SEM morphology images of S. aureus cultured on (**d**) Ti, (**e**) TiO$_2$ coating and (**f**) TiO$_2$/CuO coating after cultivating for 24 h, (He et al. 2017) (**B**) Cycling performance of AgFeNTFS for simultaneous photocatalytic disinfection of E. coli and degradation of BPA in real sewage. (He et al. 2019)

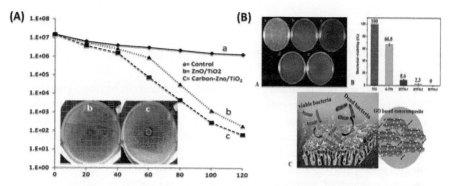

Fig. 2.7 (**A**) Antibacterial efficiency of different photocatalysts for gram negative E. coli bacteria under mild UV radiation. Insets are the respective zones of inhibition. (Pant et al. 2013), (**B**) Antibacterial properties of the membranes evaluated by the (**A**) plate colony-forming count experiments, where (**a**) is TFC, (**b**) G-TFN, (**c**) MTNF-1, (**d**) MTNF-2, and (**e**) MTN-3, (**B**) the bacterial viability (%), as a measure of the antimicrobial activity of the membranes, and (**C**) the schematic illustration of antibacterial activities of MTFN membranes. (Abadikhah et al. 2019)

light activity, thereby making it suitable for commercial application. The photocatalytic antimicrobial activity of multi walled carbon nanotubes -TiO$_2$ towards S. aureus and E. coli was found to be the highest for TC-0.5 composite (labeling as per the carbon nanotubes loading on TiO$_2$) where smaller particle size and higher surface area helped the inactivation process. Another study by Pant et al. (2013) reported the antibacterial effect of carbon nanofibers modified TiO$_2$-ZnO composite towards E. coli under light irradiation containing 10% UV light. The increased diameter of the inhibition zone in the case of carbon-TiO$_2$-ZnO is proposed for effective bactericidal killing under mild UV irradiation and Fig. 2.7A shows a comparison of bacterial inactivation process with time. Multifunctional thin film polyamide membranes embedded with rGO@TiO$_2$@Ag nanocomposites was developed by Abadikhah and coworkers (2019). A microwave irradiation assisted synthesis process was adopted for the fabrication of nanocomposites which improved the hydrophilicity, water permeation, salt rejection and antifouling properties of the membrane (Fig. 2.7B). A 100% reduction in the live E.coli population was observed when in contact for 3 hrs with the nanocomposite modified polyamide membrane. Recently Shimizu and coworkers (2019) studied the antimicrobial properties of carbon nanotubes -TiO$_2$ nanocomposite for the disinfection of E. coli. The photogenerated hydroxyl radical induced reactive oxygen species caused a physical rupture of the bacterial cell wall leading to the destruction of the microbes on the semiconductor surface. A fructose modifies titania was employed as a catalyst for photocatalytic disinfection of waste water by Rokicka-Konieczna et al. (2019). Fructose was incorporated as an inexpensive and nontoxic source of carbon on hydrothermally modified carbon. A two-step bacterial destruction by OH radials is proposed and the disinfection process began from cell wall towards intra cellular components.

2.5.2 Antimicrobial Behaviour of Zinc Oxide

ZnO is another class of wide band gap semiconductor materials having a band gap of 3.3 eV and exist in three different polymorphs say, wurtzite, rock salt and zinc blend structure. The thermodynamic stable structure is wurtzite which is widely used in applications such as gas sensors (Ahn et al. 2009), biosensors (Wei et al. 2006), piezoelectric materials (Gullapalli et al. 2010), photocatalysts for organic decomposition and photocatalytic antibacterial coating materials (Hatamie et al. 2015). The progressive research of antibacterial properties of ZnO started in 1950s.

Similar to the case of TiO$_2$, upon light irradiation, the photogenerated charge carriers move to the conduction band and valence band and participate in redox reactions. The conduction band of ZnO is situated at -0.5 V vs NHE which is more negative compared to the O_2/O^{-2} level (-0.33 V vs NHE. Whereas, the valence band positioning is 2.7 V vs NHE which is more positive compared to OH$\cdot$/H$_2$O redox position (2.53 V vs NHE). Thus, the photogenerated electrons and holes can create $O_2/O^{\cdot-}_2$ and OH$\cdot$ radical for disinfection reaction. The disinfection process by ZnO is also attributed to the reactive oxygen species assisted oxidation process. The radicals ($O_2/O^{\cdot-}_2$ and OH$\cdot$) generated by photo irradiation is oxidative enough to react and damage the cellular constituents such as DNA, lipids, proteins, carbohydrates. Three main mechanisms are proposed for the antibacterial activity of ZnO in aqueous environment.

1. The oxidative stress created by the reactive oxygen species can damage the unsaturated fatty acids of the bacterial cell membrane leading to the permanent damage of the bacterial activity (Kääriäinen et al. 2013).
2. Zn^{2+} ions released from the ZnO can transport through the cell membrane and reduce the intracellular ATP levels again leading to the improper functioning of the bacteria (Tong et al. 2015).
3. The cell membrane can damage due to the adhesion of the aggregates of ZnO nanoparticles (Li et al. 2008).

2.5.2.1 Nanostructured ZnO

Researchers have improved the photocatalytic antibacterial efficiency of ZnO by fabricating nanostructures with different morphologies such as nanoflowers, nanorods, thin films etc. For instance, 3-D flower like ZnO structures fabricated by surfactant free co precipitation method were shown to have efficient photocatalytic antibacterial property towards Enterococcus faecalis (E. faecalis) and Micrococcus luteus (M. luteus) (Quek et al. 2018). The bacteria were eradicated completely in 130 min of visible light irradiation which was detected by bacterial morphological change, K^+ ions leakage and protein leakage. The enhanced activity was attributed to the flower like morphology which increase the number of surface hydroxyls groups and photogenerated charge carriers on the ZnO surfaces leading to the formation of reactive oxygen species mainly H_2O_2 (Fig. 2.8a and b). Zhang and coworkers (2008) employed ZnO nanofluids for the disinfection of E.coli. The improved activity of the optimized ZnO sample stored for 120 days was attributed to the increased surface area and the electrostatic binding of the particle and the bacteria. A morphology dependent antibacterial activity was deduced for ZnO nanoparticle fabricated by solvothermal method (Talebian et al. 2013). The trend in the antibacterial activity of different nanostructures towards inactivation of gram-negative Escherichia coli and gram-positive Staphylococcus aureus under UV light was found as nanoflower > nanorods > spherical nanoparticle (Fig. 2.8c and d). The interstitial defect present in the nanoflowers reduced the electron-hole recombination and hence improved the light induced antibacterial activity. Raja et al. (2018) fabricated ZnO nanoparticles through an ecofriendly green synthesis route, and studied their antibacterial activity towards Salmonella paratyphi, Escherichia coli and Staphylococcus aureus. The bacterial strains were investigated and their result showed a higher antibacterial efficiency towards S.aurus and E.coli compared to S. paratyphi.

Another approach was the fabrication of ZnO nanoleaves reported by Gupta and Srivastava (2018), where disperser assisted sonochemical approach was employed as the synthesis route. The schematic diagrams of the synthesis route of the ZnO nanoleaves are shown in Fig. 2.8e. The synthesis route was claimed to be scalable in nature and the product can be employed for the disinfection of S. aureus. A green synthesis approach was employed by Bhuyan et al. (2015) for the fabrication of ZnO nanoparticles using neem leaf extract. The as prepared ZnO nanoparticles showed efficient antibacterial activity towards gram positive and gram negative bacteria (Staphylococcus aureus, Streptococcus pyogenes and Escherichia coli) by inhibiting

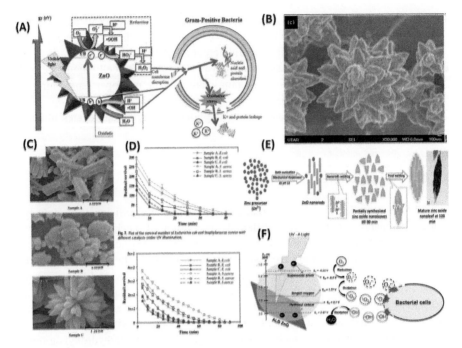

Fig. 2.8 (**a**) Proposed schematic of antimicrobial mechanism of flower-like ZnO (**b**) FESEM images of as-synthesized flower-like ZnO. (Quek et al. 2018) (**c**) SEM photographs of ZnO powders, (**d**) Plot of the survival number of Escherichia coli and Staphylococcus aureus with different catalysts under UV illumination (top) and at dark condition (bottom). (Talebian et al. 2013), (**e**) Schematic illustration showing possible growth mechanism of ZnO-NLs. Initially, after addition of base (NaOH) to the zinc precursor mixture, ZnO nanoparticles (nanorods) formation occurs which may get welded to each other in a bunch to minimize their surface energy and partially synthesized ZnO-NLs formation occurs till 90 min of the reaction which may again get welded to each other by the remaining Zn^{+2} ions in the solution to form mature leaf-shape ZnO nanoparticle at 120 min. The growth mechanism of ZnO-NLs shows, disperser assisted sonochemical method allows time and temperature dependent controlled shape and size evolution of the nanoparticles. (Gupta and Srivastava 2018). (**f**) Schematic of the antibacterial mechanisms of the ALD ZnO via photo-produced reactive oxygen species. (Park et al. 2017)

the growth of the bacteria. A similar approach was done by Anbuvannan and coworkers (2015) for the fabrication of ZnO nanoparticles using A. carnosus leaf extract. The resultant product showed improved disinfection ability for human pathogens such as S. paratyphi, V. cholerae, S. aureus, and E. coli. Park et al. (2017) fabricated ZnO thin films by atomic layer deposition route with the aim of increasing reaction surface area for microbial disinfection. For this, S. aureus was chosen as the model bacteria and a disinfection process was envisaged by the crushed morphology caused by destruction of cell wall. The reactive oxygen species generation and the cell wall destruction was the mechanism proposed and the effect of Zn^{2+} release was ruled out in this case. The schematic of the antibacterial activity of the ZnO thin films proposed by Park and coworkers is given in Fig. 2.8f.

Hydrophilic ZnO nanorods arrays were fabricated by Akhavan et al. (2009) by hydrothermal route and the effect of photoinduced antibacterial activity was observed towards E.coli. The hydrophilic behavior of the ZnO nanorods arrays is attributed to the formation of surface hydroxyl groups and is responsible for the increased antibacterial property.

Amir et al. (Hatamie et al. 2015) have developed a ZnO modified textile by a hydrothermal route and investigated its photocatalytic and antibacterial activity as a replacement for commercial silver loaded textiles. An antimicrobial packaging film was developed by Li et al. (2009) where ZnO nanoparticles were coated on polyvinyl chloride film showed better antibacterial activity towards E.Coli and S. aeurus with more resistance to S aeurus. A size dependent antibacterial activity for ZnO nanoparticles towards S. aureus under ambient light conditions was demonstrated by Raghupathi and coworkers (2011), where the antibacterial efficiency was decreasing with increase in the particle size. Reactive oxygen species generation and aggregation of nanoparticles in the cytoplasm was claimed as the main reason for the antibacterial effect of the solvo-thermally fabricated ZnO nanoparticles. Similarly shape controlled ZnO nanoparticles were fabricated by Sarika and coworkers by a soft chemical synthesis route (Singh et al. 2013). Cauliflower like ZnO nanostructures showed an effective light induced antibacterial activity towards S. aureus in contaminated water under UV irradiation. All these investigations showed that nanostructuring play an important role in the antibacterial effect of ZnO nanoparticles by improving the surface area which helps in increasing the contact area. The improved area also plays an important role in the production of reactive oxygen species responsible for photocatalytic antibacterial efficiency. Also, controlling the size and shape of the nanoparticles influences the mode of cell wall destruction.

2.5.2.2 Doped ZnO

As discussed in the previous section, ZnO can absorb only UV light due to its wide band gap nature (3.3 eV) which hinders its commercial utility. Doping of ZnO by cationic and anionic species to alter the band gap is found to be an effective approach to enhance the light absorption range (Djerdj et al. 2010; Lin et al. 2005). In this direction, Manjula and coworkers (Nair et al. 2011) developed Cobalt (Co) doped ZnO for the photocatalytic organic decontamination and microbial destruction. Very good bacterial destruction efficiency was observed by the group using Co doped ZnO for all the bacterial strains chosen for investigation (Escherichia coli, Klebsiella pneumoniae, Shigella dysenteriae, Salmonella typhi, Pseudomonas aeruginosa, Bacillus subtilis and Staphylococcus aureus). Doping noble metals such as silver, gold and platinum is an effective approach due to the improvement in the photogenerated charge separation process by scavenging the photogenerated electron by noble metals. Bechambi et al. (2015) developed Ag doped ZnO nanostructures for photocatalytic applications and observed 100% destruction of E.coli in 40 min using 1% Ag doping. The improved microbial destruction efficiency

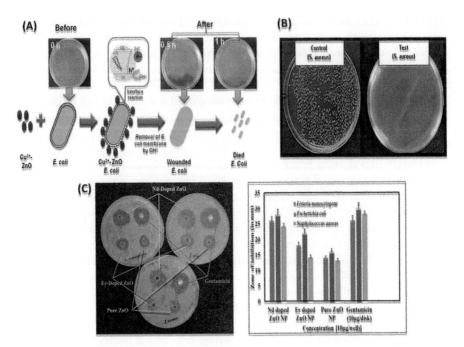

Fig. 2.9 (**A**) Pictorial illustration of the mechanism of degradation of pathogens by Cu^{2+}-modified ZnO under blue LED illumination. (Kumar et al. 2014); (**B**), Photographs of agar plates containing (**a**) Control (S. aureus), (**b**) Test sample (F doped ZnO 1:1, after exposure to visible light). (Podporska-Carroll et al. 2017); and (**C**) Bar graph showing the diameter of the zone of inhibition (in mm) produced by pure and Nd/Er-doped ZnO NPs against E. coli, S. aureus and L. monocytogene. (Raza et al. 2016)

of Ag-ZnO was attributed to the enhanced generation of reactive oxygen species such as $•O_2^-$, OH• and H_2O_2. Raju and coworkers demonstrated a similar system where Cu^{2+} doped ZnO exhibited an IFCT effect leading to increased visible light induced bactericidal activity (Kumar et al. 2014). To study this effect, the model microbe chosen was E. coli and showed an enhanced visible light bacterial killing due to IFCT and hence Cu^{2+}-ZnO induced inactivation was active only under visible light. The holes generated in the valence band under visible light irradiation and Cu^+ formation during IFCT process were responsible for bacterial inactivation (Fig. 2.9A). Antimicrobial Ce doped ZnO was developed by Karunakaran and coworkers (2010) in 2010. An efficient destruction of E. coli was achieved using 2% Ce doped ZnO. Fluorine doped ZnO photoactive catalysts were developed using a sol-gel route by Carrol et al. (Podporska-Carroll et al. 2017) and an efficient photocatalytic antimicrobial destruction was observed toward E.coli and S. aureus under visible light illumination (Fig. 2.9B). The antimicrobial effect of F doped ZnO is attributed to the synergistic effect of ZnO on microorganism by the release of Zn^{2+} ions and the formation of increased reactive oxygen species (mainly H_2O_2) due to F doping. Both the processes, lead to the cell wall destruction and decomposition of cellular materials.

The transition metals like manganese and cobalt can increase the surface defects and alter the electronic band structures leading to improvement in the visible light activity of ZnO. Rekha et al. (2010) adopted the strategy of doping different percentage of manganese in the Zn lattice of ZnO using co precipitation route. The samples showed a better antibacterial activity compared to the pristine one. Similarly, Co doped ZnO thin films fabricated by sol-gel spin coating route by Poongodi et al. (2015) displayed efficient bactericidal activity towards E. coli and S. aureus. Like other studies, the enhanced activity was attributed to the improved charge separation efficiency and reactive oxygen species generation. Surface modified metals ions on ZnO type materials can induce an interfacial charge transfer (IFCT), where photogenerated electrons from dopant levels to the surface dopant ions play a major role in enhancing microbial destruction efficiency under visible light activity.

Guo et al. fabricated Tantalum doped ZnO investigated by modified pechini type method (Guo et al. 2015). It was observed by the group that incorporation of 5% of Ta^{5+} into ZnO improved the visible light bactericidal effect towards P. aeruginosa, E. coli, and S. aureus. Hameed et al. studied the doping effect of Nd into ZnO towards the inactivation of E. coli and K. pneumonia (Hameed et al. 2016). Rare earth element doping is another approach which can improve the visible light activity and photogenerated charge separation efficiency. Bomila and coworkers developed La, a rare earth element doped ZnO nanoparticles by wet chemical route for the visible light photocatalytic applications (Bomila et al. 2018). Er/Nd doped ZnO was developed by Raza et al. using sol-gel technique (Raza et al. 2016). Improved visible light activity for doped samples was observed and exhibited improved anticancer and antimicrobial activity compared to the pristine samples (Fig. 2.9C).

2.5.2.3 Metal Loaded ZnO

Constructing metal-semiconductor hybrid structures is an effective way to tackle absorption and recombination problems in semiconductors for better photocatalytic efficiency. The noble metal loading in metal-semiconductor composite play a crucial role in extending visible light absorption capabilities of wide band gap semiconductors and increasing lifetime of photogenerated charge carriers by scavenging the photogenerated electrons thereby making photogenerated holes available for photocatalytic disinfection process. Also, the noble metals such as Ag and Au generate photogenerated electron hole pairs by a process called surface plasmon resonance (SPR).

He and coworkers (2014) fabricated Au-ZnO hybrid nanostructures by a photo-reduction method to obtain smaller nanoparticles. S. aureus and E. coli were chosen as the model microbes to study the antibacterial effect under simulated sunlight (Fig. 2.10A). The photocatalytic effect of Au nanoparticles alone showed

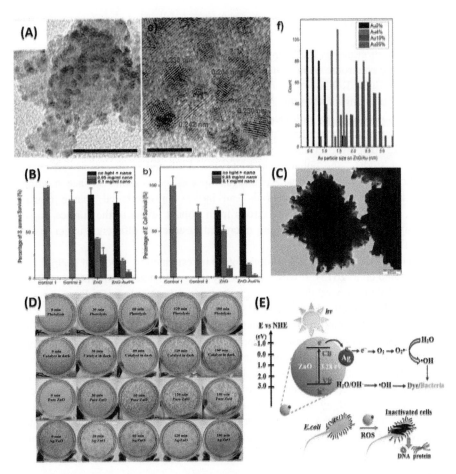

Fig. 2.10 (**A**) HRTEM image for the ZnO/Au hybrid NPs from panel d; (f) size distribution of Au particles formed on ZnO at different molar ratios for reactants. Scale bars in panels a – d are all 20 nm; scale bar in panel e is 5 nm (**B**) Ability of ZnO NPs and ZnO/Au hybrid nanostructures in killing S. aureus (**a**) and E. coli (**b**) under simulated sunlight for 10 min. Control 1 represents bacteria exposed to neither NPs nor light. Control 2 represents bacteria exposed to simulated sunlight for 10 min but without NPs. Grouped under ZnO, bacteria wasexposed to 0.1 mg/mL ZnO alone or was exposed to 10 min of solar simulated light and either 0.05 mg/mL or 0.1 mg/mL ZnO. Similarly, grouped under ZnO-Au4%, bacteria was exposed to 0.1 mg/mL ZnO/Au4% alone or was exposed to 10 min of solar simulated light and either 0.05 mg/mL or 0.1 mg/mL ZnO/Au4%. (He et al. 2014) (**C**) TEM images of 5 wt% Ag/ZnO samples (**D**) Photocatalytic antibacterial activities of photolysis, catalyst in the dark, pure ZnO and 5 wt% Ag/ZnO samples against E. coli at different irradiation times. (**E**) Schematic diagram of visible light induced photocatalytic and antibacterial mechanism of Ag/ZnO micro/nanoflowers. (Lam et al. 2018)

bactericidal activity compared to untreated control samples. Whereas in the case of Au-ZnO structures, a significant decrease in the survival of both strains with less than 10 min light exposure was evident and the effect is attributed to the increased production of reactive oxygen species as evidenced from ESR spectroscopic studies

(Fig. 2.10B). The increase in activity for Au-ZnO was about 3 times higher compared to that of ZnO nanostructures. Lu et al. (2008) employed a tyrosine assisted one pot hydrothermal synthesis of Ag-ZnO nanocomposite and demonstrated efficient light induced antibacterial efficacy towards gram positive and gram negative bacteria. Functionalization of textile fabrics using Ag-ZnO nanocomposites were carried out by Mariana and coworkers (Ibănescu et al. 2014) for effective photocatalytic and antimicrobial applications. The composites were coated onto to the fabrics by a pad-dry-cure process followed by treatment under 130 °C for 30 minutes. Antimicrobial disc diffusion tests for Ag-ZnO coated fabrics showed that the antimicrobial efficiency of the modified fabrics increased with increase in Ag loading. The results were consistent with the photocatalytic dye degradation trend for the composites. Hence, the production of reactive oxygen species due to increased surface area was attributed to improved activity of the modified fabrics.

Jin et al. (2019) developed iodine modified ZnO using reflux method to obtain higher surface oxygen vacancy and higher charge separation efficiency. Smaller grain size, cage like structure, increased surface oxygen vacancy in nanostructured I-ZnO (I-ZnO-n) led to improved charge carrier separation and generate more free radicals for bacterial disinfection. Recently Lam and coworkers (2018) investigated the bacterial inactivation ability of Ag loaded ZnO micro and nanoflowers synthesized through a surfactant free co-precipitation method followed by photodeposition route (Fig. 2.10C). The antibacterial effect of the nanoflowers was studied using E. coli under visible light irradiation. A complete bacterial destruction was obtained under 180 min visible light irradiation using 5% Ag loaded ZnO micro/nanoflowers (Fig. 2.10D). In addition, a long term antibacterial activity suppression and cytoplasmic destruction by Ag-ZnO micro/nanoflowers was confirmed by minimum inhibitory concentration and optical density studies. The photogenerated reactive oxygen species were identified to be responsible for the cytoplasmic destruction and cellular leakage of the bacteria and the mechanism proposed by the group is given in Fig. 2.10E.

2.5.2.4 Composites of ZnO

Proper band alignment of ZnO with other semiconductors favors the photocatalytic process by improving the photogenerated charge carriers and thereby ROS. Sin et al. (2018) fabricated ZnO-magnetic Fe_3O_4 composites by a surfactant free method for effective bacterial destruction. The resultant composite showed a flower like structure (Fig. 2.11a) and an effective destruction of E. coli was observed under visible light irradiation compared to pristine ZnO (Fig. 2.11b). The improved charge carrier recombination time envisaged by photoluminescence studies was attributed to enhanced effect of the composite photocatalyst. The magnetic property of the composite was an additional advantage for the efficient separation of the catalyst after use, makes it suitable for commercial. Graphene like MoS_2 sheets modified ZnO nanoflowers were developed by Awasthi and coworkers (2016) by a one pot hydrothermal route. The composites displayed effective antibacterial efficacy

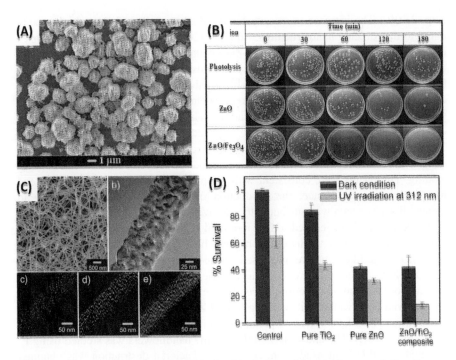

Fig. 2.11 (**a**) FESEM images of ZnO/Fe$_3$O$_4$ (**b**) Antibacterial activities toward E. coli over ZnO/Fe$_3$O$_4$ samples under visible light irradiation. (Sin et al. 2018), (**c**) FE-SEM and TEM image of the fabricated ZnO/TiO$_2$ nanofibers & EDS mapping images of the composite nanofibers with Zn element, Ti element, and Zn–Ti elements and (**d**) Graph of % survival of S. aureus after treatment with control, TiO$_2$ nanofibers, and ZnO/TiO$_2$ nanofibers in the absence and the presence of UV light irradiation at 312 nm for 30 second. The number of bacterial colonies on the untreated Petri dish surface under the dark conditions was defined as 100%. (Hwang et al. 2011)

compared to ZnO and MoS$_2$. The improvement in the bactericidal effect of these composites were attributed to the improved surface area and increased product of reactive oxygen species as observed from surface area and photoluminescence studies. Electrospun ZnO-TiO$_2$ nanofiber composite was proved to be an effective bactericidal agent towards E.coli and S. aureus under UV irradiation (Fig. 2.11c & d) (Hwang et al. 2011). A microwave assisted synthesis of CdO-NiO-ZnO, employed by Karthik et al. (2018) for optoelectronic, photocatalytic and biological application. Antibacterial activity towards gram positive and gram negative strains was investigated using the composite and bacterial cell wall destruction was observed by confocal microscopic studies.

2.5.2.5 Carbon Based Composites of ZnO

Similar to the case of TiO$_2$, it was found that a combination of carbon-based nanostructures with ZnO helped in improving the antimicrobial efficiency.

ZnO-Graphene hybrid was fabricated by Kavitha et al. (2012) by an in-situ thermal decomposition process and exhibited an excellent antibacterial effect towards E.coli. Lefatshe et al. (2017) and coworkers modified the ZnO with nanocellulose structures and demonstrated the effective antibacterial effect towards S.aureus and E.coli. A combination of Ag nanoparticle and g-C_3N_4 was utilized to modify ZnO nanoparticles by a one pot hydrothermal route for the purpose of photocatalytic disinfection towards E. coli (Adhikari et al. 2015). Synergistic effect of Ag and ZnO nanoparticles anchored on the g-C_3N_4 sheets was ascribed to the disinfection of E. coli under light and dark conditions. Carbon quantum dots, a cost effective material, are another class of carbon material which has a unique property in terms of up conversion of photoluminescence which makes the material, an efficient sunlight active photocatalyst (Li and Cao 2011). In-situ sol gel chemistry was applied for the fabrication of carbon quantum dots modified ZnO nanorods by Kuang et al. (2019), and the resultant material was able to kill ~96% bacteria under visible light irradiation even at low concentration (0.1 mgL^{-1}). The carbon quantum dots s for the carbon quantum dots -ZnO composite, was synthesized by an electrochemical route. Due to the improved charge separation efficiency of the carbon quantum dots s and the reactive oxygen species generation, antibacterial efficiency of carbon quantum dots -ZnO nanorod was three to four times higher than that of ZnO.

2.5.3 Copper (II) Oxide (CuO)

Copper (II) oxide is a p-type semiconductor with a band gap ranging from 1 to 2 eV and hence it is a visible light active photocatalyst. Due to this property it can be used as photocatalysts, photoelectrodes, antimicrobial substrate etc. CuO possess a monoclinic crystal structure, with an indirect band gap and it has a carrier diffusion length of 200 nm with an absorption depth of 500 nm. (Masudy-Panah et al. 2018). Considering photocatalytic based the bacterial disinfection property of CuO, it is cheaper than silver and can be synthesized with high surface area and interesting surface morphologies. This helps to one to tune the antimicrobial efficacy of the CuO surface and bring to a commercial scale. However due to high charge carrier recombination rates of CuO, the efficiency remains low and restricts it commercial utility.

Yousef and coworkers investigated the photocatalytic pathogenic effect of CuO/TiO_2 nanostructures fabricated by electrospinning route. The efficient charge separation in TiO_2–CuO heterojunction followed by accumulation of photogenerated electrons and holes in the conduction band and valence band respectively of CuO leads to the formation of reactive oxygen species (convert O_2 to O_2^- and OH^- to $OH^.$). These reactive oxygen species attack the outer membrane of K. pneumonia, and partially disintegrate the intact structure of the membrane. Due to the partial disintegration the surface charge of the bacterial cells

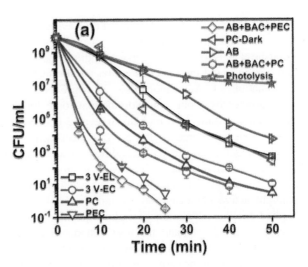

Fig. 2.12 Antibacterial activity by photocatalysis, electrolysis, electrocatalysis and photoelectrocatalysis using CuO-CSA and tetracycline (*AB* antibiotic, *BAC* bacteria, *PC* Photocatalysis, *EL* Electrolysis, *EC* Electrocatalysis, *PEC* Photoelectrocatalysis). (Reproduced with copyright from Elsevier Eswar et al. (2018b))

were negative and presence of Cu^{2+} enhanced the electrostatic force leading to complete destruction of bacterial cells. Akhavan and Kaderi developed CuO and reduced CuO (Cu) nanoparticle immobilized silica thin films for photocatalytic antibacterial applications especially E. Coli disinfection (Akhavan and Ghaderi 2010). It was observed that Cu nanoparticle with slight under layer of CuO layer showed ~63% improvement in the photocatalytic inactivation process and the improved activity is assigned the improved charge transfer process. Later Akhavan developed $CuO/Cu(OH)_2$ hierarchical nanostructures and studied the bacterial disinfection efficacy under sunlight illuminated conditions (Akhavan et al. 2011). It was observed that the chemical composition and surface density were the dominating factors which affected the bacterial disinfection under illumination. Presence of small amount of $Cu(OH)_2$ played an important role in OH^- adsorption and this boosted the photocatalytic process. Eswar and coworkers studied the photocatalytic and photoelectrocatalytic (PEC) degradation of bacteria (Eswar et al. 2018b) using photoconductive network structured CuO. As observed in the previous reports, photocatalytic process has dramatically improved the disinfection process compared to the dark conditions. When PEC was employed for the disinfection process and synergetic effect of electrocatalysis and photocatalysis was observed and as a result the time for ~100% bacterial destruction was reduced to 30 minutes. A comparison of antibacterial efficiency of CuO network structures is shown in Fig. 2.12.

A brief over view of the different semiconductor materials and it different combinations employed for photocatalytic disinfection process is shown in Table 2.1.

Table 2.1 List different semiconductor based photocatalysts employed for bacterial inactivation under different light sources

Photocatalysts	Pathogen	Light source	Reference
TiO_2	E. coli	UV	Wei et al. (1994)
TiO_2 film	G. Lamblia	UV	Lee et al. (2004)
Fe^{3+} doped TiO_2	E. coli	UV	Trapalis et al. (2003)
Sn^{4+} doped TiO_2 thin films	E. coli and S. aureus	UV	Sayılkan et al. (2009)
$TiO_{2-x-3y}N_{2y}$ film	P. Aeruginosa	Sun light and room light	Raut et al. (2012)
TiO_2-RGO	E. coli	Sunlight	Akhavan and Ghaderi (2009)
P-25 TiO_2-RGO composite electrodes	E. coli, P. Aeruginosa	Simulated light	Yin et al. (2013)
Au-TiO_2 p25-RGO composite	Fungal cells	Simulated sun light	He et al. (2013)
C-doped TiO_2	S. aureus, Shigella flexneri, and Acinetobacter baumannii	Visible light	Cheng et al. (2009)
ZnO nanofluid	E. coli	UV light	Zhang et al. (2008)
ZnO-PVC film	E. coli and S. aureus	UV light	Li et al. (2009)
ZnO nanoparticle suspension	Escherichia coli O157:H7, Listeria monocytogenes ATCL3C 7644 and Botrytis cinerea	Visible light	Kairyte et al. (2013)
ZnO/SnO_2 nanocomposite thin films	E. coli, ATCC 25922	UV light	Talebian et al. (2011)
ZnO-GO composite	Escherichia coli K-12	Visible light	Wu et al. (2015)
Sulphonated GO-ZnO-Ag	E. coli	Visible light	Gao et al. (2013)
Ag@ZnO coreshell nanostructures	Vibrio cholerae	Sunlight	Das et al. (2015)
Pd incorporated ZnO nanoparticles	E. coli	Laser beam	Khalil et al. (2011)
Fe doped ZnO	Multidrug resistant Escherichia coli	UV and solar light	Das et al. (2017)
Immobilized CuO/$CoFe_2O_4$-TiO_2 thin-film	E. coli	Simulated sunlight	Yan et al. (2011)

(continued)

Table 2.1 (continued)

Photocatalysts	Pathogen	Light source	Reference
TiO_2/ZnO/CuO	E. coli	UV light	Bai et al. (2012)
N-type Cu_2O film	E. coli	Photoelectrochemical route by visible light	Xiong et al. (2015)
Copper oxide on conducting glass	E. coli	Photoelectrochemical route by visible light	Eswar et al. (2018a)
CuO suspensions	E. coli	Mercury lamp	Paschoalino et al. (2008)
CuO/TiO_2 composite nanorods	Escherichia coli KCCM 11234 and Staphylococcus aureus KCCM 11256	Visible light	Hassan et al. (2013)
TiO_2(Eu)/CuO composite	Enterococcus species	UVA light	Michal et al. (2016)
High-surface-area CuO pretreated cotton	E. coli	Visible light	Torres et al. (2010)
Cu_xO loaded rhodium–antimony co-doped TiO_2	Salmonella typhimurium	Visible light	Dhandole et al. (2019)

2.6 Summary

In summary, the physico chemical process involved in the photocatalytic antibacterial activity of transition metal oxide based semiconductors is discussed in the present chapter. TiO_2, ZnO and CuO were chosen as the model semiconductors for discussion which are coming under the class of cost effective, abundant and stable semiconductors and can be upscaled for commercial utilization. Different strategies adopted for the enhancement of the antibacterial activity of these metal oxides are classified and discussed. The main strategies include nanostructuring, doping, composites, metal loading and modification with advanced materials. Thus morphological, electronic and optical tuning can influence the bacterial inactivation by photocatalytic means. The present chapter is indent to provide a brief idea of photo assisted antibacterial activity of transition metal oxides for graduates and entry level researchers working in this area.

References

Abadikhah H, Naderi Kalali E, Khodi S, Xu X, Agathopoulos S (2019) Multifunctional thin-film Nanofiltration membrane incorporated with reduced graphene oxide@TiO_2@Ag Nanocomposites for high desalination performance, dye retention, and antibacterial properties. ACS Appl Mater Interfaces 11:23535–23545. https://doi.org/10.1021/acsami.9b03557

Adhikari SP, Pant HR, Kim JH, Kim HJ, Park CH, Kim CS (2015) One pot synthesis and characterization of Ag-ZnO/g-C_3N_4 photocatalyst with improved photoactivity and antibacterial properties. Colloids Surf A Physicochem Eng Asp 482:477–484. https://doi.org/10.1016/j.colsurfa.2015.07.003

Ahn MW, Park KS, Heo JH, Kim DW, Choi KJ, Park JG (2009) On-chip fabrication of ZnO-nanowire gas sensor with high gas sensitivity. Sensors Actuators B Chem 138:168–173. https://doi.org/10.1016/j.snb.2009.02.008

Akhavan O, Ghaderi E (2009) Photocatalytic reduction of graphene oxide Nanosheets on TiO_2 thin film for Photoinactivation of Bacteria in solar light irradiation. J Phys Chem C 113:20214–20220. https://doi.org/10.1021/jp906325q

Akhavan O, Ghaderi E (2010) Cu and CuO nanoparticles immobilized by silica thin films as antibacterial materials and photocatalysts. Surf Coat Technol 205:219–223. https://doi.org/10.1016/j.surfcoat.2010.06.036

Akhavan O, Mehrabian M, Mirabbaszadeh K, Azimirad R (2009) Hydrothermal synthesis of ZnO nanorod arrays for photocatalytic inactivation of bacteria. J Phys D Appl Phys 42:225305. https://doi.org/10.1088/0022-3727/42/22/225305

Akhavan O, Azimirad R, Safa S, Hasani E (2011) $CuO/Cu(OH)_2$ hierarchical nanostructures as bactericidal photocatalysts. J Mater Chem 21:9634–9640. https://doi.org/10.1039/C0JM04364H

Anbuvannan M, Ramesh M, Viruthagiri G, Shanmugam N, Kannadasan N (2015) Anisochilus carnosus leaf extract mediated synthesis of zinc oxide nanoparticles for antibacterial and photocatalytic activities. Mater Sci Semicond Process 39:621–628. https://doi.org/10.1016/j.mssp.2015.06.005

Antony RP, Mathews T, Dasgupta A, Dash S, Tyagi AK, Raj B (2011) Rapid breakdown anodization technique for the synthesis of high aspect ratio and high surface area anatase TiO_2 nanotube powders. J Solid State Chem 184:624–632. https://doi.org/10.1016/j.jssc.2011.01.020

Antony RP, Mathews T, Ajikumar PK, Krishna DN, Dash S, Tyagi AK (2012a) Electrochemically synthesized visible light absorbing vertically aligned N-doped TiO_2 nanotube array films. Mater Res Bull 47:4491–4497. https://doi.org/10.1016/j.materresbull.2012.09.061

Antony RP et al (2012b) Efficient photocatalytic hydrogen generation by Pt modified TiO_2 nanotubes fabricated by rapid breakdown anodization. Int J Hydrog Energy 37:8268–8276. https://doi.org/10.1016/j.ijhydene.2012.02.089

Arenas MA et al (2013) Doped TiO_2 anodic layers of enhanced antibacterial properties. Colloids Surf B: Biointerfaces 105:106–112. https://doi.org/10.1016/j.colsurfb.2012.12.051

Awasthi GP, Adhikari SP, Ko S, Kim HJ, Park CH, Kim CS (2016) Facile synthesis of ZnO flowers modified graphene like MoS_2 sheets for enhanced visible-light-driven photocatalytic activity and antibacterial properties. J Alloys Compd 682:208–215. https://doi.org/10.1016/j.jallcom.2016.04.267

Bagchi D, Bagchi M, Hassoun EA, Stohs SJ (1993) Detection of Paraquat-Induced in vivo lipid peroxidation by gas chromatography/mass spectrometry and high-pressure liquid chromatography. J Anal Toxicol 17:411–414. https://doi.org/10.1093/jat/17.7.411

Bai H, Liu Z, Sun DD (2012) Solar-light-driven Photodegradation and antibacterial activity of hierarchical $TiO_2/ZnO/CuO$ material. ChemPlusChem 77:941–948. https://doi.org/10.1002/cplu.201200131

Bai H, Liu Z, Liu L, Sun DD (2013) Large-scale production of hierarchical TiO_2 Nanorod spheres for Photocatalytic elimination of contaminants and killing Bacteria. Chem Eur J 19:3061–3070. https://doi.org/10.1002/chem.201204013

Bechambi O, Chalbi M, Najjar W, Sayadi S (2015) Photocatalytic activity of ZnO doped with Ag on the degradation of endocrine disrupting under UV irradiation and the investigation of its antibacterial activity. Appl Surf Sci 347:414–420. https://doi.org/10.1016/j.apsusc.2015.03.049

Bhuyan T, Mishra K, Khanuja M, Prasad R, Varma A (2015) Biosynthesis of zinc oxide nanoparticles from Azadirachta indica for antibacterial and photocatalytic applications. Mater Sci Semicond Process 32:55–61. https://doi.org/10.1016/j.mssp.2014.12.053

Bomila R, Srinivasan S, Gunasekaran S, Manikandan A (2018) Enhanced Photocatalytic degradation of methylene blue dye, Opto-magnetic and antibacterial behaviour of pure and La-doped ZnO nanoparticles. J Supercond Nov Magn 31:855–864. https://doi.org/10.1007/s10948-017-4261-8

Cheng C-L et al (2009) The effects of the bacterial interaction with visible-light responsive titania photocatalyst on the bactericidal performance. J Biomed Sci 16:7. https://doi.org/10.1186/1423-0127-16-7

Chick H (1908) An investigation of the Laws of disinfection. J Hyg (Lond) 8:92–158. https://doi.org/10.1017/s0022172400006987

Cho M, Chung H, Choi W, Yoon J (2004) Linear correlation between inactivation of E. coli and OH radical concentration in TiO_2 photocatalytic disinfection. Water Res 38:1069–1077. https://doi.org/10.1016/j.watres.2003.10.029

Das S, Sinha S, Suar M, Yun S-I, Mishra A, Tripathy SK (2015) Solar-photocatalytic disinfection of Vibrio cholerae by using Ag@ZnO core–shell structure nanocomposites. J Photochem Photobiol B Biol 142:68–76. https://doi.org/10.1016/j.jphotobiol.2014.10.021

Das S et al (2017) Disinfection of multidrug resistant Escherichia coli by solar-Photocatalysis using Fe-doped ZnO nanoparticles. Sci Rep 7:104. https://doi.org/10.1038/s41598-017-00173-0

Dhandole LK, Seo Y-S, Kim S-G, Kim A, Cho M, Jang JS (2019) A mechanism study on the photocatalytic inactivation of Salmonella typhimurium bacteria by Cu_xO loaded rhodium–antimony co-doped TiO_2 nanorods. Photochem Photobiol Sci 18:1092–1100. https://doi.org/10.1039/C8PP00460A

Djerdj I, Jagličić Z, Arčon D, Niederberger M (2010) Co-doped ZnO nanoparticles: Minireview. Nanoscale 2:1096–1104. https://doi.org/10.1039/C0NR00148A

Dunlop PSM, Sheeran CP, Byrne JA, McMahon MAS, Boyle MA, McGuigan KG (2010) Inactivation of clinically relevant pathogens by photocatalytic coatings. J Photochem Photobiol A Chem 216:303–310. https://doi.org/10.1016/j.jphotochem.2010.07.004

Eswar NK, Gupta R, Ramamurthy PC, Madras G (2018a) Influence of copper oxide grown on various conducting substrates towards improved performance for photoelectrocatalytic bacterial inactivation. Mol Catal 451:161–169. https://doi.org/10.1016/j.mcat.2017.12.030

Eswar NK, Singh SA, Madras G (2018b) Photoconductive network structured copper oxide for simultaneous photoelectrocatalytic degradation of antibiotic (tetracycline) and bacteria (E.coli). Chem Eng J 332:757–774. https://doi.org/10.1016/j.cej.2017.09.117

Ganguly P, Byrne C, Breen A, Pillai SC (2018) Antimicrobial activity of photocatalysts: fundamentals, mechanisms, kinetics and recent advances. Appl Catal B Environ 225:51–75. https://doi.org/10.1016/j.apcatb.2017.11.018

Gao P, Ng K, Sun DD (2013) Sulfonated graphene oxide–ZnO–Ag photocatalyst for fast photodegradation and disinfection under visible light. J Hazard Mater 262:826–835. https://doi.org/10.1016/j.jhazmat.2013.09.055

Gong M, Xiao S, Yu X, Dong C, Ji J, Zhang D, Xing M (2019) Research progress of photocatalytic sterilization over semiconductors. RSC Adv 9:19278–19284. https://doi.org/10.1039/C9RA01826C

Goulhen-Chollet F, Josset S, Keller N, Keller V, Lett M-C (2009) Monitoring the bactericidal effect of UV-A photocatalysis: a first approach through 1D and 2D protein electrophoresis. Catal Today 147:169–172. https://doi.org/10.1016/j.cattod.2009.06.001

Gullapalli H et al (2010) Flexible Piezoelectric ZnO–Paper Nanocomposite Strain Sensor. Small 6:1641–1646. https://doi.org/10.1002/smll.201000254

Guo B-L et al (2015) The antibacterial activity of Ta-doped ZnO nanoparticles. Nanoscale Res Lett 10:336. https://doi.org/10.1186/s11671-015-1047-4

Gupta A, Srivastava R (2018) Zinc oxide nanoleaves: a scalable disperser-assisted sonochemical approach for synthesis and an antibacterial application. Ultrason Sonochem 41:47–58. https://doi.org/10.1016/j.ultsonch.2017.09.029

Hameed ASH, Karthikeyan C, Ahamed AP, Thajuddin N, Alharbi NS, Alharbi SA, Ravi G (2016) In vitro antibacterial activity of ZnO and Nd doped ZnO nanoparticles against ESBL producing Escherichia coli and Klebsiella pneumoniae. Sci Rep 6:24312. https://doi.org/10.1038/srep24312. https://www.nature.com/articles/srep24312#supplementary-information

Hassan MS, Amna T, Kim HY, Khil M-S (2013) Enhanced bactericidal effect of novel CuO/TiO$_2$ composite nanorods and a mechanism thereof. Compos Part B 45:904–910. https://doi.org/10.1016/j.compositesb.2012.09.009

Hatamie A et al (2015) Zinc oxide nanostructure modified textile and its application to biosensing, photocatalysis, and as antibacterial material. Langmuir 31:10913–10921. https://doi.org/10.1021/acs.langmuir.5b02341

He W, Huang H, Yan J, Zhu J (2013) Photocatalytic and antibacterial properties of Au-TiO$_2$ nanocomposite on monolayer graphene: from experiment to theory. J Appl Phys 114:204701. https://doi.org/10.1063/1.4836875

He W, Kim H-K, Wamer WG, Melka D, Callahan JH, Yin J-J (2014) Photogenerated charge carriers and reactive oxygen species in ZnO/Au hybrid nanostructures with enhanced Photocatalytic and antibacterial activity. J Am Chem Soc 136:750–757. https://doi.org/10.1021/ja410800y

He X et al (2017) Biocompatibility, corrosion resistance and antibacterial activity of TiO$_2$/CuO coating on titanium. Ceram Int 43:16185–16195. https://doi.org/10.1016/j.ceramint.2017.08.196

He J, Zeng X, Lan S, Lo IMC (2019) Reusable magnetic Ag/Fe, N-TiO$_2$/Fe3O4@SiO2 composite for simultaneous photocatalytic disinfection of E. coli and degradation of bisphenol A in sewage under visible light. Chemosphere 217:869–878. https://doi.org/10.1016/j.chemosphere.2018.11.072

Hom LW (1972) Kinetics of chlorine disinfection in an ecosystem. J JotSED 98:183–194

Hwang SH, Song J, Jung Y, Kweon OY, Song H, Jang J (2011) Electrospun ZnO/TiO$_2$ composite nanofibers as a bactericidal agent. Chem Commun 47:9164–9166. https://doi.org/10.1039/C1CC12872H

Ibănescu M, Muşat V, Textor T, Badilita V, Mahltig B (2014) Photocatalytic and antimicrobial Ag/ZnO nanocomposites for functionalization of textile fabrics. J Alloys Compd 610:244–249. https://doi.org/10.1016/j.jallcom.2014.04.138

Jacoby WA, Maness PC, Wolfrum EJ, Blake DM, Fennell JA (1998) Mineralization of bacterial cell mass on a photocatalytic surface in air. Environ Sci Technol 32:2650–2653. https://doi.org/10.1021/es980036f

Jin Y et al (2019) Synthesis of caged iodine-modified ZnO nanomaterials and study on their visible light photocatalytic antibacterial properties. Appl Catal B Environ 256:117873. https://doi.org/10.1016/j.apcatb.2019.117873

Joost U et al (2015) Photocatalytic antibacterial activity of nano-TiO$_2$ (anatase)-based thin films: effects on Escherichia coli cells and fatty acids. J Photochem Photobiol B Biol 142:178–185. https://doi.org/10.1016/j.jphotobiol.2014.12.010

Kääriäinen ML, Weiss CK, Ritz S, Pütz S, Cameron DC, Mailänder V, Landfester K (2013) Zinc release from atomic layer deposited zinc oxide thin films and its antibacterial effect on Escherichia coli. Appl Surf Sci 287:375–380. https://doi.org/10.1016/j.apsusc.2013.09.162

Kairyte K, Kadys A, Luksiene Z (2013) Antibacterial and antifungal activity of photoactivated ZnO nanoparticles in suspension. J Photochem Photobiol B Biol 128:78–84. https://doi.org/10.1016/j.jphotobiol.2013.07.017

Karthik K, Dhanuskodi S, Gobinath C, Prabukumar S, Sivaramakrishnan S (2018) Multifunctional properties of microwave assisted CdO–NiO–ZnO mixed metal oxide nanocomposite: enhanced photocatalytic and antibacterial activities. J Mater Sci Mater Electron 29:5459–5471. https://doi.org/10.1007/s10854-017-8513-y

Karunakaran C, Gomathisankar P, Manikandan G (2010) Preparation and characterization of antimicrobial Ce-doped ZnO nanoparticles for photocatalytic detoxification of cyanide. Mater Chem Phys 123:585–594. https://doi.org/10.1016/j.matchemphys.2010.05.019

Kavitha T, Gopalan AI, Lee K-P, Park S-Y (2012) Glucose sensing, photocatalytic and antibacterial properties of graphene–ZnO nanoparticle hybrids. Carbon 50:2994–3000. https://doi.org/10.1016/j.carbon.2012.02.082

Khalil A, Gondal MA, Dastageer MA (2011) Augmented photocatalytic activity of palladium incorporated ZnO nanoparticles in the disinfection of Escherichia coli microorganism from water. Appl Catal A Gen 402:162–167. https://doi.org/10.1016/j.apcata.2011.05.041

Khraisheh M, Wu L, Al-Muhtaseb AH, Al-Ghouti MA (2015) Photocatalytic disinfection of Escherichia coli using TiO_2 P25 and Cu-doped TiO_2. J Ind Eng Chem 28:369–376. https://doi.org/10.1016/j.jiec.2015.02.023

Kiwi J, Nadtochenko V (2005) Evidence for the mechanism of Photocatalytic degradation of the bacterial wall membrane at the TiO_2 Interface by ATR-FTIR and laser kinetic spectroscopy. Langmuir 21:4631–4641. https://doi.org/10.1021/la0469831

Koli VB, Dhodamani AG, Raut AV, Thorat ND, Pawar SH, Delekar SD (2016) Visible light photo-induced antibacterial activity of TiO_2-MWCNTs nanocomposites with varying the contents of MWCNTs. J Photochem Photobiol A Chem 328:50–58. https://doi.org/10.1016/j.jphotochem.2016.05.016

Kramer A, Schwebke I, Kampf G (2006) How long do nosocomial pathogens persist on inanimate surfaces? A systematic review. BMC Infect Dis 6:130. https://doi.org/10.1186/1471-2334-6-130

Kuang W et al (2019) Antibacterial nanorods made of carbon quantum dots-ZnO under visible light irradiation. J Nanosci Nanotechnol 19:3982–3990. https://doi.org/10.1166/jnn.2019.16320

Kumar R, Anandan S, Hembram K, Narasinga Rao T (2014) Efficient ZnO-based visible-light-driven Photocatalyst for antibacterial applications. ACS Appl Mater Interfaces 6:13138–13148. https://doi.org/10.1021/am502915v

Lam S-M, Quek J-A, Sin J-C (2018) Mechanistic investigation of visible light responsive Ag/ZnO micro/nanoflowers for enhanced photocatalytic performance and antibacterial activity. J Photochem Photobiol A Chem 353:171–184. https://doi.org/10.1016/j.jphotochem.2017.11.021

Laxma Reddy PV, Kavitha B, Kumar Reddy PA, Kim K-H (2017) TiO_2-based photocatalytic disinfection of microbes in aqueous media: a review. Environ Res 154:296–303. https://doi.org/10.1016/j.envres.2017.01.018

Lee JH et al (2004) The preparation of TiO_2 nanometer photocatalyst film by a hydrothermal method and its sterilization performance for Giardia lamblia. Water Res 38:713–719. https://doi.org/10.1016/j.watres.2003.10.011

Lefatshe K, Muiva CM, Kebaabetswe LP (2017) Extraction of nanocellulose and in-situ casting of ZnO/cellulose nanocomposite with enhanced photocatalytic and antibacterial activity. Carbohydr Polym 164:301–308. https://doi.org/10.1016/j.carbpol.2017.02.020

Li B, Cao H (2011) ZnO@graphene composite with enhanced performance for the removal of dye from water. J Mater Chem 21:3346–3349. https://doi.org/10.1039/C0JM03253K

Li Q, Mahendra S, Lyon DY, Brunet L, Liga MV, Li D, Alvarez PJJ (2008) Antimicrobial nanomaterials for water disinfection and microbial control: potential applications and implications. Water Res 42:4591–4602. https://doi.org/10.1016/j.watres.2008.08.015

Li X, Xing Y, Jiang Y, Ding Y, Li W (2009) Antimicrobial activities of ZnO powder-coated PVC film to inactivate food pathogens. Int J Food Sci Technol 44:2161–2168. https://doi.org/10.1111/j.1365-2621.2009.02055.x

Li J, Liu X, Qiao Y, Zhu H, Ding C (2014) Antimicrobial activity and cytocompatibility of Ag plasma-modified hierarchical TiO_2 film on titanium surface. Colloids Surf B: Biointerfaces 113:134–145. https://doi.org/10.1016/j.colsurfb.2013.08.030

Lin H-F, Liao S-C, Hung S-W (2005) The dc thermal plasma synthesis of ZnO nanoparticles for visible-light photocatalyst. J Photochem Photobiol A Chem 174:82–87. https://doi.org/10.1016/j.jphotochem.2005.02.015

Liu N et al (2019) Superior disinfection effect of Escherichia coli by hydrothermal synthesized TiO_2-based composite photocatalyst under LED irradiation: influence of environmental factors and disinfection mechanism. Environ Pollut 247:847–856. https://doi.org/10.1016/j.envpol.2019.01.082

Lu W, Liu G, Gao S, Xing S, Wang J (2008) Tyrosine-assisted preparation of Ag/ZnO nanocomposites with enhanced photocatalytic performance and synergistic antibacterial activities. Nanotechnology 19:445711. https://doi.org/10.1088/0957-4484/19/44/445711

Maness PC, Smolinski S, Blake DM, Huang Z, Wolfrum EJ, Jacoby WA (1999) Bactericidal activity of photocatalytic TiO(2) reaction: toward an understanding of its killing mechanism. Appl Environ Microbiol 65:4094–4098

Masudy-Panah S, Zhuk S, Tan HR, Gong X, Dalapati GK (2018) Palladium nanostructure incorporated cupric oxide thin film with strong optical absorption, compatible charge collection and low recombination loss for low cost solar cell applications. Nano Energy 46:158–167. https://doi.org/10.1016/j.nanoen.2018.01.050

Matai I, Sachdev A, Dubey P, Uday Kumar S, Bhushan B, Gopinath P (2014) Antibacterial activity and mechanism of Ag–ZnO nanocomposite on S. aureus and GFP-expressing antibiotic resistant E. coli. Colloids Surf B: Biointerfaces 115:359–367. https://doi.org/10.1016/j.colsurfb.2013.12.005

Matsunaga T, Tomoda R, Nakajima T, Wake H (1985) Photoelectrochemical sterilization of microbial cells by semiconductor powders. FEMS Microbiol Lett 29:211–214

Michal R, Dworniczek E, Caplovicova M, Monfort O, Lianos P, Caplovic L, Plesch G (2016) Photocatalytic properties and selective antimicrobial activity of TiO_2(Eu)/CuO nanocomposite. Appl Surf Sci 371:538–546. https://doi.org/10.1016/j.apsusc.2016.03.003

Nair MG, Nirmala M, Rekha K, Anukaliani A (2011) Structural, optical, photo catalytic and antibacterial activity of ZnO and Co doped ZnO nanoparticles. Mater Lett 65:1797–1800. https://doi.org/10.1016/j.matlet.2011.03.079

Pant B, Pant HR, Barakat NAM, Park M, Jeon K, Choi Y, Kim H-Y (2013) Carbon nanofibers decorated with binary semiconductor (TiO_2/ZnO) nanocomposites for the effective removal of organic pollutants and the enhancement of antibacterial activities. Ceram Int 39:7029–7035. https://doi.org/10.1016/j.ceramint.2013.02.041

Panthi G, Yousef A, Barakat NAM, Abdelrazek Khalil K, Akhter S, Ri Choi Y, Kim HY (2013) Mn_2O_3/TiO_2 nanofibers with broad-spectrum antibiotics effect and photocatalytic activity for preliminary stage of water desalination. Ceram Int 39:2239–2246. https://doi.org/10.1016/j.ceramint.2012.08.068

Park K-H, Han GD, Neoh KC, Kim T-S, Shim JH, Park H-D (2017) Antibacterial activity of the thin ZnO film formed by atomic layer deposition under UV-A light. Chem Eng J 328:988–996. https://doi.org/10.1016/j.cej.2017.07.112

Paschoalino M, Guedes NC, Jardim W, Mielczarski E, Mielczarski JA, Bowen P, Kiwi J (2008) Inactivation of E. coli mediated by high surface area CuO accelerated by light irradiation >360nm. J Photochem Photobiol A Chem 199:105–111. https://doi.org/10.1016/j.jphotochem.2008.05.010

Podporska-Carroll J, Panaitescu E, Quilty B, Wang L, Menon L, Pillai SC (2015) Antimicrobial properties of highly efficient photocatalytic TiO_2 nanotubes. Appl Catal B Environ 176-177:70–75. https://doi.org/10.1016/j.apcatb.2015.03.029

Podporska-Carroll J et al (2017) Antibacterial properties of F-doped ZnO visible light photocatalyst. J Hazard Mater 324:39–47. https://doi.org/10.1016/j.jhazmat.2015.12.038

Poongodi G, Anandan P, Kumar RM, Jayavel R (2015) Studies on visible light photocatalytic and antibacterial activities of nanostructured cobalt doped ZnO thin films prepared by sol–gel spin coating method. Spectrochim Acta A Mol Biomol Spectrosc 148:237–243. https://doi.org/10.1016/j.saa.2015.03.134

Preethi LK, Mathews T (2019) Electrochemical tuning of heterojunctions in TiO_2 nanotubes for efficient solar water splitting. Cat Sci Technol 9:5425–5432. https://doi.org/10.1039/C9CY01216H

Preethi LK, Antony RP, Mathews T, Loo SCJ, Wong LH, Dash S, Tyagi AK (2016) Nitrogen doped anatase-rutile heterostructured nanotubes for enhanced photocatalytic hydrogen production: promising structure for sustainable fuel production. Int J Hydrog Energy 41:5865–5877. https://doi.org/10.1016/j.ijhydene.2016.02.125

Preethi LK, Antony RP, Mathews T, Walczak L, Gopinath CS (2017a) A study on doped Heterojunctions in TiO_2 nanotubes: An efficient Photocatalyst for solar water splitting. Sci Rep 7:14314. https://doi.org/10.1038/s41598-017-14463-0

Preethi LK, Mathews T, Nand M, Jha SN, Gopinath CS, Dash S (2017b) Band alignment and charge transfer pathway in three phase anatase-rutile-brookite TiO_2 nanotubes: An efficient photocatalyst for water splitting. Appl Catal B Environ 218:9–19. https://doi.org/10.1016/j.apcatb.2017.06.033

Quek J-A, Lam S-M, Sin J-C, Mohamed AR (2018) Visible light responsive flower-like ZnO in photocatalytic antibacterial mechanism towards Enterococcus faecalis and Micrococcus luteus. J Photochem Photobiol B Biol 187:66–75. https://doi.org/10.1016/j.jphotobiol.2018.07.030

Raghupathi KR, Koodali RT, Manna AC (2011) Size-dependent bacterial growth inhibition and mechanism of antibacterial activity of zinc oxide nanoparticles. Langmuir 27:4020–4028. https://doi.org/10.1021/la104825u

Raja A et al (2018) Eco-friendly preparation of zinc oxide nanoparticles using Tabernaemontana divaricata and its photocatalytic and antimicrobial activity. J Photochem Photobiol B Biol 181:53–58. https://doi.org/10.1016/j.jphotobiol.2018.02.011

Raut NC, Mathews T, Ajikumar PK, George RP, Dash S, Tyagi AK (2012) Sunlight active antibacterial nanostructured N-doped TiO_2 thin films synthesized by an ultrasonic spray pyrolysis technique. RSC Adv 2:10639–10647. https://doi.org/10.1039/C2RA21024J

Raza W, Faisal SM, Owais M, Bahnemann D, Muneer M (2016) Facile fabrication of highly efficient modified ZnO photocatalyst with enhanced photocatalytic, antibacterial and anticancer activity. RSC Adv 6:78335–78350. https://doi.org/10.1039/C6RA06774C

Rekha K, Nirmala M, Nair MG, Anukaliani A (2010) Structural, optical, photocatalytic and antibacterial activity of zinc oxide and manganese doped zinc oxide nanoparticles. Phys B Condens Matter 405:3180–3185. https://doi.org/10.1016/j.physb.2010.04.042

Rokicka-Konieczna P, Markowska-Szczupak A, Kusiak-Nejman E, Morawski AW (2019) Photocatalytic water disinfection under the artificial solar light by fructose-modified TiO_2. Chem Eng J 372:203–215. https://doi.org/10.1016/j.cej.2019.04.113

Saito T, Iwase T, Horie J, Morioka T (1992) Mode of photocatalytic bactericidal action of powdered semiconductor TiO_2 on mutans streptococci. J Photochem Photobiol B Biol 14:369–379. https://doi.org/10.1016/1011-1344(92)85115-B

Sayılkan F, Asiltürk M, Kiraz N, Burunkaya E, Arpaç E, Sayılkan H (2009) Photocatalytic antibacterial performance of Sn^{4+}-doped TiO_2 thin films on glass substrate. J Hazard Mater 162:1309–1316. https://doi.org/10.1016/j.jhazmat.2008.06.043

Shimizu Y, Ateia M, Wang M, Awfa D, Yoshimura C (2019) Disinfection mechanism of E. coli by CNT-TiO_2 composites: photocatalytic inactivation vs. physical separation. Chemosphere 235:1041–1049. https://doi.org/10.1016/j.chemosphere.2019.07.006

Sin J-C, Tan S-Q, Quek J-A, Lam S-M, Mohamed AR (2018) Facile fabrication of hierarchical porous ZnO/Fe_3O_4 composites with enhanced magnetic, photocatalytic and antibacterial properties. Mater Lett 228:207–211. https://doi.org/10.1016/j.matlet.2018.06.027

Singh S, Barick KC, Bahadur D (2013) Shape-controlled hierarchical ZnO architectures: photocatalytic and antibacterial activities. CrystEngComm 15:4631–4639. https://doi.org/10.1039/C3CE27084J

Sökmen M, Candan F, Sümer Z (2001) Disinfection of E. coli by the Ag-TiO_2/UV system: lipidperoxidation. J Photochem Photobiol A Chem 143:241–244. https://doi.org/10.1016/S1010-6030(01)00497-X

Talebian N, Nilforoushan MR, Zargar EB (2011) Enhanced antibacterial performance of hybrid semiconductor nanomaterials: ZnO/SnO$_2$ nanocomposite thin films. Appl Surf Sci 258:547–555. https://doi.org/10.1016/j.apsusc.2011.08.070

Talebian N, Amininezhad SM, Doudi M (2013) Controllable synthesis of ZnO nanoparticles and their morphology-dependent antibacterial and optical properties. J Photochem Photobiol B Biol 120:66–73. https://doi.org/10.1016/j.jphotobiol.2013.01.004

Tong T, Wilke CM, Wu J, Binh CTT, Kelly JJ, Gaillard J-F, Gray KA (2015) Combined toxicity of Nano-ZnO and Nano-TiO$_2$: from single- to multinanomaterial systems. Environ Sci Technol 49:8113–8123. https://doi.org/10.1021/acs.est.5b02148

Torres A, Ruales C, Pulgarin C, Aimable A, Bowen P, Sarria V, Kiwi J (2010) Innovative high-surface-area CuO pretreated cotton effective in bacterial inactivation under visible light. ACS Appl Mater Interfaces 2:2547–2552. https://doi.org/10.1021/am100370y

Trapalis CC, Keivanidis P, Kordas G, Zaharescu M, Crisan M, Szatvanyi A, Gartner M (2003) TiO$_2$(Fe3+) nanostructured thin films with antibacterial properties. Thin Solid Films 433:186–190. https://doi.org/10.1016/S0040-6090(03)00331-6

Wang W, Zhang L, An T, Li G, Yip H-Y, Wong P-K (2011) Comparative study of visible-light-driven photocatalytic mechanisms of dye decolorization and bacterial disinfection by B–Ni-codoped TiO$_2$ microspheres: the role of different reactive species. Appl Catal B Environ 108-109:108–116. https://doi.org/10.1016/j.apcatb.2011.08.015

Watson HE (1908) A note on the variation of the rate of disinfection with change in the concentration of the disinfectant. J Hyg (Lond) 8:536–542. https://doi.org/10.1017/s0022172400015928

Wei C et al (1994) Bactericidal activity of TiO$_2$ photocatalyst in aqueous media: toward a solar-assisted water disinfection system. Environ Sci Technol 28:934–938. https://doi.org/10.1021/es00054a027

Wei A et al (2006) Enzymatic glucose biosensor based on ZnO nanorod array grown by hydrothermal decomposition. Appl Phys Lett 89:123902. https://doi.org/10.1063/1.2356307

Wu D et al (2015) Mechanistic study of the visible-light-driven photocatalytic inactivation of bacteria by graphene oxide–zinc oxide composite. Appl Surf Sci 358:137–145. https://doi.org/10.1016/j.apsusc.2015.08.033

Xiong L et al (2015) N-type Cu$_2$O film for Photocatalytic and Photoelectrocatalytic processes: its stability and inactivation of E. coli. Electrochim Acta 153:583–593. https://doi.org/10.1016/j.electacta.2014.11.169

Yan J, Chen H, Zhang L, Jiang J (2011) Inactivation of Escherichia coli on immobilized CuO/CoFe$_2$O$_4$-TiO$_2$ thin-film under simulated sunlight irradiation. Chin J Chem 29:1133–1138. https://doi.org/10.1002/cjoc.201190212

Yin S et al (2013) Functional free-standing graphene honeycomb films. Adv Funct Mater 23:2972–2978. https://doi.org/10.1002/adfm.201203491

Zhang L, Ding Y, Povey M, York D (2008) ZnO nanofluids – a potential antibacterial agent. Prog Nat Sci 18:939–944. https://doi.org/10.1016/j.pnsc.2008.01.026

Chapter 3
Metal and Metal Oxide Nanomaterials for Wastewater Decontamination

Mohd. Tauqeer, Mohammad Ehtisham Khan, Radhe Shyam Ji, Prafful Bansal, and Akbar Mohammad

Abstract There is an unremarkable development in human life by industrialization and modernization which consequences the harmful contamination in the fresh water. This contamination is rapidly growing by the use of a variety of organic dyes, inorganic and other harmful pollutants like chemicals, pesticides, soil erosion, etc. which causes the quality of fresh and clean water continuously degraded. To get an access of clean water removal of contaminants is a major challenge for the water industries. It is a high demand to develop useful material in water treatment with high separation capacity, low-priced, porosity, and recyclability. Adsorption, coagulation, oxidative-reductive degradation and membrane separation, etc. have been developed for water/waste water treatment. Adsorption is found to be one of the most widely used method in water treatment. Variety of different adsorbents and their hybrids are used for the removal of pollutants from water. Recently, investigation on remediation of water involves the use of nanomaterials with high purifying capability. Metal and metal oxide nanomaterials are one of the important and excellent reagents for the decontamination of water/waste water treatment. The present chapter is a compilation of various adsorbents based on metal and metal oxide nanomaterials for decontamination of water/waste water available till date.

Keywords Metal-metal oxide nanomaterials · Pollutants · Waste-water decontamination

---

M. Tauqeer (✉) · P. Bansal
Department of Chemistry, Aligarh Muslim University, Aligarh, Uttar Pradesh, India
e-mail: mohdtauqeer.ch@amu.ac.in

M. E. Khan
Department of Chemical Engineering Technology, College of Applied Industrial Technology (CAIT), Jazan University, Jazan, Kingdom of Saudi Arabia

R. S. Ji
Discipline of Chemistry, Indian Institute of Technology Indore, Indore, Madhya Pradesh, India

A. Mohammad
School of Chemical Engineering, Yeungnam University, Gyeongsan-si, Gyeongbuk, South Korea

© The Author(s), under exclusive license to Springer Nature Switzerland AG 2021
S. Rajendran et al. (eds.), *Metal, Metal-Oxides and Metal-Organic Frameworks for Environmental Remediation*, Environmental Chemistry for a Sustainable World 64, https://doi.org/10.1007/978-3-030-68976-6_3

3.1 Introduction

Amongst the abundant natural resources on earth, water is found to be affecting all section of human life and it is found that about 1% of water is used in human consumption (Grey et al. 2013; Adeleye et al. 2016). However, contamination of fresh water is rapidly growing due to use of various organic/inorganic pollutants (Schwarzenbach et al. 2006; Ferroudj et al. 2013). It is mentioned in reports that around 1.1 billion people do not get clean water all over the world which is continuously decreases with increase in population and contamination of fresh water by variety of organic/inorganic pollutants (Adeleye et al. 2016; Leonard et al. 2003; Ashbolt 2004; Hutton et al. 2007; Wigginton et al. 2012; Ritter et al. 2002; Fawell and Nieuwenhuijsen 2003; Rodriguez-Mozaz et al. 2004; Falconer and Humpage 2005; Wang et al. 2009a, b). The inorganic/organic pollutants come from different industries mainly paper, textiles, leather cosmetics, printing, dye and metal processing industries. Contamination of these pollutants potentially affects the access to get clean water because removal of these contaminants is very difficult as these are highly stable and affect the aquatic life (Liu et al. 2012; Tian et al. 2013; Jia et al. 2013; Pereira and Alves 2012; Gajda 1996; Ivanov 1996; Kabdaşli et al. 1996; Bensalah et al. 2009; Wróbel et al. 2001; Dawood et al. 2014; Wong et al. 2004; Hoffmann et al. 1995; Robles et al. 2013; Clarke and Anliker 1980). Dangerous results have already seen due to these contaminations of pollutant in water although the exact amount of pollutants is not known (Robles et al. 2013; Clarke and Anliker 1980; Mishra and Tripathy 1993; Banat 1996; Gupta et al. 1990; Malaviya and Singh 2011; Bhatnagar and Sillanpää 2009). Production of high-quality drinking water has been developed by the different water industries which still needed improvement due to emergence of other contaminants like pharmaceutical products, viruses etc.

Development of an appropriate material to be used for the purification of water is achieved when the material posses high separation capacity, cheap, porosity, and recyclability (Zelmanov and Semiat 2008; Anjum et al. 2016). Water treatment has already been applied biologically but these are slow and limited process which sometime causes toxicity to microorganism due to toxic agents (Daraei et al. 2013). Remediation of water has been adapted by different methods including adsorption, coagulation, oxidative-reductive degradation and membrane separation and so on which are used to remove the various pollutants (Dutta et al. 2001; Moghaddam et al. 2010; Gao et al. 2012; Wang et al. 2006a, b). Adsorption among them is most widely used method in water treatment process. Variety of different adsorbents and their hybrids are used for the decontamination of water and these material surfaces have high adsorption capacity (Wang et al. 2006a, b; Gandhi et al. 2016; Upadhyay et al. 2014; Perreault et al. 2015). A big achievement has been made in water treatment process by enhancing the properties of the catalyst, electro-catalyst, photo electro-catalyst, and disinfection and desalination agent (Chatterjee and Dasgupta 2005; Moo et al. 2014; Wang et al. 2013a, b; Mahata et al. 2007; Chang and Wu 2013; Daer et al. 2015).

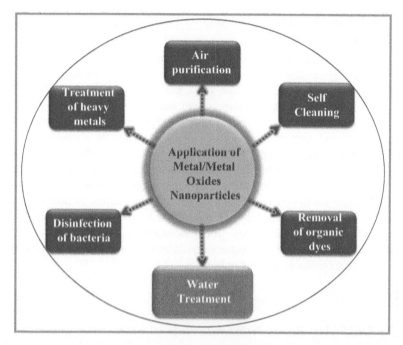

Fig. 3.1 Metal/metal oxide nanoparticles for various applications including water treatment

In recent years, Nanotechnology has emerged as an important area for providing materials that are useful for water decontamination treatment process by optimizing their properties like their hydrophobic and hydrophilic character, high toughness, high strength and porosity (Liu et al. 2014a, b; Ray and Shipley 2015; Tauqeer et al. 2020; Mohammad et al. 2018a, b). Metal or metal oxide-based nanomaterials are broad area of inorganic nanomaterials that are useful in removing different pollutants from water/waste water. Various metals-based nanoparticles/nanomaterials like silver nanoparticles, gold nanoparticles, iron nanoparticles, etc., and metal oxide-based nanoparticles/nanomaterials like, ferric oxides, titanium oxides, magnesium oxides, zinc oxides, cerium oxides, etc. are useful in removing different pollutants. Several applicable categories of metal and metal-oxide nanoparticles are used in various applications such as, disinfection of bacteria, dye sensitized solar cells, heavy metals decontamination, purification of air, self-cleaning and the most important is water treatment (Fig. 3.1) (Khan et al. 2020; Yunus et al. 2012; Yaqoob et al. 2020; Mohammad et al. 2018a, b).

This chapter reviews and highlights the recent findings on the use of metal and metal oxide based nanoadsorbents for water/wastewater treatment. However, this chapter represents comprehensive review of the important field using limited number of examples. Variety of adsorbents based on metal and metal oxide nanomaterials have been listed to the best of our efforts in a controllable manner and their recycling by focusing on the challenges of future research.

3.2 Metal Based Nanomaterials in Water Treatment

Modernization and industrialization cause serious and harmful effect on human society mainly by contaminating the water. Different pollutants used for the contamination are heavy metal ions like Pb^{2+}, Cd^{2+}, Zn^{2+}, Ni^{2+} and Hg^2, organic compounds like, phenols, dyes, benzene and bacteria, viruses, and so on. These pollutants are mostly eliminated by adsorption technology because of its low cost, high efficiency, and easy removal of pollutants. To enhance the absorption process, improvement in adsorption capacity of nanomaterials is taken into consideration by using various doping phenomenon which ultimately are used to remove various contaminants from wastewater. Metal based nanomaterials emerge as one of the beneficial agents for the effective waste water decontamination of pollutants. This part briefly explains different types of metal based nano adsorbents mainly gold nanomaterials, silver nanomaterials, iron nanomaterials, palladium nanomaterial, platinum nanomaterials and zinc nanomaterials along with their hybrid adsorbents for the decontamination of various pollutants.

3.2.1 Gold (Au) Based Nanomaterials

Among the various metal-based nanomaterials, gold nanomaterials are one of the most widely used nanomaterials for the decontamination of different pollutant from waste water (Tiwari and Paul 2015; Pantapasis and Grumezescu 2017). Interest in gold nanoparticles (AuNPs) to be used as nanoprobes for the different pollutants like heavy metals, organic pollutants and other pathogenic pollutants is continuously increasing because of their high selectivity and high binding capacity (Koedrith et al. 2015; Tamer et al. 2013). Jimoh et al. (2015) reported the selectivity for iron ion (Fe^{3+}) using nanogold-Schiff base chemosensor (l-thiolated Schiff base coating on gold nanoparticles) in a mixture of water. Determination of cobalt ion (Co^{2+}) in aqueous solution is achieved using gold nanoparticles stabilized with $S_2O_3^{2-}$ and ethylenediamine (en) (Zhang et al. 2012a, b).

Gold nanomaterials for the decontamination of organic compounds are important because of high toxicity of organic pollutants which affects the human badly. Phenols are considered to be the most toxic organic contaminants in the environments (CDC 2011). Gold nanoparticles is used to reduce the p-nitrophenol to 4-aminophenol by excess of borohydride (Lin et al. 2013) and another study (Lerma-García et al. 2014) uses Au/Al_2O_3 for this reduction in presence of $NaBH_4$.

Gold nanoparticles (AuNPs) are known to exhibit good antibacterial effect against both gram positive and gram-negative bacteria. As we have seen in literature the biosynthesized gold nanoparticles are found to inhibit the growth of Staphylococcus aureus, Basilus subtilus and E.*coli* (Khan et al. 2016). Without external excitations, amine-functionalized gold and titanium dioxide nanoparticles have shown excellent antibacterial excellent antibacterial effect for waste water

3 Metal and Metal Oxide Nanomaterials for Wastewater Decontamination

Table 3.1 Different types of gold-based nanomaterial adsorbents and pollutants

S. No.	Adsorbent	Polluted species	References
1	AuNPs functionalized by l-cysteine	Cu^{2+}	Wenrong et al. (2007)
2	Ag/AuNPs	Cu^{2+}	Lou et al. (2011)
3	Cit stabilized AuNPs	Co^{2+}	Raghav and Srivastava (2015)
4	AuNPs (green synthesis)	Co^{2+}, Ni^{2+}	Annadhasan et al. (2015)
5	AuNPs using as stabilizer l-dopa, (l-3,4-dihydroxyphenylalanine)	Mn^{2+}	Narayanan and Park (2014)
6	AuNPs in presence of $NaBH_4$	p-nitrophenol	Lerma-García et al. (2014)

treatment with presence of ammonium salts as a surfactant (Wan and Yeow 2012). Polydopamine-functionalized poly (vinyl alcohol) microporous supported gold nanoparticles are extreme efficient for waste water treatment to remove organic pollutant (Li et al. 2019). Chitosan functionalized activated coke for gold nanoparticles anchoring has been shown tremendous results towards pollutant treatment of nitrophenol and azo dyes (Fu et al. 2019). Intercalating gold nanoparticles and graphitic carbon nitride into graphene oxide (GNPs/g-C_3N_4/GO) membrane is useful for removal of environmental pollutants (Qu et al. 2017). Rao and co-worker reported Silver (Ag) and Gold (Au) nanoparticles (NPs), which is using an aqueous extract of *A. elaeagnoidea* flower for the decontamination of Methylene Blue, Congo Red and p-Nitrophenol (Manjari et al. 2017). Some of the gold-based adsorbents are listed in Table 3.1 along with polluted species.

3.2.2 Silver (Ag) Based Nanomaterials

Silver nanoparticles (Ag-NPs) are now-a-days rapidly used in water filtration because large numbers of reports mention the effects of silver nanoparticles in the aquatic environment (Lu et al. 2016). Silver nanoparticles are considered to be good antimicrobial agents in different aqueous solutions. Prevention of bio fueling is achieved by killing and self-cleaning of membrane surface followed by incorporation of antibacterial silver nanoparticles and negatively charged carboxylic and amine functional groups (Prince et al. 2014). Size control silver nanoparticle when uniformly loaded onto the hierarchical nano columnar structures of zinc oxide. It is showing very good photo degradation of formaldehyde in water (Liu et al. 2019). Extraction of *Ives* cultivar pomace produces silver nanoparticles which is used as auxiliaries in wastewater purification and shows 47% reduction in the bacterial count of Escherichia coli on the disinfection of the wastewater (Raota et al. 2019). Rezayi et al. reported the Bacillus sp. and Amaranthus sp. as an efficient natural source for creation of biologically active silver nanoparticles (Bahrami-Teimoori et al. 2019). Silver nanoparticles exhibit antibacterial activity towards suspension of *Escherichia Coli* and *Enterococcus Faecalis*in water treatment. Ceramic material supported silver nanoparticles also showed better filtration for waste water in house hold (Ren and Smith 2013; Kallman et al. 2011; Oyanedel-Craver and Smith 2008). Degradation rate of polluting dyes, methyl orange, methylene blue and eosin Y by $NaBH_4$ showing size effect of silver nanoparticles have been observed for removal of pollutants from waste water (Vidhu and Philip 2014).

3.2.3 Iron (Fe) Based Nanomaterials

Recently, iron-based nanomaterials have drawn attention because of its ability to reduce easily, high adsorption capacity, cheap, large surface area (Rivero-Huguet and Marshall 2009) and are presented it as the most studied zero-valent metal nanomaterials for the decontamination of various pollutants from water wastewater (Matheson and Tratnyek 1994). Iron based nanomaterials are useful in removing various contaminants from water, for example, chromium (VI) ion is removed by iron hydroxide ($Fe(OH)_3$) (Wang et al. 2014a, b), hydrogen peroxide (H_2O_2) and iron(II) ion (Fu et al. 2014): Other contaminants' which is removed by using zero valent iron nanomaterials are halogenated compounds (Liang et al. 2014), nitrogen containing aromatic compounds (Xiong et al. 2015), dyes (Hoag et al. 2009), phenols (Wang et al. 2013a, b), metals (Arancibia et al. 2016), phosphates (Marková et al. 2013), nitrates (Muradova et al. 2016), metalloids (Ling et al. 2015),and trace elements (Ling et al. 2015). Recently, iron-based nanomaterials are also used in soil treatment (Gueye et al. 2016).

Apart from many usefulness, iron-based nanomaterials have the limitations to aggregates, oxidizes and difficulty in separation which can be improved by modifying the adsorbents by doping with other metal (Liou et al. 2005), matrix encapsulation and emulsification (Stefaniuk et al. 2016). Iron based nanomaterials are supposed to enhance its adsorbent properties by doping with other metals (Singh and Misra 2015; Chen et al. 2011; Li et al. 2016; Lv et al. 2014; Berge and Ramsburg 2009). Biogenic iron and Nickel(Ni)-Iron(Fe) nanoparticles prepared using *Terminalia bellirica* extracts from water and these nanoparticle played role for polyphenolsin the plant extract as a reducing and capping agent (Yang et al. 2005; Kumpiene et al. 2008; Mueller et al. 2012; El-Temsah et al. 2013; Gunarani et al. 2019). Few of the iron-based nanomaterials as adsorbents are shown in Table 3.2.

3.2.4 Palladium (Pd) Based Nanomaterials

Altamimi research group have reported the photocatalytic degradation of Rhodamine B in an aqueous medium over these photocatalysts enabled disclosures showing the effect of palladium nanoparticles nature on the photocatalytic activity (Alshammari et al. 2019). Polyethyleneimine stabilized palladium nanoparticles

Table 3.2 Different types of iron-based nanomaterial adsorbents and pollutants

S. No.	Adsorbent	Polluted species	References
1	Nanoscale zero-valent iron with polyphenols	Bromothymol blue	Hoag et al. (2009)
2	Pd-Fe NPs modified with PMMA, PAA and CTAB	2,4-Dichorophenol	Wang et al. (2013a, b)
3	Zeolite and montmorillonite functionalized with nanoscale zero valent iron	Pb^{2+}	Arancibia et al. (2016)
4	Magnetic bimetallic Fe-Ag nanoparticles	Microorganism and phosphorus	Marková et al. (2013)
5	Iron based nanomaterials and Fe/Cu nanomaterial	Nitrate ion$^+$	Muradova et al. (2016)
6	Iron based nanomaterials	Uranium$^+$	Ling and Zhang (2015)

(Pd NPs) have shown great result as catalysts for water treatment for removal of 4-nitrophenol (Cui et al. 2019).

3.2.5 Platinum (Pt) Based Nanomaterials

Platinum nanoparticles measure the concentration of ammonia in artificial and real waste water samples showing its excellent selectivity, sensitivity, and capacity (Ning et al. 2017) Granger research group shows the physico-chemical properties of platinum nanoparticles, by dispersing homogenously on alumina framework and acting as catalytic active in purification of water (Sekhar et al. 2018).

3.2.6 Zinc (Zn) Based Nanomaterials

Zinc based nanomaterial has also been considered as an alternative for iron nanomaterials. Dehalogenation reactions are considered to be the main application of nano-zero-valent zinc. The nano-zero-valent zinc has been utilized for CCl_4 and octachlorodibenzo-p-dioxin removal from waste water treatment via dehalogenation process (Tratnyek et al. 2010). Owija and co-worker have reported zinc nanoparticles were used for the decontamination of acid red dye in real wastewater sample, and results showed the high removal efficiency (Bokare et al. 2013). The bio carrier containing zinc nanoparticles in bio film reactor have performed better for textile waste water treatment (Wang et al. 2018; Salam et al. 2019).

3.3 Waste Water Decontamination by Metal Oxide Nanomaterials

Development of nanotechnology in the field of waste water decontamination continuously emerges as one of important area of research. The main pollutants namely toxic metal ions, organic pollutants and microorganisms in water treatment process can be removed by different nanomaterials in particular by metal oxide nanomaterials. Recently, metal oxide nanomaterials are on its way to develop an efficient and economical reagent to clean the waste water pollutants (Fei and Li 2010). Remediation of water by metal-oxide nanomaterials is due to their variable properties and stable valences, high surface area and variable electronic configuration. This chapter includes the current advances on the use of metal-oxide based nanomaterials (TiO_2, Fe_2O_3, ZnO, CeO_2 and Al_2O_3) in waste water decontamination. Some of the important metal oxide nanomaterials and their derivatives as adsorbents have been discussed below with the improving adsorbent properties in detail.

3.3.1 Titanium Dioxide (TiO_2) Nanomaterials

Titanium dioxide (TiO_2) is one of the important and widely used photocatalyst for water treatment (Nolan et al. 2009; Hu et al. 2013; Oh et al. 2003). Titanium dioxide is important decontaminating agent for the degradation and removal of organic dyes and other organic contaminants because of its high photo-catalytic activity, cheap, non-toxicity and high stability (Zhou et al. 2011; Akpan and Hameed 2009; Lydakis-Simantiris et al. 2010; Shinde et al. 2017). Among the three different polymorph anatase, rutile, and brookite, the Degussa P- 25 photocatalyst (mixture of anatase and rutile titanium dioxide) exhibited high photocatalytic activity for the degradation of organic dyes compared to other forms of titanium dioxide, uses commercially (Zhou et al. 2012a, b; Hou et al. 2015). Titanium dioxide nanoparticles exhibits low selectivity and are able to degrade various pollutants, like chlorinated compounds (Ohsaka et al. 2008), polycyclic aromatic hydrocarbons (Guo et al. 2015), dyes (Lee et al. 2008), phenolic compounds (Nguyen et al. 2016), pesticides (Alalm et al. 2015), arsenic (Moon et al. 2014), cyanide (Kim et al. 2016), and heavy metals (Chen and Guoetal 2016).

Recently, efficiency of photocatalytic activity of titanium dioxide nanomaterials for waste water treatment is continuously improving either by modifying the morphology with increase in surface area and porosity or by modifying chemically with the incorporation of various semiconductor, doping of metal ion/non-metal ion (Jiang et al. 2006), co-doping with foreign ion (Li et al. 2012) and noble metal deposition and combination with electron acceptor materials (Pelaez et al. 2012). Degradation of phenol is achieved by Fe(III)-doped titanium dioxide nanoparticles under solar light irradiation (Nahar et al. 2006). Noble metals like Ag, Au, Pt and Pd and Si, Fe are used to modify the titanium dioxide nanomaterial that enhances its use as waste water decontaminating agent. For example, removal of 2–chlorophenol in aqueous phase was achieved by Co-doped titanium dioxide nanomaterial catalyst (Barakat et al. 2005). Rare earth Pr-doped titanium dioxide nanomaterials shows good photocatalytic activity to degrade the phenol (Chiou and Juang 2007) Shah et al. (2002) investigated the metalloorganic chemical vapor deposition method to synthesize pure titanium dioxide and Pd^{2+}, Nd^{3+}, Pt^{4+} and Fe^{3+}-doped titanium dioxide nanoparticles and their photocatalytic activity to degrade-chlorophenol in UV light suggesting the position of dopants in the nanomaterials.

The photo degrading properties of titanium dioxide nanomaterials are used to remove various microorganism like bacteria (Gram-negative and Gram-positive), fungi, algae, protozoa, and viruses (Foster et al. 2011). Disinfection of wastewater can be achieved when titanium dioxide nanoparticles is used as dopant. For example, sulphur-doped titanium dioxide exhibits visible -light-induced antibacterial effect (Yu et al. 2005), while Fe- doped titanium dioxide sol-gel electrode is shown to exhibit higher photo electrocatalytic disinfection of *E. coli* compared to the disinfection by the corresponding undoped electrode (Egerton et al. 2006). Different adsorbents using titanium dioxide for the removal of different contaminants are listed in Table 3.3.

3.3.2 Iron Oxide (Fe_2O_3) Nanomaterials

Another important metal oxide nanomaterial for the decontamination of water is iron oxide nanomaterial due to good sorption capacity, low cost, high removal capability and easy isolation (Li et al. 2003; Oliveira et al. 2004). Recent investigation on iron-based nanomaterials clearly showed its excellent adsorption capacity for decontamination of metals, inorganic and organic pollutants (Hai and Chen 2001; Onyango et al. 2003; Oliveira et al. 2004; Herrera et al. 2001; Wu et al. 2004, 2005) Fe_2O_3 and Fe_3O_4 is the most common iron nanomaterial used as an adsorbent for the decontamination of water (Takafuji et al. 2004; Wu et al. 2005). Various parameters have been dealt with iron oxide nanomaterial for the decontamination of metal ions (Takafuji et al. 2004; Cornell and Schwertmann 2003). For example, Shen et al. (2009) reported the adsorption efficiency of Ni^{2+}, Cu^{2+}, Cd^{2+} and Cr^{6+} ions by Fe_3O_4 nanoparticles and shows strong dependency of different parameters like pH, temperature, the adsorbent species and the incubation time.

As far as removal mechanism is concerned for the decontamination of pollutants different mechanism is used because of variable oxidation state of these nanomaterials (Tang and Lo 2013). For instance, phosphate isolation in aqueous phase has been investigated using iron oxide nanomaterial (Yoon et al. 2014). E. Petala et al. (2017) report the mechanism for the isolation of arsenic using nanocomposite of magnetic carbon nanocages (iron oxide based). Another study by Cao et al. (2012) explained the mechanism for the decontamination of As(V) and

Table 3.3 Different types of titanium oxide-based nanomaterial adsorbents and pollutants

S. No.	Adsorbent	Polluted species	References
1	Ag-doped TiO_2	Toluene	Li et al. (2010)
2	Nd doped TiO_2	Methylene blue	Yang et al. (2012a, b)
3	10 wt% MgO doped TiO_2	4-chlorophenol	Pozan and Kambur (2013)
4	Bi & B, co-doped TiO_2	AO7, 2, 4-DCP	Bagwasi et al. (2013)
5	Gd doped TiO_2	Methyl orange	Lv et al. (2011)
6	PANI/TiO_2	Methylene blue/ rhodamine B	Radoičić et al. (2013)
7	C-doped TiO_2 at 200 °C	Toluene	Dong et al. (2011)
8	C-doped TiO_2 at 500 °C	Toluene	Dong et al. (2011)
9	C-self-doped TiO_2 sheets	Methylene blue	Wu et al. (2009)
10	N-TiO_2 at 500 °C	Methylene blue/ 4-chlorophenol	Wu et al. (2009)
11	N-doped TiO_2	Methylene blue, phenol, Lindane	Nolan et al. (2012), Diker et al. (2011), Cheng et al. (2012), Senthilnathan and Philip (2010)
12	C–N co-doped rod-like TiO_2	Methylene blue	Yu et al. (2011)
13	B-doped TiO_2	Methylene blue	Zheng et al. (2011)
14	C, S, N and Fe doped TiO_2	Rhodamine B	Yang et al. (2009)
15	N, S-TiO_2, 500 °C	Methyl orange	Ju et al. (2013)

Cr(VI) using α-Fe_2O_3 nanostructures which is thought to start with electrostatic attraction of α-Fe_2O_3 and As(V),/Cr(VI) ions for surface bonding.

Decontamination of organic pollutants have also been used by iron oxide nanomaterials because of efficient and easy isolation by external magnetic field, cheap, high removal capacity, insensitive to toxic contaminants (Xu et al. 2012; Nizamuddin et al. 2019) and is thought to proceed by the interaction with surface followed by diffusion of pollutant species on active sites of the functional group (Hu et al. 2011; Zhao et al. 2010). Previously, use of iron oxide based nanomaterials for the decontamination of organic pollutants like carbon tetrachloride (CCl_4), trichlorobenzene ($C_6H_3Cl_3$), 1,1-dichloro ethane ($C_2H_2Cl_2$), hexachlorobenzene (C_6Cl_6), lindane ($C_6H_6Cl_6$), DDT ($C_{14}H_9Cl_5$), orange II ($C_{16}H_{11}N_2NaO_4S$), tropaeolin O ($C_{12}H_9N_2NaO_5S$), n-nitro sodium ethylamine ($C_4H_{10}N_2O$), chrysoidin ($C_{12}H_{13}C_lN_4$), TNT ($C_7H_5N_3O_6$) etc. has been observed (Li et al. 2006a, b; Zhang 2003). Microbial decontamination of pathogenic bacteria by the use of iron oxide nanomaterial have been demonstrated by a magnetic nanoparticle comprises of iron oxide/titania ($Fe_3O_4@TiO_2$) core/shell (Chen et al. 2008). The mechanism may involve the adsorbent capacity to inhibit the cell growth targeted under low power UV irradiation with a short period of time. Different types of pollutants and the iron oxide-based adsorbents are shown in Table 3.4.

3.3.3 Zinc Oxide (ZnO) Nanomaterials

Zinc oxide nanomaterials is found another important and efficient metal oxide-based nanomaterials in water treatment process due to their useful properties like environment friendly, variable oxidation state and efficient photocatalytic properties (similar to titanium oxide nanoparticles because of almost same band gap energy) with the advantage of its cheapness (Janotti and Van deWalle 2009; Reynolds et al. 1999; Chen et al. 1998; Schmidt-Mende and MacManus-Driscoll 2007; Daneshvar et al. 2004; Behnajady et al. 2006; Gomez-Solís et al. 2015; Singh et al. 2013a, b, c).

Various strategies have been used for the improvement of zinc oxide nanomaterials for the decontamination of water/waste water such as metal doping including anionic, cationic, rare earth etc. (Lee et al. 2016), coupling with other semiconductor like CdO (Samadi et al. 2014), CeO_2 (Liu et al. 2014a, b), SnO_2 (Uddin et al. 2012), TiO_2 (Pant et al. 2012), graphene oxide (Dai et al. 2014) and reduced graphene oxide (Zhou et al. 2012a, b). For example, use of transition metal/non-metal doping (Lu et al. 2011; Hsu and Chang 2014), and loading with noble metal (Yin et al. 2012) have been summarized to modify the band gap electrically conductive support is used to improve the photocatalytic efficiency of Zinc oxide. Ba-Abbad et al. (2013) reported the preparation of Fe doped Zinc oxide matrix by sol-gel method with improved photocatalytic activity by introducing the ferromagnetism to the host material ad changed the lattice constant and optical property. Co and Mn as dopants have been reported by (Ekambaram et al. 2007) to increase the mobility of charge carrier. Graphene based nanocomposites is used to enhance the photocatalytic activity of zinc oxide nanoparticles in water treatment by limiting electron/hole recombination and by preventing corrosion and leaching of metal oxide nanoparticle into water (Xu et al. 2011; Mohammad et al. 2018a, b). Other important adsorbents of zinc oxide nanomaterials for the remediation of different pollutants have been listed in Table 3.5.

3.3.4 Aluminium Oxide (Al_2O_3) Nanomaterials

Aluminium oxide nanomaterials have been used for the decontamination of heavy metals, defluorination, nitrate removal and dyes (Ahmad et al. 2017). Carbon nanotubes (CNTs) on micro-sized Al_2O_3 adsorbent is used in the decontamination of Pb^{2+}, Cu^{2+}, and Cd^{2+} from water following adsorption order as: $Cd^{2+} < Cu^{2+} < Pb^{2+}$ (Hsieh and Horng 2007). Afkhami et al. (2010) shows the decontamination of Cd(II),

Table 3.4 Different types of iron oxide-based nanomaterial adsorbents and pollutants

S. No.	Adsorbents	Polluted species	References
1	Magnetic iron–nickel oxide or (MnO_2/Fe_3O_4/MWCNT)	Cr(VI)	Wei et al. (2009), Luo et al. (2013)
2	Fe_3O_4-humic acid nanoparticles	Pb(II), Hg(II), Cu(II) Cd(II)	Liu et al. (2008a)
3	Fe_2O_3/biochar or Iron oxide–based carbon aerogel	As(V)	Zhang et al. (2013), Lin and Chen (2014)
4	Magnetic chitosan	Pb(II)	Liu et al. (2008a, b)
5	Mesoporous iron oxide carbon encapsulates	As(V)	Wu et al. (2012)
6	Ascorbic acid-Fe_3O_4	As^{3+}, As^{5+}	Feng et al. (2012)
7	DMSA-Fe_3O_4	Co^{2+}, Cu^{2+}, Cr^{3+}, Ni^{2+}, Cd^{2+}, Pb^{2+}, As^{3+}	Singh et al. (2011a, b)
8	Fe_3O_4-ZnO nanoparticle/nanocomposite/polymer modified nanocomposite	Ni^{2+}, Cu^{2+}, Cd^{2+}, Co^{2+}, Pb^{2+}, As^{3+} Hg^{2+}	Feng et al. (2012), Singh et al. (2013a, b, c)
9	Fe_3O_4@APS@AA-coCA Fe_3O_4	Pb^{2+}, Cd^{2+}, Zn^{2+} Cu^{2+}	Singh et al. (2011a, b)
10	Fe_3O_4–SiO_2-poly(1,2diaminobenzene)	Cu^{2+}, As^{3+}, Cr^{3+}	Zhang et al. (2012a, b)
11	EDTA-γ-Fe_3O_4@SiO_2 / Thiol-γ-Fe_3O_4@SiO_2	Cd^{2+}, Pb^{2+}, Hg^{2+}, As^{3+}	Sinha and Jana (2012)
12	Fe_3O_4/SiO_2/Schiff base	Pb^{2+}, Cd^{2+}, Cu^{2+}	Bagheri et al. (2012)
13	Iron oxide nanostructures	Orange dye II	Zhong et al. (2006)
14	Mesoporous carbon iron oxide nanocomposite	2,4-Dichloro phenoxy acetic acid	Tang et al. (2015)
15	Amine/Fe_3O_4 functionalized with resin	Methyl orange, reactive brilliant red K-2BP, and acid red 18	Song et al. (2016)
16	Fe_3O_4-loaded coffee waste hydrochar	Acid red 17	Khataee et al. (2017)
17	Magnetic carbon composite	Malachite green	Song et al. (2016)
18	Graphene/iron oxide	1-Naphthol and 1-naphthylamine	Zhu et al. (2014)
19	Activated carbon trapped on iron oxide/graphene oxide/reduced grapheme oxide-iron oxide hybrid	Methyl blue and methyl orange	Luo and Xang (2009), Yang et al. (2012a, b)
20	Magnetic Fe_3O_4/C	Methylene blue and cresol red	Zhang and Kong (2011)

Cr(III) and Pb(II) ions from waters through nano alumina modified with 2, 4-dinitrophenylhydrazine. Another study reports that the iron oxide–alumina mixed nanocomposite fibre is used for the decontamination of Cu^{2+}, Pb^{2+}, Ni^{2+} and

Table 3.5 Different types of zinc oxide-based nanomaterial adsorbents and pollutants

S. No.	Adsorbents	Polluted species	References
1	ZnO Nano-assembly	Cu^{2+}, Pb^{2+}, Co^{2+}, Hg^{2+}, Cd^{2+}, As^{3+}, Ni^{2+}	Singh et al. (2011a, b)
2	Mesoporous ZnO nanorods	Pb^{2+}, Cd^{2+}	Kumar et al. (2013)
3	ZnO Nanorods	Methyl Orange, Rh6G	Ma et al. (2011), Kim and Huh (2011)
5	Flower like ZnO	Phenol	Xu et al. (2009)
6	ZnO (Nano-disks)	Methyl Orange	Zhai et al. (2012)
7	ZnO (porous octahedron)	Methyl Orange	Zheng et al. (2009)
8	ZnO (flower like assembly)	MB, RhB	Singh et al. (2013a, b, c)
9	1% Ag-doped ZnO	Methyl Orange	Yildirim et al. (2013)
10	Cu-doped ZnO	Resazurin	Mohan et al. (2012)
11	Mn doped ZnO Nano-assembly	MB	Barick et al. (2010)
12	Co doped ZnO	RhB	Qiu et al. (2008)

Hg^{2+} (Mahapatra et al. 2013). Activated Al_2O_3 loaded with alum at pH: 6.5 (Tripathy et al. 2006) and nano-AlO(OH) at pH: 7 ha been used for the removal of fluoride ion from water (Wang et al. 2009a, b). Further study by Bhatnagara et al. (2010) showed the decontamination of nitrate from waters using alumina nanomaterials with the sorption capacity 4.0 mg/g at optimum pH: 4.4.

Recently, functionalized Al_2O_3 membranes have been reported for water filtration processes. For example, DeFriend et al. studied the fabrication of alumina ultrafilter (UF) membranes which is showing high selectivity toward dyes. Removal of methyl violet and malachite green from water is achieved by nano aluminium hydroxide (Kerebo et al. 2016) and nano Al_2O_3 particle (Pathania et al. 2016). Few aluminium oxide nanomaterials as adsorbents have been listed along with polluted species in Table 3.6.

3.3.5 Cerium Oxide (CeO_2) Nanomaterials

Rare earth oxides emerge as other adsorbents for the decontamination of arsenate, fluoride and phosphate anions as well as for removal of arsenate, fluoride and phosphate anions (Hideaki et al. 1987; Tokunaga et al. 1995). Ceria among them is considered to be important due to its unique properties like ease in conduction of oxygen ion and increase in capacity to store oxygen (Trovarelli 1996; Inaba and Tagawa 1996). However, use of cerium oxide nanomaterials in water treatment is rare, due to their costs involved. Although few adsorbents are used to remove

Table 3.6 Different types of aluminium oxide-based nanomaterial adsorbents and pollutants

S. No.	Adsorbents	Polluted species	References
1	Alumina ultrafilter membrane	Synthetic dyes	DeFriend et al. (2003)
2	Alumina Nanofilter membrane	Ca^{2+}, Mg^{2+}, Cl^-/SO_4^{2-}	Stanton et al. (2003)
3	Pd-Cu/ γ –alumina	NO_3^-	Chaplin et al. (2006)
4	Alumina-polyelectrolytes and citrate-stabilized gold nanoparticles	4- nitrophenol	Dotzauer et al. (2006)

various contaminants using cerium-based nanomaterials like Carbon nanotubes (CeO_2-CNT) (Peng et al. 2005) or on aligned carbon nanotubes (CeO_2-ACNTs) supported by cerium oxide nanomaterials are effective in removing As(V) (Di et al. 2006).

3.4 Mechanism of Metal/Metal Oxide-Based Nanomaterials Used in Waste Water Decontamination

Various methods such as coagulation, oxidation, adsorption electrochemical methods, and bioremediation etc. have been used for the effective removal of pollutants from water. Adsorption among them is one of the important and widely used method for water/waste water treatment (Dutta et al. 2001; Moghaddam et al. 2010; Gao et al. 2012; Wang et al. 2006a, b; Gandhi et al. 2016; Upadhyay et al. 2014; Perreault et al. 2015). There are various nano sized metal and metal oxide adsorbents for the effective removal of different contaminants from water (Yang et al. 2019). Silver nanoparticles when added with ceramic filters have emerged as the effective removal of *Escherichia coli* with high porosity (Kallman et al. 2011). Gold nanomaterial is one of the important adsorbents for the decontamination of mercury due to easy formation of AuHg, $AuHg_3$, and Au_3Hg (Zhang et al. 2016).

The basic mechanism of photocatalysis using metal and metal oxide nanoparticles is shown in Fig. 3.2. The photocatalysis may involve the light absorption, electron-hole pair generation, and the separation and free charge carrier-induced redox reactions. Previous reports suggest that the nano-based photocatalysts can increase the oxidation ability by producing oxidizing agents at material surface which helps in degradation of pollutants in water (Yang et al. 2019; Shinde et al. 2017; Khan et al. 2016; Moon et al. 2014; O'Carroll et al. 2013).

Zero-valent iron is one of the most extensively studied metal-based nanomaterials for water treatment processes and discussed briefly with possible mechanism in this part. Nanoscale zero valent iron (free nanoscale zero valent iron size less than 100 nm diameter) is a composite formed by Fe (0) and ferric oxide coating (Fig. 3.3) (O'Carroll et al. 2013). Nanoscale zero valent iron particles reacted with water and oxygen giving an iron (hydroxide layer which may lead to the formation

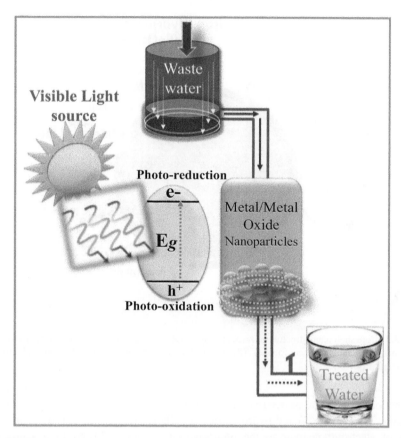

Fig. 3.2 Basic Mechanism of metal and metal-oxide mediated photocatalysis under visible-light irradiation

of core shell structure in aqueous solution (Nurmi et al. 2005; Li and Zhang 2007; Sun et al. 2006). The oxide layer may allow the electron transfer from the metal to reduce the contaminants. The electron transfer process may be achieved indirectly by various bands (conduction, impurity, localized), or directly through defect processes or by sorption or structural Fe^{2+} (Li et al. 2006a, b). The outer (hydroxide layer may also function as an adsorbent for various contaminants,

Metal oxide based nonmaterial's is found to be another important adsorbent used for the decontamination of water/waste water due to their high removal efficiency and selectivity. Manganese oxide-based nanomaterials with its high surface area contributes high adsorption capacity and $M-O^{\delta+}$ and $M-O^{\delta-}$ units on manganese oxide surface enhances the sorption of pollutants (Mukherjee et al. 2013).

Recently, degradation of pollutants by photocatalysis has emerged as one of the important technologies in waste water treatment. The possible principle involve in photocatalysis is that photocatalyst absorb a photon of light energy equal to or greater than the photocatalyst band gap energy which causes electron/hole pair

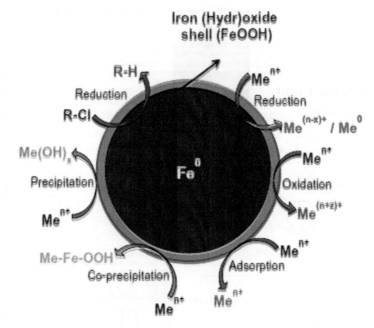

Fig. 3.3 Core shell nano-structure of zerovalent iron (nZVI) describing several mechanisms for the elimination of chlorinated compounds. (Reproduced with permission from (O'Carroll et al. 2013). Copyright Elsevier, 2013)

formation. Generation of reactive molecules like hydrogen peroxide (H_2O_2), superoxide ion radical (O^{2-}) and hydroxyl radical ($OH^{\cdot}$) may be achieved by the migration of electron/hole pair to the photocatalyst surface followed by reaction of absorbed pollutants (Gaya and Abdullah 2008; Thongsuriwong et al. 2012; Pare et al. 2009). It is believed that photocatalysis may involve the following steps; (i) formation of hole by the promotion of electron from valence band to conduction band (ii) Migration of excited electron and hole towards the surface (Wang et al. 2014a, b). (iii) Reaction of electron and hole with the electron donor and electron acceptor due to the fact that the chemical potential of electron is $+0.5$ to -1.5 V w.r.t. NHE which exhibit a strong reductive potential while the chemical potential of hole is $+1.0$ to $+3$ V w.r.t. NHE which exhibit strong oxidative potential. The species formed are extremely reactive and may involve in degradation of different pollutants into harmless products. To better increase the photocatalysis of various adsorbents separation of electron/hole pair should be higher along with ease of the charge transfer.

Recent investigation from the last few decades suggested that metal oxide or sulphide semiconductors are the important photocatalysts among which titanium dioxide has been widely studied. Titanium dioxide photocatalyst having the large band gap energy of 3.2 eV excited to induce charge separation within the particles under ultraviolet (UV) region. Titanium dioxide under UV irradiation leads to the generation of reactive oxygen species that can degrade the contaminants completely

Fig. 3.4 Plausible mechanism of titanium dioxide photocatalytic process. (Reproduced with permission from (Moon et al. 2014). Copyright American Chemical Society, 2014)

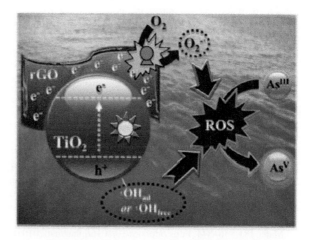

in very short reaction time (Fig. 3.4) (Moon et al. 2014). On the other hand, hydroxyl radicals which may be formed in the photocatalytic process enable titanium dioxide nanoparticles to affect various cells (Mills and Le Hunte 1997).

3.5 Conclusion and Future Perspective

Present chapter describes the water treatment process which still needed development of more advanced materials to get clean water. Adsorbents process is found to be the effective process in water treatment. In this regards, various advanced adsorbents have been developed in water treatment process. Remediation of water is mainly depending on material properties and its types which can be used to enhance the removal efficiency. Nano based adsorbents is emerged as the important candidate for the decontamination of water and various nanomaterials based adsorbents have already been reported in water treatment process. Metal based nanomaterial and metal oxide-based nanomaterials are used now a day with increased purifying capability, cheapness and selectivity. Various metal and metal oxide-based nanomaterials as adsorbents have been known and discussed for the efficient decontamination of different pollutants from water/ waste water and emerged as one of the most advanced and useful material for the purification of water. However, some limitations are found like these adsorbents may contaminate the water / waste water during sample treatment processes which can seriously affect the environment and human being. In conclusion, advanced materials are still needed to develop in water treatment process with less toxicity, cheap and maximum removal capability of pollutants.

Acknowledgements We would like to kindly acknowledge all the work taken into consideration. M. Tauqeer gratefully acknowledges the Department of Chemistry, A.M.U.,

Aligarh. P. Bansal is grateful to the Department of Chemistry and Department of Industrial Chemistry, A.M.U. Aligarh., RSJ gratefully acknowledges the SIC and Department of Chemistry, IIT Indore. Mohammad Ehtisham Khan is indebted to the deanship research program of Jazan University, Kingdom of Saudi Arabia and A. Mohammad is grateful to IIT Indore for Institute Postdoctoral fellowship. There is no conflict of interest for the author.

References

Adeleye AS, Conway JR, Garner K, Huang Y, Su Y, Keller AA (2016) Engineered nanomaterials for water treatment and remediation: costs, benefits, and applicability. Chem Eng J 286:640–662. https://doi.org/10.1016/j.cexcXcXXvccvccvj.2015.10.105

Afkhami A, Saber-Tehrani M, Bagheri H (2010) Simultaneous removal of heavy-metal ions in wastewater samples using nano-alumina modified with 2,4-dinitrophenylhydrazine. J Hazard Mater 181:836–844. https://doi.org/10.1016/j.jhazmat.2010.05.089

Ahmad IZ, Ahmad A, Tabassum H, Kuddus M (2017) Applications of nanoparticles in the treatment of wastewater. In: Martínez LMT et al (eds) Handbook of ecomaterials. Springer, Cham. https://doi.org/10.1007/978-3-319-48281-1_37-1

Akpan UG, Hameed BH (2009) Parameters affecting the photocatalytic degradation of dyes using TiO_2 based photocatalysts: a review. J Hazard Mater 170:520–529. https://doi.org/10.1016/j.jhazmat.2009.05.039

Alalm MG, Tawfik A, Ookawara S (2015) Comparison of solar TiO_2 photocatalysis and solar photo-Fenton for treatment of pesticides industry wastewater: operational conditions, kinetics, and costs. J Water Process Eng 8:55–63. https://doi.org/10.1016/j.jwpe.2015.09.007

Alshammari AS, Bagabas A, Alarifi N, Altamimi R (2019) Effect of the nature of metal nanoparticles on the photocatalytic degradation of rhodamine B. Top Catal 62:7–11. https://doi.org/10.1007/s11244-019-01180-3

Anjum M, Miandad R, Waqas M, Gehnay F, Barakat MA (2016) Remediation of wastewater using various nanomaterials. Arab J Chem 12:4897. https://doi.org/10.1016/j.arabjc.2016.10.004

Annadhasan M, Kasthuri J, Rajendiran N (2015) Green synthesis of gold nanoparticles under sunlight irradiation and their colorimetric detection of Ni^{2+} and Co^{2+} ions. RSC Adv 5(15):11458–11468. https://doi.org/10.1039/C4RA14034F

Arancibia MN, Baltazar SE, García A (2016) Nanoscale zero valent supported by zeolite and montmorillonite: template effect of the removal of lead ion from an aqueous solution. J Hazard Mater 301:371–380. https://doi.org/10.1016/j.jhazmat.2015.09.007

Ashbolt NJ (2004) Microbial contamination of drinking water and disease outcomes in developing regions. Toxicology 198:229–238. https://doi.org/10.1016/j.tox.2004.01.030

Ba-Abbad MM, Kadhum H, Mohamad AA, Takriff AB, Sopian MS (2013) Visible light activity of Fe^{3+} doped ZnO nanoparticle prepared viasol-gel technique. Chemosphere 91:1604–1611. https://doi.org/10.1016/j.chemosphere.2012.12.055

Bagheri H, Afkhami A, Saber-Tehrani M, Khoshsafar H (2012) Preparation and characterization of magnetic nanocomposite of Schiff base/silica/magnetite as a preconcentration phase for the trace determination of heavy metal ions in water, food and biological samples using atomic absorption spectrometry. Talanta 97:87–95. https://doi.org/10.1016/j.talanta.2012.03.066

Bagwasi S, Tian B, Zhang J, Nasir M (2013) Synthesis, characterization and application of bismuth and boron Co-doped TiO_2: a visible light active photocatalyst. Chem Eng J 217:108–118. https://doi.org/10.1016/j.cej.2012.11.080

Bahrami-Teimoori B, Pourianfar HR, Akhlaghi M, Abbas T, Rezayi M (2019) Biosynthesis and antibiotic activity of silver nanoparticles using different sources: glass industrial sewage-adapted Bacillus sp. and herbaceous Amaranthus sp. Biotechnol Appl Biochem 66:900. https://doi.org/10.1002/bab.1803

Banat IM (1996) Microbial decolorization of textile-dye containing effluents: a review. Bioresour Technol 58:217–227. https://doi.org/10.1016/S0960-8524(96)00113-7

Barakat MA, Schaeffer H, Hayes G, Ismat-Shah S (2005) Photocatalytic degradation of 2-chlorophenol by Co-doped TiO_2 nanoparticles. Appl Catal B Environ 57:23–30. https://doi.org/10.1016/j.apcatb.2004.10.001

Barick KC, Singh S, Aslam M, Bahadur D (2010) Porosity and photocatalytic studies of transition metal doped ZnO nanoclusters. Microporous Mesoporous Mater 134:195–202. https://doi.org/10.1016/j.micromeso.2010.05.026

Behnajady MA, Modirshahla N, Hamzavi R (2006) Kinetic study on photocatalytic degradation of C.I. acid yellow 23 by ZnO photocatalyst. J Hazard Mater 133(1–3):226–232. https://doi.org/10.1016/j.jhazmat.2005.10.022

Bensalah N, Alfaro M, Martínez-Huitle C (2009) Electrochemical treatment of synthetic wastewaters containing Alphazurine A dye. Chem Eng J 149:348–352. https://doi.org/10.1016/j.cej.2008.11.031

Berge ND, Ramsburg CA (2009) Oil-in-water emulsions for encapsulated delivery of reactive iron particles. Environ Sci Technol 43(13):5060–5066. https://doi.org/10.1021/es900358p

Bhatnagar A, Sillanpää M (2009) Applications of chitin- and chitosan derivatives for the detoxification of water and wastewater — a short review. Adv Colloid Interf Sci 152:26–38. https://doi.org/10.1016/j.cis.2009.09.003

Bhatnagara A, Kumar E, Sillanpaa M (2010) Nitrate removal from water by nano-alumina: characterization and sorption studies. Chem Eng J 163:317. https://doi.org/10.1016/j.cej.2010.08.008

Bokare V, Jung JL, Chang Y, Chang YS (2013) Reductive dechlorination of octachlorodibenzo-p-dioxin by nanosized zero-valent zinc: modeling of rate kinetics and congener profile. J Hazard Mater 250–251:397–402. https://doi.org/10.1016/j.jhazmat.2013.02.020

Cao CY, Qu J, Yan WS, Zhu JF, Wu ZY, Song WG (2012) Low-cost synthesis of flowerlike α-Fe_2O_3 nanostructures for heavy metal ion removal: adsorption property and mechanism. Langmuir 28:4573–4579. https://doi.org/10.1021/la300097y

CDC (2011) Phenol. NIOSH pocket guide to chemical hazards. https://www.cdc.gov/niosh/npg/npgd0493.html

Chang H, Wu H (2013) Graphene-based nanocomposites: preparation, functionalization, and energy and environmental application. Energy Environ Sci 6:3483–3507. https://doi.org/10.1039/C3EE42518E

Chaplin BP, Roundy E, Guy KA, Shaply JR, Werth CJ (2006) Effects of natural water ions and humic acid on catalytic nitrate reduction kinetics using an alumina supported Pd-Cu catalyst. Environ Sci Technol 40:3075–3081. https://doi.org/10.1021/es0525298

Chatterjee D, Dasgupta SJ (2005) Visible light induced photocatalytic degradation of organic pollutants. Photobiol C: Photochem Rev 6(2005):186–205. https://doi.org/10.1016/j.jphotochemrev.2005.09.001

Chen ZP, Guoetal YLM (2016) One-pot synthesis of Mn-doped TiO_2 grown on graphene and the mechanism for removal of Cr(VI) and Cr(III). J Hazard Mater 310:188–198. https://doi.org/10.1016/j.jhazmat.2016.02.034

Chen Y, Bagnall DM, Koh HJ et al (1998) Plasma assisted molecular beam epitaxy of ZnO on c-plane sapphire: growth and characterization. J Appl Phys 84(7):3912–3918. https://doi.org/10.1063/1.368595

Chen WJ, Tsai PJ, Chen YJ (2008) Functional Fe_3O_4/TiO_2 core/shell magnetic nanoparticles as photokilling agents for pathogenic bacteria. Small 4:485–491. https://doi.org/10.1002/smll.200701164

Chen ZX, Jin XY, Chen Z, Megharaj M, Naidu R (2011) Removal of methyl orange from aqueous solution using bentonite-supported nanoscale zero-valent iron. J Colloid Interface Sci 363(2):601–607. https://doi.org/10.1016/j.jcis.2011.07.057

Cheng X, Yu X, Xing Z, Wan J (2012) Enhanced photocatalytic activity of nitrogen doped TiO_2 anatase nano-particle under simulated sunlight irradiation. Energy Procedia 16:598–605. https://doi.org/10.1016/j.egypro.2012.01.096

Chiou C-H, Juang RS (2007) Photocatalytic degradation of phenol in aqueous solutions by Pr-doped TiO_2 nanoparticles. J Hazard Mater 149:1–7. https://doi.org/10.1016/j.jhazmat.2007.03.035

Clarke E, Anliker R (1980) Organic dyes and pigments. In: Handbook of environmental chemistry, vol 3, pp 181–215. https://doi.org/10.1007/978-3-540-38522-6_7

Cornell RM, Schwertmann U (2003) The iron oxides: structure, properties, reactions, occurrences and uses. Wiley, Hoboken. https://doi.org/10.1002/3527602097

Cui Y, Liang B, Zhang J, Wang R, Sun H, Wang L, Gao D (2019) Polyethyleneimine-stabilized palladium nanoparticles for reduction of 4-nitrophenol. Transit Met Chem 44:655–662. https://doi.org/10.1007/s11243-019-00330-6

Daer S, Kharraz D, Giwa A, Hasan SA (2015) Recent applications of nanomaterials in water desalination: a critical review and future opportunities. Desalination 367:37–48. https://doi.org/10.1016/j.desal.2015.03.030

Dai K, Lu L, Liang C et al (2014) Graphene oxide modified ZnO nanorods hybrid with high reusable photocatalytic activity under UV-LED irradiation. Mater Chem Phys 143(3):1410–1416. https://doi.org/10.1016/j.matchemphys.2013.11.055

Daneshvar N, Salari D, Khataee AR (2004) Photocatalytic degradation of azo dye acid red 14 in water on ZnO as an alternative catalyst to TiO_2. J Photochem Photobiol A Chem 162(2–3):317–322. https://doi.org/10.1016/S1010-6030(03)00378-2

Daraei P, Madaeni SS, Salehi E, Ghaemi N, Ghari HS, Khadivi MA, Rostami E (2013) Novel thin film composite membrane fabricated by mixed matrix nanoclay/chitosan on PVDF microfiltration support: preparation, characterization and performance in dye remova. J Membr Sci 436:97–108. https://doi.org/10.1016/j.memsci.2013.02.031

Dawood S, Sen TK, Phan C (2014) Synthesis and characterization of novel-activated carbon from waste biomass pine cone and its application in the removal of Congo red dye from aqueous solution by adsorption. Water Air Soil Pollut 225:1–16. https://doi.org/10.1007/s11270-013-1818-4

DeFriend KA, Wiesner MR, Barron AR (2003) Alumina and aluminate ultrafiltration membranes derived from alumina nanoparticles. J Membr Sci 224:11–28. https://doi.org/10.1016/S0376-7388(03)00344-2

Di ZC, Ding J, Peng XJ, Li YH, Luan ZK, Liang J (2006) Chromium adsorption by aligned carbon nanotubes supported ceria nanoparticles. Chemosphere 62:861–865. https://doi.org/10.1016/j.chemosphere.2004.06.044

Diker H, Varlikli C, Mizrak K, Dana A (2011) Characterizations and photocatalytic activity comparisons of N-doped nc-TiO_2 depending on synthetic conditions and structural differences of amine sources. Energy 36:1243–1254. https://doi.org/10.1016/j.energy.2010.11.020

Dong F, Guo S, Wang H, Li X, Wu Z (2011) Enhancement of the visible light photocatalytic activity of C-doped TiO_2 nanomaterials prepared by a green synthetic approach. J Phys Chem C 115:13285–13292

Dotzauer DM, Dai J, Sun L, Bruening ML (2006) Catalytic membrane prepared using layer – by-layer adsorption of polyelectrolyte/metal nanoparticle films in porous supports. Nano Lett 6:2268–2272. https://doi.org/10.1021/nl061700q

Dutta K, Mukhopadhyay S, Bhattacharjee S, Chaudhuri B (2001) Chemical oxidation of methylene blue using a Fenton-like reaction. J Hazard Mater 84:57–71. https://doi.org/10.1016/s0304-3894(01)00202-3

Egerton TA, Kosa SA, Christensen PA (2006) Photoelectrocatalytic disinfection of E. coli suspensions by iron doped TiO_2. Phys Chem Chem Phys: PCCP 8:398–406. https://doi.org/10.1039/B507516E

Ekambaram S, Iikubo Y, Kudo A (2007) Combustion synthesis and photocatalytic activity of transition metal incorporated ZnO. J Alloys Compd 433:237–240. https://doi.org/10.1016/j.jallcom.2006.06.045

El-Temsah YS, Oughton DH, Joner EJ (2013) Effects of nano-sized zero-valent iron on DDT degradation and residual toxicity in soil: a column experiment. Plant Soil 368(1/2):189–200. https://doi.org/10.1007/s11104-012-1509-8

Falconer IR, Humpage AR (2005) Health risk assessment of cyanobacterial (blue-green algal) toxins in drinking water. Int J Environ Res Public Health 2:43–50. https://doi.org/10.3390/ijerph2005010043

Fawell J, Nieuwenhuijsen MJ (2003) Contaminants in drinking water. Br Med Bull 68:199–208. https://doi.org/10.1093/bmb/ldg027

Fei J, Li J (2010) Metal oxide nanomaterials for water treatment. In: Kumar CS (ed) Nanotechnologies for the life sciences. Wiley, Weinheim. https://doi.org/10.1002/9783527610419.ntls0145

Feng L, Cao M, Ma X, Zhu Y, Hu C (2012) Superparamagnetic high-surface-area Fe_3O_4 nanoparticles as adsorbents for arsenic removal. J Hazard Mater 217–218:439–446. https://doi.org/10.1016/j.jhazmat.2012.03.073

Ferroudj N, Nzimoto J, Davidson A, Talbot D, Briot E, Dupuis V, Abramson S (2013) Maghemite nanoparticles and maghemite/silica nanocomposite microspheres as magnetic Fenton catalysts for the removal of water pollutants. Appl Catal B Environ 136:9–18. https://doi.org/10.1016/j.apcatb.2013.01.046

Foster HA, Ditta IB, Varghese S, Steele A (2011) Photocatalytic disinfection using titanium dioxide: spectrum and mechanism of antimicrobial activity. Appl Microbiol Biotechnol 90(6):1847–1868. https://doi.org/10.1007/s00253-011-3213-7

Fu F, Dionysiou DD, Liu H (2014) The use of zero-valent iron for groundwater remediation and wastewater treatment: a review. J Hazard Mater 267:194–205. https://doi.org/10.1016/j.jhazmat.2013.12.062

Fu Y, iQin L, Huang DL, Zeng G, Lai C, Li B, He JF, Yi H, Zhang M, Cheng M, Wen XF (2019) Chitosan functionalized activated coke for Au nanoparticles anchoring: green synthesis and catalytic activities in hydrogenation of nitrophenols and azo dyes. Appl Catal B Environ 255:117740. https://doi.org/10.1016/j.apcatb.2019.05.042

Gajda S (1996) Synthetic dyes based on environmental considerations. Dyes Pigments 30:1–20. https://doi.org/10.1016/0143-7208(95)00048-8

Gandhi MR, Vasudevan S, Shibayama A, Yamada M (2016) Graphene and graphene-based composites: a rising star in water purification – a comprehensive overview. ChemistrySelect 1:4358–4385. https://doi.org/10.1002/slct.201600693

Gao Y, Pu X, Zhang D, Ding G, Shao X, Ma J (2012) Combustion synthesis of graphene oxide–TiO_2 hybrid materials for photodegradation of methyl orange. Carbon 50:4093–4101. https://doi.org/10.1016/j.carbon.2012.04.057

Gaya UI, Abdullah AH (2008) Heterogeneous photocatalytic degradation of organic contaminants over TiO_2. J Photochem Photobiol C: Photochem Rev 9:1–12. https://doi.org/10.1016/j.jphotochemrev.2007.12.003

Gomez-Solís C, Ballesteros JC, Torres-Martínez LM et al (2015) Rapid synthesis of ZnO nano-corncobs from Nital solution and its application in the photodegradation of methyl orange. J Photochem Photobiol A Chem 298:49–54. https://doi.org/10.1016/j.jphotochem.2014.10.012

Grey D, Garrick D, Blackmore D, Kelman J, Muller M, Sadof C (2013) Water security in one blue planet: twenty-first century policy challenges for science. Philos Trans R Soc Lond A Math Phys Eng Sci 371:20120406. https://doi.org/10.1098/rsta.2012.0406

Gueye MT, Palma LD, Allahverdeyeva G (2016) Hexavalent chromium reduction by nano zero valent iron in soil. Chem Eng Trans 47:289–294. https://doi.org/10.3303/CET1647049

Gunarani GI, Raman AB, Kumar JD, Natarajan S, Jegadeesan GB (2019) Biogenic synthesis of Fe and NiFe nanoparticles using Terminalia bellirica extracts for water treatment applications. Mater Lett 247:90–94. https://doi.org/10.1016/j.matlet.2019.03.104

Guo M, Song W, Wang T, Li Y, Wang X, Du X (2015) Phenyl functionalization of titanium dioxide-nanosheets coating fabricated on a titanium wire for selective solid-phase microextraction of polycyclic aromatic hydrocarbons from environment watersamples. Talanta 144:998–1006. https://doi.org/10.1016/j.talanta.2015.07.064

Gupta G, Prasad G, Singh V (1990) Removal of chrome dye from aqueous solutions by mixed adsorbents: fly ash and coal. Water Res 24:45–50. https://doi.org/10.1016/0043-1354(90)90063-c

Hai CH, Chen CY (2001) Removal of metal ions and humic acid from water by iron-coated filter media. Chemosphere 44:1177–1184. https://doi.org/10.1016/s0045-6535(00)00307-6

Herrera F, Lopez A, Mascolo G, Albers E, Kiwi J (2001) Catalytic combustion of Orange II on hematite: surface species responsible for the dye degradation. Appl Catal B 29:147–162. https://doi.org/10.1016/S0926-3373(00)00198-3

Hideaki I, Junji N, Yuzuru I, Tokuzo K (1987) Anion adsorption behavior of rare earth oxide hydrates. Chem Soc Jpn 5:807–813. https://doi.org/10.1246/nikkashi.1987.807

Hoag GE, Collins JB, Holcomb JL, Hoag JR, Nadagouda MN, Varma RS (2009) Degradation of bromothymol blue by 'greener' nano-scale zero-valent iron synthesized using tea polyphenols. J Mater Chem 19(45):8671–8677. https://doi.org/10.1039/B909148C

Hoffmann MR, Martin ST, Choi W, Bahnemann DW (1995) Environmental applications of semiconductor photocatalysis. Chem Rev 95:69–96. https://doi.org/10.1021/cr00033a004

Hou H, Shang M, Wang L, Li W, Tang B, Yang W (2015) Efficient photocatalytic activities of TiO_2 hollow fibers with mixed phases and mesoporous walls. Sci Rep 5:15228. https://doi.org/10.1038/srep15228

Hsieh S-H, Horng J-J (2007) Adsorption behavior of heavy metal ions by carbon nanotubes grown on microsized Al_2O_3 particles. J Univ Sci Technol B 14(1):77–84. https://doi.org/10.1016/S1005-8850(07)60016-4

Hsu MH, Chang CJ (2014) S-doped ZnO nanorods on stainless-steel mesh as immobilized hierarchical photocatalysts for photocatalytic H_2 production. Int J Hydrog Energy 39:16524–16533. https://doi.org/10.1016/j.ijhydene.2014.02.110

Hu J, Shao D, Chen C, Sheng G, Ren X, Wang X (2011) Removal of 1-naphthylamine from aqueous solution by multiwall carbon nanotubes/iron oxides/cyclodextrin composite. J Hazard Mater 185:463–471. https://doi.org/10.1016/j.jhazmat.2010.09.055

Hu A, Liang R, Zhang X, Kurdi S, Luong D, Huang H, Peng P, Marzbanrad E, Oakes KD, Zhou Y, Servos MR (2013) Enhanced photocatalytic degradation of dyes by TiO_2 nanobelts with hierarchical structures. J Photochem Photobiol A 256:7–15. https://doi.org/10.1016/j.jphotochem.2013.01.015

Rivero-Huguet M, Marshall WD (2009) Reduction of hexavalentchromium mediated by micron- and nano-scale zerovalentmetallic particles. J Environ Monit 11(5):1072–1079. https://doi.org/10.1039/B819279K

Hutton G, Haller L, Bartram J (2007) Economic and health effects of increasing coverage of low-cost household drinking water supply and sanitation interventions. World Health Organization WHO (World Health Organization) Drinking-water: Fact sheet no. 391. http://www.who.int/mediacentre/factsheets/fs391/en/

Inaba H, Tagawa H (1996) Ceria-based solid electrolytes. Solid State Ionics 83:1–16. https://doi.org/10.1016/0167-2738(95)00229-4

Ivanov K (1996) Possibilities of using zeolite as filler and carrier for dyestuffs in paper. Papier-Zeitschrift fur die Erzeugung von HolzstoffZellstoff Papier und Pappe 50: 456–460. ISSN:0031-1340

Janotti A, Van deWalle CG (2009) Fundamentals of zinc oxide as a semiconductor. Rep Prog Phys 72(12):126501. https://doi.org/10.1088/0034-4885/72/12/126501

Jia WN, Wu X, Jia BX, Qu FY, Fan HJ (2013) Self-assembled porous ZnS nanospheres with high photocatalytic performance. Sci Adv Mater 5:1329–1336. https://doi.org/10.1002/adma.200401839

Jiang F, Zheng Z, Xu Z, Zheng S, Guo Z, Chen L (2006) Aqueous Cr(VI) photo-reduction catalyzed by TiO$_2$ and sulphated TiO$_2$. J Hazard Mater 134:94–103. https://doi.org/10.1016/j.jhazmat.2005.10.041

Jimoh AA, Helal A, Shaikh MN, Aziz MA, Yamani ZH, Ahmed AA, Kim JP (2015) Schiff base ligand coated gold nanoparticles for the chemical sensing of Fe(III) ions. J Nanomater 2015:1–7. https://doi.org/10.1155/2015/101694

Ju J, Chen X, Shi Y, Miao J, Wu D (2013) Hydrothermal preparation and photocatalytic performance of N, S-doped nanometer TiO$_2$ under sunshine irradiation. Powder Technol 237:616–622. https://doi.org/10.1016/j.powtec.2012.12.048

Kabdaşli I, Tünay O, Orhon D (1996) Wastewater control and management in a leather tanning district. Water Sci Technol 40:261–267. https://doi.org/10.1016/S0273-1223(99)00393-5

Kallman EN, Oyanedel-Craver VA, Smith JA (2011) Ceramic filters impregnated with silver nanoparticles for point-of-use water treatment in rural Guatemala. J Environ Eng 137(6): 407–415. https://doi.org/10.1061/(ASCE)EE.1943-7870.0000330

Kerebo A, Desta A, Duraisamy R (2016) Removal of methyl violet from synthetic wastewater using nano aluminium oxyhydroxide. Int J Eng Res Dev 12(8):2.0efda360239c114de0c86ec53d3174cbde34. https://doi.org/10.1515/nano.0038.2016-2586

Khan AU, Yuan Q, Wei Y, Khan GM, Khan ZUH, Khan S, Ali F, Tahir K, Ahmad A, Khan FU (2016) Photocatalytic and antibacterial response of biosynthesized gold nanoparticles. J Photochem Photobiol B 162:273–277. https://doi.org/10.1016/j.jphotobiol.2016.06.055

Khan ME, Mohammad A, Cho MH (2020) Nanoparticles based surface plasmon enhanced photocatalysis. In: Green Photocatalysts. Springer, Cham, pp 133–143. https://doi.org/10.1007/978-3-030-15608-4_5

Khataee A, Kayan B, Kalderis D, Karimi A, Akay S, Konsolakis M (2017) Ultrasound-assisted removal of acid red 17 using nanosized Fe$_3$O$_4$-loaded coffee waste hydrochar. Ultrason Sonochem 35:72–80. https://doi.org/10.1016/j.ultsonch.2016.09.004

Kim D, Huh Y-D (2011) Morphology-dependent photocatalytic activities of hierarchical microstructures of ZnO. J Mater Lett 65:2100–2103. https://doi.org/10.1016/j.matlet.2011.04.074

Kim SH, Lee SW, Lee GM, Lee BT, Yun ST, Kim SO (2016) Monitoring of TiO$_2$-catalytic UV-LED photo-oxidation of cyanide contained in mine waste water and leachate. Chemosphere 143:106–114. https://doi.org/10.1016/j.chemosphere.2015.07.006

Koedrith P, Thasiphu T, Weon J, Boonprasert R, Tuitemwong K, Tuitemwong P (2015) Recent trends in rapid environmental monitoring of pathogens and toxicants: potential of nanoparticle-based biosensor and applications. Sci World J 2015:1–12. https://doi.org/10.1155/2015/510982

Kumar KY, Muralidhara HB, Arthoba Nayaka Y, Balasubramanyam J, Hanumanthappa H (2013) Hierarchically assembled mesoporous ZnO nanorods for the removal of lead and cadmium by using differential pulse anodic stripping voltammetric method. Powder Technol 239:208–216. https://doi.org/10.1016/j.powtec.2013.02.009

Kumpiene J, Lagerkvist A, Maurice C (2008) Stabilization of As, Cr, Cu, Pb and Zn in soil using amendments – a review. Waste Manag 28(1):215–222. https://doi.org/10.1016/j.wasman.2006.12.012

Lee Y, Kim S, Venkateswaran P, Jang J, Kim H, Kim J (2008) Anion co-doped Titania for solar photocatalytic degradation of dyes. Carbon Lett 9(2):131–136

Lee KM, Lai CW, Ngai KS, Juan JC (2016) Recent developments of zinc oxide based photocatalyst in water treatment technology: a review. Water Res 88:428–448. https://doi.org/10.1016/j.watres.2015.09.045

Leonard P, Hearty S, Brennan J (2003) Advances in biosensors for detection of pathogens in food and water. Enzym Microb Technol 32:3–13. https://doi.org/10.1016/S0141-0229(02)00232-6

Lerma-García MJ, Ávila M, Alfonso EFS, Ríos Á, Zougagh M (2014) Synthesis of gold nanoparticles using phenolic acids and its application in catalysis. J Mater Environ Sci 5(6):1919–1926

Li XQ, Zhang WX (2007) Sequestration of metal cations with zerovalent iron nanoparticles–a study with high resolution X-ray photoelectron spectroscopy (HR-XPS). J Phys Chem C 111:6939–6946. https://doi.org/10.1021/jp0702189

Li P, Miser DE, Babier S, Yadav RT, Hajaligol MR (2003) The removal of carbon monoxide by iron oxide nanoparticles. Appl Catal B 43:151–162. https://doi.org/10.1016/S0926-3373(02)00297-7

Li XQ, Elliott DW, Zhang WX (2006a) Zero-valent iron nanoparticles for abatement of environmental pollutants: materials and engineering aspects. Crit Rev Solid State Mater Sci 31:111–122. https://doi.org/10.1080/10408430601057611

Li XQ, Elliot DW, Zhang WX (2006b) Zero-valent iron nanoparticles for abatement of environmental pollutants: materials and engineering aspects. Crit Rev Solid State Mater Sci 31:111–122. https://doi.org/10.1080/10408430601057611

Li X, Wang L, Lu X (2010) Preparation of silvermodified TiO_2 via microwave-assisted method and its photocatalytic activity for toluene degradation. J Hazard Mater 177:639–647. https://doi.org/10.1016/j.jhazmat.2009.12.080

Li LH, Lu J, Wang ZS, Yang L, Zhou XF, Han L (2012) Fabrication of the C–N co-doped rod-like TiO_2 photocatalyst with visible-light responsive photocatalytic activity. Mater Res Bull 47:1508–1512. https://doi.org/10.1016/j.materresbull.2012.02.032

Li XY, Ai LH, Jiang J (2016) Nanoscale zerovalent iron decorated on graphene nanosheets for Cr (VI) removal from aqueous solution: surface corrosion retard induced the enhanced performance. Chem Eng J 288:789–797. https://doi.org/10.1016/j.cej.2015.12.022

Li Y, Lu H, Wang Y, Li X (2019) Deposition of Au nanoparticles on PDA-functionalized PVA beads as a recyclable catalyst for degradation of organic pollutants with $NaBH_4$ in aqueous solution. J Alloys Compd 793:115–126. https://doi.org/10.1016/j.jallcom.2019.04.148

Liang DW, Yang YH, Xu WW, Peng SK, Lu SF, Xiang Y (2014) Non-ionic surfactant greatly enhances the reductive debromination of polybrominated diphenyl ethers by nanoscale zero-valent iron: mechanism and kinetics. J Hazard Mater 278:592–596. https://doi.org/10.1016/j.jhazmat.2014.06.030

Lin YF, Chen JL (2014) Magnetic mesoporous Fe/carbon aerogel structures with enhanced arsenic removal efficiency. J Colloid Interface Sci 420:74–79. https://doi.org/10.1016/j.jcis.2014.01.008

Lin C, Tao K, Hua D, Ma Z, Zhou S (2013) Size effect of gold nanoparticles in catalytic reduction of p-nitrophenol with $NaBH_4$. Molecules 18(10):12609. https://doi.org/10.3390/molecules181012609

Ling L, Zhang WX (2015) Enrichment and encapsulation of uranium with iron nanoparticle. J Am Chem Soc 137(8):2788–2791. https://doi.org/10.1021/ja510488r

Ling L, Pan B, Zhang WX (2015) Removal of selenium from water with nanoscale zero-valent iron: mechanisms of intraparticle reduction of Se(IV). Water Res 71:274–281. https://doi.org/10.1016/j.watres.2015.01.002

Liou YH, Lo SL, Lin CJ, Kuan WH, Weng SC (2005) Chemical reduction of an unbuffered nitrate solution using catalyzed and uncatalyzed nanoscale iron particles. J Hazard Mater 12(1–3):02–110. https://doi.org/10.1016/j.jhazmat.2005.06.029

Liu JF, Zhao ZS, Jiang GB (2008a) Coating Fe_3O_4 magnetic nanoparticles with humic acid for high efficient removal of heavy metals in water. Environ Sci Technol 42:6949–6954. https://doi.org/10.1021/es800924c

Liu X, Hu Q, Fang Z, Zhang X, Zhang B (2008b) Magnetic chitosan nanocomposites: a useful recyclable tool for heavy metal ion removal. Langmuir 25:3–8. https://doi.org/10.1021/la802754t

Liu B, Xu J, Ran SH, Wang ZR, Chen D, Shen GZ (2012) High-performance photodetectors, photocatalysts, and gas sensors based on polyol reflux synthesized porous ZnO nanosheets. CrystEngComm 14:4582–4588. https://doi.org/10.1039/C2CE25278C

Liu Y, Tourbin M, Lachaize S, Guiraud P (2014a) Nanoparticles in wastewaters: hazards, fate and remediation. Powder Technol 255:249–156. https://doi.org/10.1016/j.powtec.2013.08.025

Liu IT, Hon MH, Teoh LG (2014b) The preparation, characterization and photocatalytic activity of radical-shaped CeO_2/ZnO microstructures. Ceram Int 40(3):4019–4024. https://doi.org/10.1016/j.ceramint.2013.08.053

Liu S, Shan Y, Chen L, Boury B, Huang L, Xiao H (2019) Probing nanocolumnar silver nanoparticle/zinc oxide hierarchical assemblies with advanced surface plasmon resonance and their enhanced photocatalytic performance for formaldehyde removal. Appl Organomet Chem 33:e5209. https://doi.org/10.1002/aoc.5209

Lou T, Chen L, Chen Z, Wang Y, Chen L, Li J (2011) Colorimetric detection of trace copper ions based on catalytic leaching of silver-coated gold nanoparticles. ACS Appl Mater Interfaces 3(11):4215–4220. https://doi.org/10.1021/am2008486

Lu YC, Lin YH, Wang DJ, Wang LL, Xie TF, Jiang TF (2011) A high performance cobalt doped ZnO visible light photocatalyst and its photogenerated charge transfer properties. Nano Res 4:1144–1152. https://doi.org/10.1007/s12274-011-0163-4

Lu H, Wang J, Stoller M, Wang T, Bao Y, Hao H (2016) An overview of nanomaterials for water and wastewater treatment. Adv Mater Sci Eng 2016(4964828):1–10. https://doi.org/10.1155/2016/4964828

Luo X, Xang L (2009) High effective adsorption of organic dyes on magnetic cellulose beads entrapping activated carbon. J Hazard Mater 171(1–3):340–347. https://doi.org/10.1016/j.jhazmat.2009.06.009

Luo C, Tian Z, Yang B, Zhang L, Yan S (2013) Manganese dioxide/iron oxide/acid oxidized multi-walled carbon nanotube magnetic nanocomposite for enhanced hexavalent chromium removal. Chem Eng J 234:256–265. https://doi.org/10.1016/j.cej.2013.08.084

Lv C, Zhou Y, Li H, Dang M, Guo C, Ou Y, Xiao B (2011) Synthesis and characterisation of Gd^{3+} doped mesoporous TiO_2 materials. Appl Surf Sci 257:5104–5108. https://doi.org/10.1016/j.apsusc.2011.01.029

Lv XS, Xue XQ, Jiang GM (2014) Nanoscale zero-valent iron (nZVI) assembled on magnetic Fe_3O_4/graphene for chromium(VI) removal from aqueous solution. J Colloid Interface Sci 417:51–59. https://doi.org/10.1016/j.jcis.2013.11.044

Lydakis-Simantiris N, Riga D, Katsivela E, Mantzavinos D, Xekoukoulotakis NP (2010) Disinfection of spring water and secondary treated municipal wastewater by TiO_2 photocatalysis. Desalination 250:351–355. https://doi.org/10.1016/j.desal.2009.09.055

Ma S, Li R, Lv C, Xu W, Gou X (2011) Facile synthesis of ZnO nanorods arrays and hierarchical nanostructures for photocatalysis and gas sensor applications. J Hazard Mater 192:730–740. https://doi.org/10.1016/j.jhazmat.2011.05.082

Mahapatra A, Mishra BG, Hota G (2013) Electrospun Fe_2O_3-Al_2O_3 nanocomposite fibers as efficient adsorbent for removal of heavy metal ions from aqueous solution. J Hazard Mater 116:258–259. https://doi.org/10.1016/j.jhazmat.2013.04.045

Mahata P, Aarthi T, Madras G, Natarajan S (2007) Photocatalytic degradation of dyes and organics with nanosized $GdCoO_3$. J Phys Chem C 111:1665–1674. https://doi.org/10.1021/jp066302q

Malaviya P, Singh A (2011) Physicochemical technologies for remediation of chromium-containing waters and wastewaters. Crit Rev Environ Sci Technol 41:1111–1172. https://doi.org/10.1080/10643380903392817

Manjari G, Saran S, Arun T, Devipriya SP, Rao AVB (2017) Facile Aglaia elaeagnoidea mediated synthesis of silver and gold nanoparticles: antioxidant and catalysis properties. J Clust Sci 28(4):2041–2056

Marková Z, Šišková KM, Filip J (2013) Air stable magnetic bimetallic Fe-Ag nanoparticles for advanced antimicrobial treatment and phosphorus removal. Environ Sci Technol 47(10):5285–5293. https://doi.org/10.1021/es304693g

Matheson LJ, Tratnyek PG (1994) Reductive dehalogenation of chlorinated methanes by iron metal. Environ Sci Technol 28(12):2045–2053. https://doi.org/10.1021/es00061a012

Mills A, Le Hunte S (1997) An overview of semiconductor photocatalysis. J Photochem Photobiol A Chem 108(1):1–35. https://doi.org/10.1016/S1010-6030(97)00118-

Mishra G, Tripathy MA (1993) Critical review of the treatments for decolourization of textile effluent. Colourage 40:35–35

Moghaddam SS, Moghaddam MA, Arami M (2010) Coagulation/flocculation process for dye removal using sludge from water treatment plant: optimization through response surface methodology. J Hazard Mater 175:651–657. https://doi.org/10.1016/j.jhazmat.2009.10.058

Mohammad A, Ansari SN, Chaudhary A, Ahmad K, Rajak R, Tauqeer M, Mobin SM (2018a) Enthralling adsorption of different dye and metal contaminants from aqueous systems by cobalt/cobalt oxide nanocomposites derived from single-source molecular precursors. ChemistrySelect 3:5733–5741. https://doi.org/10.1002/slct.201703169

Mohammad A, Ahmad K, Qureshi A, Tauqeer M, Mobin SM (2018b) Zinc oxide-graphitic carbon nitride nanohybrid as an efficient electrochemical sensor and photocatalyst. Sensors Actuators B Chem 277(20):467–476. https://doi.org/10.1016/j.snb.2018.07.086

Mohan R, Krishnamoorthy K, Kim S-J (2012) Enhanced photocatalytic activity of Cu-doped ZnO nanorods. Solid State Commun 152:375–380. https://doi.org/10.1016/j.ssc.2011.12.008

Moo JGS, Khezri B, Webster RD, Pumera M (2014) Graphene oxides prepared by Hummers', Hofmann's, and Staudenmaier's methods: dramatic influences on heavy-metal-ion adsorption. ChemPhysChem 15:2922–2929. https://doi.org/10.1002/cphc.201402279

Moon G, Kim D, Kim H, Bokare AD, Choi W (2014) Platinum-like behaviour of reduced grapheme oxide as a co-catalyst on TiO_2 for the efficient photocatalytic oxidation of arsenite. Environ Sci Technol Lett 1(2):185–190. https://doi.org/10.1021/ez5000012

Mueller NC, Braun J, Bruns J, Černík M, Rissing P, Rickerby D, Nowack B (2012) Application of nanoscale zero valent iron (NZVI) for groundwater remediation in Europe. Environ Sci Pollut Res 19(2):550–558. https://doi.org/10.1007/s11356-011-0576-3

Mukherjee J, Ramkumar J, Shukla R, Tyagi AK (2013) Sorption characteristics of nano manganese oxide: efficient sorbent for removal of metal ions from aqueous streams. J Radioanal Nucl Chem 297:49–57. https://doi.org/10.1007/s10967-012-2393-7

Muradova GG, Gadjieva SR, Palma LD, Vilardi G (2016) Nitrates removal by bimetallic nanoparticles in water. Chem Eng Trans 47:205–210. https://doi.org/10.3303/CET1647035

Nahar S, Hasegawa K, Kagaya S (2006) Photocatalytic degradation of phenol by visible light-responsive iron- doped TiO_2 and spontaneous sedimentation of the TiO_2 particles. Chemosphere 65:1976–1982. https://doi.org/10.1016/j.chemosphere.2006.07.002

Narayanan KB, Park HH (2014) Colorimetric detection of manganese(II) ions using gold/dopa nanoparticles. Spectrochim Acta Mol Biomol Spectrosc 131:132–137. https://doi.org/10.1016/j.saa.2014.04.081

Nguyen AT, Hsieh CT, Juang RS (2016) Substituent effects on photodegradation of phenols in binary mixtures by hybrid H_2O_2 and TiO_2 suspensions under UV irradiation. J Taiwan Inst Chem Eng 62:68–75. https://doi.org/10.1016/j.jtice.2016.01.012

Ning YF, Yan P, Chen YP, Guo JS, Shen Y, Fang F, Tang Y, Gao X (2017) Development of a Pt modified microelectrode aimed for the monitoring of ammonium in solution. Int J Environ Anal Chem 97(1):85–98. https://doi.org/10.1080/03067319.2016.1277994

Nizamuddin S, Siddiqui MTH, Mubarak NM, Baloch HA, Abdullah EC, Mazari SA, Griffin GJ, Srinivasan MP, Tanksale A (2019) For the removal of heavy metals and dyes from wastewater. In: Nanoscale materials in water purification. Elsevier, Amsterdam, pp 447–472. https://doi.org/10.1016/B978-0-12-813926-4.00023-9

Nolan NT, Seery MK, Pillai SC (2009) Spectroscopic investigation of the anatase to rutile transformation of sol-gel synthesisized TiO_2 photocatalyst. J Phys Chem C 113:16151–16157. https://doi.org/10.1021/jp904358g

Nolan NT, Synnott DW, Seery MK, Hinderc SJ, Wassenhovend AV, Pillai SC (2012) Effect of N doping on the photocatalytic activity of sol–gel TiO_2. J Hazard Mater 211–212:88–94. https://doi.org/10.1016/j.jhazmat.2011.08.074

Nurmi JT, Tratnyek PG, Sarathy V, Baer DR, Amonette JE, Pecher K (2005) Characterization and properties of metallic iron nanoparticles: spectroscopy, electrochemistry, and kinetics. Environ Sci Technol 39:1221–1230. https://doi.org/10.1021/es049190u

O'Carroll D, Sleep B, Krol M, Boparai H, Kocur C (2013) Nanoscale zero valent iron and bimetallic particles for contaminated site remediation. Adv Water Resour 51:104–122. https://doi.org/10.1016/j.advwatres.2012.02.005

Oh SM, Kim SS, Lee JE, Ishigaki T, Park DW (2003) Effect of additives on photocatalytic activity of titanium dioxide powders synthesized by thermal plasma. Thin Solid Films 435:252–258. https://doi.org/10.1016/S0040-6090(03)00388-2

Ohsaka T, Shinozaki K, Tsuruta K, Hirano K (2008) Photo electrochemical degradation of some chlorinated organic compounds on n-TiO_2 electrode. Chemosphere 73(8):1279–1283. https://doi.org/10.1016/j.chemosphere.2008.07.016

Oliveira LCA, Petkowicz DI, Smaniotto A, Pergher SBC (2004) Magnetic zeolites: a new adsorbent for removal of metallic contaminants from water. Water Res 38:699–704. https://doi.org/10.1016/j.watres.2004.06.008

Onyango MS, Kojima Y, Matsuda H, Ochieng A (2003) Adsorption kinetics of arsenic removal from groundwater by iron-modified zeolite. J Chem Eng Jpn 36:1516–1522. https://doi.org/10.1252/jcej.36.1516

Oyanedel-Craver VA, Smith JA (2008) Sustainable colloidal silver-impregnated ceramic filter for point-of-use water treatment. Environ Sci Technol 42(3):927–933. https://doi.org/10.1021/es071268u

Pant HR, Park CH, Pant B, Tijing LD, Kim HY, Kim CS (2012) Synthesis, characterization, and photocatalytic properties of ZnO nano-flower containing TiO_2 NPs. Ceram Int 38(4):2943–2950. https://doi.org/10.1016/j.ceramint.2011.11.071

Pantapasis K, Grumezescu AM (2017) 13-Gold nanoparticles: advances in water purification process. In: Water purification, pp 447–477. https://doi.org/10.1016/B978-0-12-804300-4.00013-7

Pare B, Singh P, Jonnalgadda SB (2009) Degradation and mineralization of Victoria blue B dye in a slurry photoreactor using advanced oxidation process. J Sci Ind Res 68:724–729

Pathania D, Katwal R, Kaur H (2016) Enhanced photocatalytic activity of electrochemically synthesized aluminium oxide nanoparticles. Int J Miner Metall Mater 23(3):358. https://doi.org/10.1007/s12613-016-1245-9

Pelaez M, Nolan NT, Pillai SC, Seery MK, Falaras P, Kontos AG et al (2012) A review on the visible light active TiO_2 photocatalysts for environmental applications. Appl Catal B Environ 125:331–349. https://doi.org/10.1016/j.apcatb.2012.05.036

Peng XJ, Luan ZK, Ding J, Di ZC, Li YH, Tian BH (2005) Ceria nanoparticles supported nanotubes for the removal of arsenate from water. Mater Lett 59:399–403. https://doi.org/10.1016/j.matlet.2004.05.090

Pereira L, Alves M (2012) In: Malik A, Grohmann E (eds) Environmental protection strategies for sustainable development, vol 4. Springer, New York, pp 111–162. ISBN 978-94-007-1591-2

Perreault F, Faria FDA, Elimelech M (2015) Environmental applications of graphene- based nanomaterials. Chem Soc Rev 44:5861–5896. https://doi.org/10.1039/C5CS00021A

Petala E, Georgiou Y, Kostas V, Dimos K, Karakassides MA, Deligiannakis Y, Aparicio C, Zboril JTR (2017) Magnetic carbon nanocages: an advanced architecture with surface-and morphology-enhanced removal capacity for arsenites. ACS Sustain Chem Eng 5(7):5782–5792. https://doi.org/10.1021/acssuschemeng.7b00394

Pozan GS, Kambur A (2013) Removal of 4-chlorophenol from wastewater: preparation, characterization and photocatalytic activity of alkaline earth oxide doped TiO_2. Appl Catal B Environ 129:409–415. https://doi.org/10.1016/j.apcatb.2012.09.050

Prince JA, Bhuvana S, Anbharasi V, Ayyanar N, Boodhoo KVK, Singh G (2014) Self-cleaning Metal Organic Framework (MOF) based ultra-filtration membranes – a solution to bio-fouling in membrane separation processes. Sci Rep 4:6555–6555. https://doi.org/10.1038/srep06555

Qiu X, Li G, Sun X, Li L, Fu X (2008) Doping effects of Co(2+) ions on ZnO nanorods and their photocatalytic properties. Nanotechnology 19:215703–215710. https://doi.org/10.1088/0957-4484/19/21/215703

Qu L, Wang N, Xu H, Wang W, Liu Y, Kuo L, Yadav TP, Wu J, Joyner J, Song Y, Li H, Lou J, Vajtai R, Ajayan PM (2017) Gold nanoparticles and g-C_3N_4-intercalated graphene oxide membrane for recyclable surface enhanced raman scattering. Adv Funct Mater 27(31):1701714

Radoičić M, Šaponjić Z, Janković IA, ĆirićMarjanović G, Ahrenkiel SP, Čomor MI (2013) Improvements to the photocatalytic efficiency of polyaniline modified TiO_2 nanoparticles. Appl Catal B Environ 136–137:133–139. https://doi.org/10.1016/j.apcatb.2013.01.007

Raghav R, Srivastava S (2015) Gold nanoparticles based colorimetric detection of cobalt (II) ions. Sens Lett 13(3):254–258. https://doi.org/10.1166/sl.2015.3426

Raota CS, Cerbaro AF, Salvador M, Delamare APL, Sergio E, da Silva Crespo J, da Silva TB, Giovanela M (2019) Green synthesis of silver nanoparticles using an extract of Ives cultivar (Vitis labrusca) pomace: characterization and application in wastewater disinfection. J Environ Chem Eng 7(5):103383

Ray PZ, Shipley HJ (2015) Inorganic nano-adsorbents for the removal of heavy metals and arsenic: a review. RSC Adv 5:29885–29907. https://doi.org/10.1039/C5RA02714D

Ren D, Smith JA (2013) Retention and transport of silver nanoparticles in a ceramic porous medium used for point-ofuse water treatment. Environ Sci Technol 47(8):3825–3832. https://doi.org/10.1021/es4000752

Reynolds DC, Look DC, Jogai B, Litton CW, Cantwell G, Harsch WC (1999) Valence-band ordering in ZnO. Phys Rev B Condens Matter Mater Phys 60(4):2340–2344. https://doi.org/10.1103/PhysRevB.60.2340

Ritter L, Solomon K, Sibley P (2002) Sources, pathways, and relative risks of contaminants in surface water and ground water: a perspective prepared for the Walkerton inquiry. J Toxicol Environ Health 65:1–142. https://doi.org/10.1080/152873902753338572

Robles AC, Martinez E, Alcantar IR, Frontana C, Gutierrez LG (2013) Development of an activated carbon-packed microbial bioelectrochemical system for azo dye degradation. Bioresour Technol 127:37–43. https://doi.org/10.1016/j.biortech.2012.09.066

Rodriguez-Mozaz S, De Alda MJL, Barceló D (2004) Monitoring of estrogens, pesticides and bisphenol A in natural waters and drinking water treatment plants by solid-phase extraction-liquid chromatography-mass spectrometry. J Chromatogr A 1045:85–92. https://doi.org/10.1016/j.chroma.2004.06.040

Salam MA, Kosa SA, Al-Nahdi NA, Owija NY (2019) Removal of acid red dye from aqueous solution using zero-valent copper and zero-valent zinc nanoparticles. Desalin Water Treat 141:310–320. https://doi.org/10.5004/dwt.2019.23303

Samadi M, Pourjavadi A, Moshfegh AZ (2014) Role of CdO addition on the growth and photocatalytic activity of electrospun ZnO nanofibers: UV vs. visible light. Appl Surf Sci 298:147–154. https://doi.org/10.1016/j.apsusc.2014.01.146

Schmidt-Mende L, MacManus-Driscoll JL (2007) ZnO-nanostructures, defects, and devices. Mater Today 10:40–48. https://doi.org/10.1016/S1369-7021(07)70078-0

Schwarzenbach RP, Escher BI, Fenner K, Hofstetter TB, Johnson CA, Von Gunten U, Wehrli B (2006) The challenge of micropollutants in aquatic systems. Science 313:1072–1077. https://doi.org/10.1126/science.1127291

Sekhar ACS, Zaki A, Tronce S, Casale S, Vinod CP, Dacquin JP, Granger P (2018) Enhanced selectivity of 3-D ordered macroporous Pt/Al_2O_3 catalysts in nitrites removal from water. Appl Catal A Gen 564:26–32. https://doi.org/10.1016/j.apcata.2018.07.014

Senthilnathan J, Philip L (2010) Photocatalytic degradation of lindane under UV and visible light using N-doped TiO_2. Chem Eng J 161:83–92. https://doi.org/10.1016/j.cej.2010.04.034

Shah SI, Li W, Huang CP, Jung O, Ni C (2002) Study of Nd^{3+}, Pd^{2+}, Pt^{4+} and Fe^{3+} dopant effect on photoreactivity of TiO_2 nanoparticles. Proc Natl Acad Sci U S A 99:6482–6486. https://doi.org/10.1073/pnas.052518299

Shen YF, Tang J, Nie ZH, Wang YD, Ren Y, Zuo L (2009) Preparation and application of magnetic Fe_3O_4 nanoparticles for wastewater purification. Sep Purif Technol 25:312–319. https://doi.org/10.1016/j.seppur.2009.05.020

Shinde DR, Tambade PS, Chaskar MG, Gadave KM (2017) Photocatalytic degradation of dyes in water by analytical reagent grades ZnO, TiO2 and SnO2: a comparative study. Drink Water Eng Sci 10:109–117. https://doi.org/10.5194/dwes-10-109-2017

Singh R, Misra V (2015) Stabilization of zero-valent iron nanoparticles: role of polymers and surfactants. In: Handbook of nanoparticles. Springer, New York, pp 1–19. https://doi.org/10.1007/978-3-319-13188-7_44-1

Singh S, Barick KC, Bahadur D (2011a) Surface engineered magnetic nanoparticles for removal of toxic metal ions and bacterial pathogens. J Hazard Mater 192:1539–1547. https://doi.org/10.1016/j.jhazmat.2011.06.074

Singh S, Barick KC, Bahadur D (2011b) Novel and efficient three dimensional mesoporous ZnO nanoassemblies for environmental remediation. Int J Nanosci 10.1001–1005. https://doi.org/10.1142/S0219581X11008654

Singh S, Barick KC, Bahadur D (2013a) Fe_3O_4 embedded ZnO nanocomposites for the removal of toxic metal ions, organic dyes and bacterial pathogens. J Mater Chem A 1:3325–3333. https://doi.org/10.1039/C2TA01045C

Singh S, Barick KC, Bahadur D (2013b) Functional oxide nanomaterials and nanocomposites for the removal of heavy metals and dyes. Nanomater Nanotechnol 3(20):1–19. https://doi.org/10.5772/57237

Singh S, Barick KC, Bahadur D (2013c) Shape controlled hierarchical ZnO architectures: photocatalytic and antibacterial activities. CrystEngComm 15:4631–4639. https://doi.org/10.1039/C3CE27084J

Sinha A, Jana NR (2012) Functional, mesoporous, superparamagnetic colloidal sorbents for efficient removal of toxic metals. Chem Commun 48:9272–9274. https://doi.org/10.1039/C2CC33893A

Song W, Gao B, Xu X, Xing L, Han S, Duan P, Song W, Jia R (2016) Adsorption–desorption behavior of magnetic amine/Fe_3O_4 functionalized biopolymer resin towards anionic dyes from wastewater. Bioresour Technol 210:123–130. https://doi.org/10.1016/j.biortech.2016.01.078

Stanton BW, Harris JJ, Miller MD, Bruening ML (2003) Ultrathin, multilayered polyelectrolyte films as nanofiltration membranes. Langmuir 19:7038–7042. https://doi.org/10.1021/la034603a

Stefaniuk M, Oleszczuk P, Ok YS (2016) Review on nano zerovalent iron (nZVI): from synthesis to environmental applications. Chem Eng J 287:618–632. https://doi.org/10.1016/j.cej.2015.11.046

Sun YP, Li XQ, Cao JS, Zhang WX, Wang HP (2006) Characterization of zero-valent iron nanoparticles. Adv Colloid Interf Sci 120:47–56. https://doi.org/10.1016/j.cis.2006.03.001

Takafuji M, Ide S, Ihara H, Xu Z (2004) Preparation of poly (1-vinylimidazole)-grafted magnetic nanoparticles and their application for removal of metal ions. Chem Mater 16:1977–1983. https://doi.org/10.1021/cm030334y

Tamer U, Cetin D, Suludere Z, Boyaci IH, Temiz HT, Yegenoglu H, Daniel P, Dinçer İ, Elerman Y (2013) Gold-coated iron composite nanospheres targeted the detection of Escherichia coli. Int J Mol Sci 14(3):6223–6240. https://doi.org/10.3390/ijms14036223

Tang SC, Lo IM (2013) Magnetic nanoparticles: essential factors for sustainable environmental applications. Water Res 47:2613–2632. https://doi.org/10.1016/j.watres.2013.02.039

Tang L, Zhang S, Zeng GM, Zhang Y, Yang GD, Chen J, Wang JJ, Wang JJ, Zhou YY, Deng YC (2015) Rapid adsorption of 2,4-dichlorophenoxyacetic acid by iron oxide nanoparticles-doped carboxylic ordered mesoporous carbon. J Colloid Interface Sci 445:1–8. https://doi.org/10.1016/j.jcis.2014.12.074

Tauqeer M, Ahmad MS, Siraj M, Mohammad A, Ansari O, Baig MT (2020) Nanocomposite materials for wastewater decontamination. In: Modern age waste water problems. Springer, Cham, pp 23–46. https://doi.org/10.1007/978-3-030-08283-3_2

Thongsuriwong K, Amornpitoksuk P, Suwanboon S (2012) Photocatalytic and antibacterial of Ag doped ZnO thin films prepared by sol-gel dip coating method. J Sol-Gel Sci Technol 62:304–312. https://doi.org/10.1007/s10971-012-2725-7

Tian J, Sang YH, Yu GW, Jiang HD, Mu XN, Liu H (2013) Asymmetric supercapacitors based on graphene/MnO$_2$ nanospheres and graphene/MoO$_3$ nanosheets with high energy density. Adv Mater 25:5074–5083. https://doi.org/10.1002/adfm201301851

Tiwari A, Paul B (2015) A brief review on the application of gold nanoparticles as sensors in multi-dimensional aspects. J Environ Sci Toxicol Food Technol 1(4):1–7

Tokunaga S, Hardon MJ, Wasay SA (1995) Removal of flouride ions from aqueous solution by multivalent metal compounds. Int J Environ Stud 48:17–28. https://doi.org/10.1080/00207239508710973

Tratnyek PG, Salter AJ, Nurmi JT, Sarathy V (2010) Environmental applications of zerovalent metals: iron vs. zinc. Nanoscale materials in chemistry: environmental applications 1045(9): 165–178 of ACS Symposium Series. https://doi.org/10.1021/bk-2010-1045.ch009

Tripathy SS, Bersillon J-L, Gopal K (2006) Removal of fluoride from drinking water by adsorption onto alum-impregnated activated alumina. Sep Purif Technol 50:310–317. https://doi.org/10.1016/j.seppur.2005.11.036

Trovarelli A (1996) Catalytic properties of ceria and CeO$_2$-containing materials. Catal Rev Sci Eng 38:439–520. https://doi.org/10.1080/01614949608006464

Uddin MT, Nicolas Y, Olivier C et al (2012) Nanostructured SnO$_2$-ZnO heterojunction photocatalysts showing enhanced photocatalytic activity for the degradation of organic dyes. Inorg Chem 51(14):7764–7773. https://doi.org/10.1021/ic300794j

Upadhyay RK, Soin N, Roy SS (2014) Role of graphene/metal oxide composites as photocatalysts, adsorbents and disinfectants in water treatment: a review. RSC Adv 4:3823–3851. https://doi.org/10.1039/C3RA45013A

Vidhu VK, Philip D (2014) Catalytic degradation of organic dyes using biosynthesized silver nanoparticles. Micron 56:54. https://doi.org/10.1016/j.micron.2013.10.006

Wan W, Yeow JT (2012) Antibacterial properties of poly(quaternary ammonium) modified gold and titanium dioxide nanoparticles. J Nanosci Nanotechnol 12(6):4601–4606. https://doi.org/10.1166/jnn.2012.6147

Wang S, Li H, Xu L (2006a) Application of zeolite MCM-22 for basic dye removal from wastewater. J Colloid Interface Sci 295:71–78. https://doi.org/10.1016/j.jcis.2005.08.006

Wang S, Li H, Xie S, Liu S, Xu L (2006b) Physical and chemical regeneration of zeolitic adsorbents for dye removal in wastewater treatment. Chemosphere 65:82–87. https://doi.org/10.1016/j.chemosphere.2006.02.043

Wang Y, Tang XW, Chen YM, Zhan LT, Li ZZ, Tang Q (2009a) Adsorption behavior and mechanism of Cd(II) on loess soil from China. J Hazard Mater 172:30–37. https://doi.org/10.1016/j.jhazmat.2009.06.121

Wang S-G, Ma Y, Shi Y-J, Gong W-X (2009b) Defluoridation performance and mechanism of nano-scale aluminum oxide hydroxide in aqueous solution. J Chem Technol Biotechnol 84:1043. https://doi.org/10.1002/jctb.2131

Wang S, Sun H, Ang HM, Tad MO (2013a) Adsorptive remediation of environmental pollutants using novel graphene-based nanomaterials. Chem Eng J 226:336–347. https://doi.org/10.1016/j.cej.2013.04.070

Wang XY, Zhu MP, Liu HL, Ma J, Li F (2013b) Modification of Pd–Fe nanoparticles for catalytic dechlorination of 2,4- dichlorophenol. Sci Total Environ 449:157–167. https://doi.org/10.1016/j.scitotenv.2013.01.008

Wang Y, Fang Z, Kang Y, Tsang EP (2014a) Immobilization and phytotoxicity of chromium in contaminated soil remediated by CMC-stabilized nZVI. J Hazard Mater 275:230–237. https://doi.org/10.1016/j.jhazmat.2014.04.056

Wang H, Zhang L, Chen Z, Hu J, Li S, Wang Z, Liu J (2014b) Semiconductor heterojunction photocatalyst: design, construction and photocatalytic performance. Chem Soc Rev 43:5234–5244. https://doi.org/10.1039/C4CS00126E

Wang F, Zhou L, Zhao JQ (2018) The performance of biocarrier containing zinc nanoparticles in biofilm reactor for treating textile wastewater. Process Biochem (Oxford, U K) 74:125–131. https://doi.org/10.1016/j.procbio.2018.08.022

Wei L, Yang G, Wang R, Ma W (2009) Selective adsorption and separation of chromium (VI) on the magnetic iron–nickel oxide from waste nickel liquid. J Hazard Mater 164:1159–1163. https://doi.org/10.1016/j.jhazmat.2008.09.016

Wenrong Y, Justin GJ, Zhicong H, Qiong L, Guonan C (2007) Fast colorimetric detection of copper ions using l-cysteine functionalized gold nanoparticles. J Nanosci Nanotechnol 7(2):712–716. https://doi.org/10.1166/jnn.2007.116

Wigginton NY, Eston N, Malakoff JD (2012) More treasure than trash. Science 337:662–663. https://doi.org/10.1126/science.337.6095.662

Wong Y Szeto YS, Cheung WH, McKay G (2004) Adsorption of acid dyes on chitosan— equilibrium isotherm analyses. Process Biochem 39:695–704. https://doi.org/10.1016/S0032-9592(03)00152-3

Wróbel D, Boguta A, Ion RM (2001) Mixtures of synthetic organic dyes in a photoelectrochemical cell. J Photochem Photobio A Chem 138:7–22. https://doi.org/10.1016/S1010-6030(00)00377-4

Wu RC, Qu HH, He H, Yu YB (2004) Removal of azo-dye acid red B (ARB) by adsorption and catalytic combustion using magnetic $CuFe_2O_4$ powder. Appl Catal B 48:49–56. https://doi.org/10.1016/j.apcatb.2003.09.006

Wu RC, Qu JH, Chen YS (2005) Magnetic powder $MnO–Fe_2O_3$ composite –a novel material for the removal of azo-dye from water. Water Res 39:630–638. https://doi.org/10.1016/j.watres.2004.11.005

Wu ZB, Dong F, Zhao WR, Wang HQ, Liu Y, Guan BH (2009) The fabrication and characterization of novel carbon doped TiO_2 nanotubes, nanowires and nanorods with high visible light photocatalytic activity. Nanotechnology 20:235701–235709. https://doi.org/10.1088/0957-4484/20/23/235701

Wu Z, Li W, Webley PA, Zhao D (2012) General and controllable synthesis of novel mesoporous magnetic iron oxide@ carbon encapsulates for efficient arsenic removal. Adv Mater 24:485–491. https://doi.org/10.1002/adma.201103789

Xiong Z, Lai B, Yang P, Zhou Y, Wang J, Fang S (2015) Comparative study on the reactivity of Fe/Cu bimetallic particles and zero valent iron (ZVI) under different conditions of N_2 air or without aeration. J Hazard Mater 297:261–268. https://doi.org/10.1016/j.jhazmat.2015.05.006

Xu L, Yan-Ling H, Pelligra C, Chun-Hu C, Jin L, Huang H, Sithambaram S, Aindow M, Joesten R, Suib SL (2009) ZnO with different morphologies synthesized by solvothermal methods for enhanced photocatalytic activity. Chem Mater 21:2875–2885. https://doi.org/10.1021/cm900608d

Xu T, Zhang L, Cheng H, Zhu Y (2011) Significantly enhanced photocatalytic performance of ZnO viagraphene hybridization and the mechanism study. Appl Catal B Environ 101:382–387. https://doi.org/10.1016/j.apcatb.2010.10.007

Xu P, Zeng GM, Huang DL, Feng CL, Hu S, Zhao MH, Lai C, Wei Z, Huang C, Xie GX (2012) Use of iron oxide nanomaterials in wastewater treatment: a review. Sci Total Environ 424:1–10. https://doi.org/10.1016/j.scitotenv.2012.02.023

Yang X, Feng Y, He Z, Stoffella PJ (2005) Molecular mechanisms of heavy metal hyperaccumulation and phytoremediation. J Trace Elem Med Biol 18(4):339–353. https://doi.org/10.1016/j.jtemb.2005.02.007

Yang X, Cao C, Erickson L, Hohn K, Maghirang R, Klabunde K (2009) Photo-catalytic degradation of rhodamine B on C, S, N, and Fe-doped TiO_2 under visible-light irradiation. Appl Catal B Environ 91:657–662. https://doi.org/10.1016/j.apcatb.2009.07.006

Yang L, Liu P, Xi L, Li S (2012a) The photocatalytic activities of neodymium and fluorine doped TiO_2 nanoparticles. Ceram Int 38:4791–4796. https://doi.org/10.1016/j.ceramint.2012.02.067

Yang X, Chen C, Li J, Zhao G, Ren X, Wang X (2012b) Graphene oxide-iron oxide and reduced graphene oxide iron oxide hybrid materials for the removal of organic and inorganic pollutants. RSC Adv 2:8821–8826. https://doi.org/10.1039/C2RA20885G

Yang J, Hou B, Wang J, Tian B, Bi J, Wang N, Li X, Huang X (2019) Nanomaterials for the removal of heavy metals from wastewater. Nanomaterials 9:424. https://doi.org/10.3390/nano9030424

Yaqoob AA, Parveen T, Umar K, Ibrahim MNM (2020) Role of nanomaterials in the treatment of wastewater: a review. Water 12(2):495. https://doi.org/10.3390/w12020495

Yildirim OA, Unalan HE, Durucan C (2013) Highly efficient room temperature synthesis of silver-doped zinc oxide (ZnO:Ag) nanoparticles: structural, optical, and photocatalytic properties. J Am Ceram Soc 96:766–773. https://doi.org/10.1111/jace.12218

Yin XT, Que WX, Fei D, Shen FY, Guo QS (2012) Ag nanoparticle/ZnO nanorods nanocomposites derived by a seed-mediated method and their photocatalytic properties. J Alloys Compd 524:13–21. https://doi.org/10.1016/j.jallcom.2012.02.052

Yoon SY, Lee CG, Park JA, Kim JH, Kim SB, Lee SH, Choi JW (2014) Kinetic, equilibrium and thermodynamic studies for phosphate adsorption to magnetic iron oxide nanoparticles. Chem Eng J 236:341–347. https://doi.org/10.1016/j.cej.2013.09.053

Yu JC, Ho W, Yu J, Yip H, Wong PK, Zhao J (2005) Efficient visible light – induced photocatalytic disinfection on sulfur-doped nanocrystalline titania. Environ Sci Technol 39:1175–1179. https://doi.org/10.1021/es035374h

Yu J, Dai G, Xiang Q, Jaroniec M (2011) Fabrication and enhanced visible-light photocatalytic activity of carbon self-doped TiO_2 sheets with exposed {001} facets. J Mater Chem 21:1049–1057. https://doi.org/10.1039/C0JM02217A

Yunus IS, Harwin, Kurniawan A, Adityawarma D, Indarto A (2012) Nanotechnologies in water and air pollution treatment. Environ Technol Rev 1(1):136–148. https://doi.org/10.1080/21622515.2012.733966

Zelmanov G, Semiat R (2008) Phenol oxidation kinetics in water solution using iron (3)-oxide-based nano-catalysts. Water Res 42:3848–3856. https://doi.org/10.1016/j.watres.2008.05.009

Zhai T, Xie S, Zhao Y, Sun X, Lu X, Yu M, Xu M, Xiao F, Tong Y (2012) Controllable synthesis of hierarchical ZnO nanodisks for highly photocatalytic activity. CrystEngComm 14:1850–1855. https://doi.org/10.1039/C1CE06013A

Zhang X (2003) Nanoscale iron particles for environmental remediation: an overview. J Nanopart Res 5:323–332. https://doi.org/10.1023/A:1025520116015

Zhang Z, Kong J (2011) Novel magnetic Fe_3O_4@ C nanoparticles as adsorbents for removal of organic dyes from aqueous solution. J Hazard Mater 193:325–329. https://doi.org/10.1016/j.jhazmat.2011.07.033

Zhang Z, Zhang J, Lou T, Pan D, Chen L, Qu C, Chen Z (2012a) Label-free colorimetric sensing of cobalt (II) based on inducing aggregation of thiosulfate stabilized gold nanoparticles in the presence of ethylenediamine. Analyst 137(2):400–405. https://doi.org/10.1039/c1an15888k

Zhang F, Lan J, Zhao Z, Yang Y, Tan R, Song W (2012b) Removal of heavy metal ions from aqueous solution using Fe_3O_4–SiO_2-poly (1,2-diaminobenzene) core–shell sub-micron particles. J Colloid Interface Sci 387:205–212. https://doi.org/10.1016/j.jcis.2012.07.066

Zhang M, Gao B, Varnoosfaderani S, Hebard A, Yao Y, Inyang M (2013) Preparation and characterization of a novel magnetic biochar for arsenic removal. Bioresour Technol 130:457–462. https://doi.org/10.1016/j.biortech.2012.11.132

Zhang Y, Bing W, Hui X, Hui L, Wang M, He Y, Pan B (2016) Nanomaterials-enabled water and waste water treatment. Nanoimpact 3–4:22–39. https://doi.org/10.1016/j.impact.2016.09.004

Zhao X, Wang J, Wu F, Wang T, Cai Y, Shi Y, Jiang G (2010) Removal of fluoride from aqueous media by Fe_3O_4@ $Al(OH)_3$ magnetic nanoparticles. J Hazard Mater 173:102–109. https://doi.org/10.1016/j.jhazmat.2009.08.054

Zheng J, Jiang Z-Y, Kuang Q, Xie Z-X, Huang R-B, Zheng L-S (2009) Shape-controlled fabrication of porous ZnO architectures and their photocatalytic properties. J Solid State Chem 182:115–121. https://doi.org/10.1016/j.jssc.2008.10.009

Zheng J, Liu Z, Liu X, Yan X, Li D, Chu W (2011) Facile hydrothermal synthesis and characteristics of B-doped TiO_2 hybrid hollow microspheres with higher photo-catalytic activity. J Alloys Compd 509:3771–3776. https://doi.org/10.1016/j.jallcom.2010.12.152

Zhong LS, Hu JS, Liang HP, Cao AM, Song WG, Wan LJ (2006) Self-assembled 3D flower like iron oxide nanostructures and their application in water treatment. Adv Mater 18:2426–2431. https://doi.org/10.1002/adma.200600504

Zhou W, Du G, Hu P, Li G, Wang D, Liu H, Wang J, Boughton R, Liu D, Jiang H (2011) Nanoheterostructures on TiO_2 nanobelts achieved by acid hydrothermal method with enhanced photocatalyic and gas sensitive performance. J Mater Chem 21:7937–7945. https://doi.org/10.1039/C1JM10588D

Zhou XT, Ji HB, Huang XJ (2012a) Photocatalytic degradation of methyl orange over metalloporphyrins supported on TiO_2 degussa P25. Molecules 17:1149–1158. https://doi.org/10.3390/molecules17021149

Zhou X, Shi T, Zhou H (2012b) Hydrothermal preparation of ZnO-reduced graphene oxide hybrid with high performance in photocatalytic degradation. Appl Surf Sci 258(17):6204–6211. https://doi.org/10.1016/j.apsusc.2012.02.131

Zhu X, Liu Y, Zhou C, Zhang S, Chen J (2014) Novel and high-performance magnetic carbon composite prepared from waste hydrochar for dye removal. ACS Sustain Chem Eng 2:969–977. https://doi.org/10.1021/sc400547y

Chapter 4
MoS₂ Based Nanocomposites for Treatment of Industrial Effluents

Manjot Kaur, Unni Krishnan, and Akshay Kumar

Abstract Water pollution due to various industrial pollutants is a major threat for safe environment. Conventional methodology for removal of these pollutants is hindered by various obstacles. However, photocatalysis has been considered by many industries as solution for removal of these pollutants because of its advantageous features. Semiconductor transition metal dichalcogenides with unique properties are gaining lot of interest as photocatalyst. Molybdenum disulfide (MoS_2) from metal dichalcogenide family is promising for photocatalysis due to existence of direct band gap but it possesses some limitations such as lack of emission, high recombination and stacking faults. MoS_2 based nanocomposites overcome these limitations and has a great degradation efficiency for degrading all major pollutants and dyes.

This chapter reviews the emerging trends and various combinations of MoS_2 based nanocomposites utilised for wastewater treatment. The major pollutants are the industrial wastes of azo and nitroaromatic based compounds. MoS_2 based nanocomposites like $SnO_2/Ag/MoS_2$, MoS_2/GO and MoS_2 /TiO_2 have exhibited degradation efficiency of ~ 90% against 2,4-diclorophenol (DCP), 99% against methylene blue and 99.35% reduction in 4-nitrophenol respectively. A comprehensive review of mechanism followed by nanocomposites for degradation is also elucidated in this chapter. Most of their degradation mechanism followed a Z-scheme or heterojunction formation, which showed high results. Additionally, PANI@MoS_2 nanocomposites has good removal rate against heavy metals like Cr (VI). This chapter also discusses various factors which effect the photocatalytic property. Degradation efficiencies are dependent on factors like bandgap, phases and morphology. The photocatalytic efficacy is attributed to OH• and •O_2^- radicals, which affects reusability and efficiency. Finally, the chapter also provides an insight into constraints in photocatalysis and the advantages of MoS_2 based nanocomposites to overcome the same.

M. Kaur · U. Krishnan · A. Kumar (✉)
Advanced Functional Materials Laboratory, Department of Nanotechnology, Sri Guru Granth Sahib World University, Fatehgarh Sahib, Punjab, India

© The Author(s), under exclusive license to Springer Nature Switzerland AG 2021
S. Rajendran et al. (eds.), *Metal, Metal-Oxides and Metal-Organic Frameworks for Environmental Remediation*, Environmental Chemistry for a Sustainable World 64,
https://doi.org/10.1007/978-3-030-68976-6_4

Keywords MoS$_2$ · Nanocomposites · Industrial dyes · Photocatalysis · Degradation · Wastewater treatment

4.1 Introduction

Increase of consumption capacity and population of world has escalated energy demands and amount of toxics and pollutants released into the environment. Textile industry expels out more than 10,000 different types of dyes and chemicals accounting to 700,000 tonnes globally. Water is used in every segment of its utilisation and cleaning which is later disposed as wastewater. Various methodologies are adhered to treat the outlet from various industries into general water and air medium. Dyes and chemicals in wastewater from textile industry, tanning industry and paint industry should be treated before expelling it into open source. This wastewater is treated by methods like adsorption (ONG et al. 2008), biodegradation (Casas et al. 2007) and photocatalytic oxidation (Hu et al. 2014). Various method used for removal of pollutants are shown in Fig. 4.1. The cost effectiveness, reusability, ease of operation, and non-toxicity of bi-products / resultants are few parameters which decide on selection of methodology for wastewater treatment process. Photocatalysis procedure which has all the above advantages along with usage of free source of energy i.e. sunlight has always been the favourite for waste water

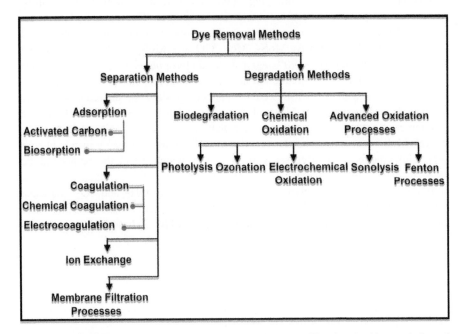

Fig. 4.1 Methodologies for removal of industrial pollutants. ('Reprinted with permission of Elsevier' from Reference Nidheesh et al. 2018)

treatment. Additionally, catalytic behaviour of these photocatalytic materials also assisted in resolving energy crises by assisting in H_2 production. Semiconductors which has the ability to transfer electrons from valency layer to conduction layer by absorbing small amount of photons, are established to be the favourite. Thus, photocatalytic materials were explored to obtain maximum degradation efficiency for multiple chemicals. Transitional metal dichalcogenides consisting of molybdenum disulfide (MoS_2), tungsten disulfide etc., which has excellent semiconductor characteristics, are investigated extensively in this decade, because of their promising results and ease of preparation.

Unlike graphene, MoS_2 is a prominent member of transitional metal dichalcogenides which is expected to replace graphene as it has an ability to transform from indirect bandgap to direct band gap on change of bulk to nanostructure. MoS_2 has already been a successful as photocatalyst in degradation of polluted water of textile industry, desulphurisation and decomposition of pollutants from refineries and degradation of toluene, nitric oxide (NO) removal for air purification. Thus, MoS_2 is deemed to be a solution for wastewater treatment and energy production (Abinaya et al. 2018). Its other specific features like porous structure, enlarged surface area with increased active sites, charge separation features in two dimensional format has been the reason for its success as good photocatalyst.

However, MoS_2 has a few drawbacks which hinder its full utilisation and extracting the complete effectiveness. Few of its limitations are as listed below:

- The poor electronic conductivity of MoS_2 effects the employing MoS_2 as a photocatalytic material (Li et al. 2014).
- The large interlayer spacing with an weak interlayer bonding will promote MoS_2 nanosheets into a thick form which is unfavourable for photocatalysis (Zeng et al. 2015).
- High electron hole recombination rate.
- Additionally, the formed MoS_2 multi-layer nanostructures tend to agglomeration or stacked multilayers on a substrate. This causes the reduction in efficiency due to migration of charge carrier and photon absorption.

Numerous strategies have been formulated to enhance the photocatalytic efficiency of MoS_2. The successful methodologies included defect engineering, doping with other active metals, deposition of noble metals and formation of semiconductor heterostructures. However, formation of semiconductor heterostructure by coupling with other semiconductors is evaluated to be highly effective in enhancing the photocatalytic efficiency by effective transfer of charge pair and prohibiting the recombination of photogenerated electro-hole pairs.

4.2 Classification of Pollutants

Pollutants which pollute water can be classified into natural and synthetic dyes as shown in Fig. 4.2. The natural dyes are produced from animals and plants. Synthetic dyes are man-made and can be further classified into non-Azo dyes and Azo dyes,

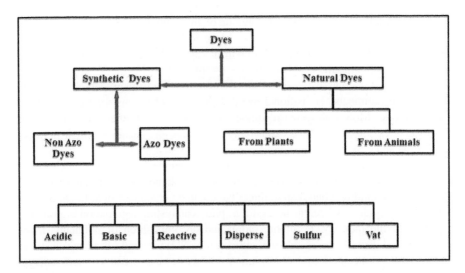

Fig. 4.2 Classification of various dyes. In this, azo dyes are extensively used in food, cosmetic, pharmaceutical, textile and leather industries which account for nearly 60–70% of all dyes used commercially

which are major contributor for contamination. This class is further divided as per their basic nature. Nitroaromatic compounds are extensively utilised in fungicides, pesticides, plasticizers, and dyes. It is a major water pollutant in industrial effluents. 4-Nitrophenol (4-NP)— listed as a "priority pollutant" by the US Environmental Protection Agency, makes substantial health and environmental risks, resulting in toxicity and carcinogenicity. It is difficult to be removed by natural degradation and is highly soluble in water.

Bulk MoS_2 has a large size and small bandgap, which is unfavourable for production of free hydroxyl radicals, on the contrary, nano sized MoS_2 due to its larger surface area displace excellent photo catalytic property. Additionally, modifying MoS_2 to a lesser number of layers is also reported to improve photocatalytic property. The photocatalytic performance of MoS_2 is a function of morphology, particle size and bandgap. Photocatalytic property of MoS_2 is influenced by its optical and structural properties as it decides on the active sites and its defect state. MoS_2 is an active photocatalytic material in the visible range. Many researches have reported its high degradation efficiency against various chemicals and dyes. Photocatalytic property of MoS_2 depends on type and number of elements in its configuration, which can be classified into three forms namely pure MoS_2, Bi-functional elements and Ternary hybrid combination with MoS_2. Further to enhance its capabilities MoS_2 nanocomposites were fabricated along with other successful materials. Various examples of MoS_2 based nanocomposites for photocatalytic degradation are shown in Table 4.1. Few of these elements forming nanocomposites with MoS_2 are zinc oxide, carbon nitride, graphene etc. The success rate of photocatalytic materials is governed by function that rate of recombination should be slower and lesser than the electron capturing rate.

Table 4.1 MoS$_2$ nanocomposites for removal of industrial effluents

Nanocomposite	Morphology	Removal efficiency (%)	References
Coupling between the MoS$_2$ and GO with g-C$_3$N$_4$	MoS$_2$ nanosheets dispersed in the CN-M-G ternary hybrid.	89.5% after 2.5 h Rhodamine B (RhB) under visible light irradiation	Hu et al. (2014)
MoS$_2$–GO	Homogeneous dispersion of MoS$_2$ in graphene hydrogel	99% removal of methylene blue (MB) in 60 min under solar light	Ding et al. (2015)
TiO$_2$/ MoS$_2$@zeolite	TiO$_2$/MoS$_2$ coated on surface of zeolite	90% of methyl orange (MO) in 60 min in visible light	W. Zhang et al. (2015a)
TiO$_2$/MoS$_2$	Hybrid structure	90% of MO in 10 min in xenon lamp simulating solar irradiation	W. Zhang et al. (2015b)
MoS$_2$/g-C$_3$N$_4$	MoS$_2$ particles dispersed on surface of layered g-C$_3$N$_4$	51.67%, higher rate of NO nitric oxide removal than g-C$_3$N$_4$ nanoparticles	Wen et al. (2016)
MoS$_2$ nanosheets with ZnO–g-C$_3$N$_4$	ZnO nanospheres embedded in co-stacked g-C$_3$N$_4$/MoS$_2$ nanosheets	99.50% of MB in 30 min of visible light irradiation,	Jo et al. (2016)
g-C$_3$N$_4$/Ag/MoS$_2$	Flowerlike architecture	9.43-fold increase in degradation in visible light	Fang (2017)
MoS$_2$ /TiO$_2$	MoS$_2$ nanosheet encapsulating TiO$_2$ hollow spheres	99.35%, reduction of 4-nitrophenol simulated sun light irradiation	Guo et al. (2017)
MoS$_2$/ZnO	Deposition of ZnO on MoS$_2$	80% of Novacron red hunts-man (NRH) dye after 80 min of light irradiation	Krishnan et al. (2019a)
MoS$_2$/Ag	Accumulation of Ag nanoparticles on MoS$_2$	77% of Novacron red hunts-man dye after 80 min of light irradiation	Krishnan et al. (2019a)
SnO$_2$/Ag/MoS$_2$	Flower like nanosheets of MoS$_2$ with even distribution of Ag and SnO$_2$	90% of 2,4-diclorophenol (DCP) in 20 min. 100% of MB and MO within 12 min. RhB 100% within 14 min	Khan et al. (2019)
GO/MoS$_2$/g-C$_3$N$_4$	MoS$_2$ well-dispersed in the ternary system	96.7% towards RhB with good photostability	Yan et al. (2019)

4.3 Degradation Mechanism Followed by MoS$_2$ Based Composites

Degradation mechanism proposed is a function of material to be degraded and co-catalyst forming nanocomposite. Few dyes self-degraded on absorption of energy. Schematic diagrams depicting photocatalytic mechanism are shown in Figs. 4.3 and 4.4. The absorption ability of dye can be attributed to large surface area which creates many active zones and improved electron-hole pair separation

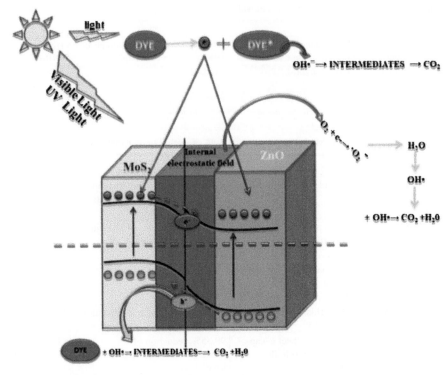

Fig. 4.3 Degradation mechanism with reactive dye. ('Reprinted with permission of Elsevier' from Reference Krishnan et al. 2019a)

capability of nanocomposites (Saha et al. 2015). The modus operandi for degradation of dyes may follow two different routes to achieve dye removal. In route-1, with absorption of energy, the dye molecules transforms into activated dye molecule (ADM) (Li and Cao 2011; Sacco et al. 2012) by transferring its electrons into the conduction band of MoS_2 and ZnO (Ji et al. 2009). These molecules flow into transition state of electrostatic field formed at the heterojunction. However, there is a likely probability that electron would be energized and again assists in the formation of more ADM. The likely mechanism may be as follows:

$$h^+ + OH^- \rightarrow OH^\bullet \quad (4.1)$$

$$Dye + OH^\bullet \rightarrow INTERMEDIATES \rightarrow CO_2 \quad (4.2)$$

$$DYE + e^- \rightarrow (ADM)^{\bullet -} \quad (4.3)$$

$$(ADM)^{\bullet -} \rightarrow DYE + (ADM)^{\bullet -} \quad (4.4)$$

$$(ADM)^{\bullet -} + OH^\bullet \rightarrow INTERMEDIATES \rightarrow CO_2 \quad (4.5)$$

In route-2, the heterojunction formed improves electron-hole pair separation and enhances photocatalytic and degradation process (Tan et al. 2014). The UV light will

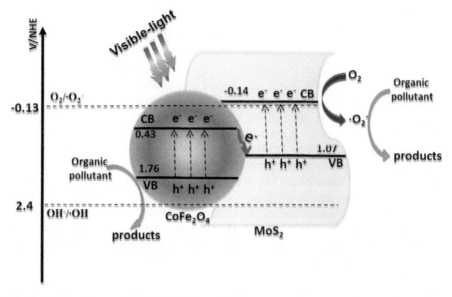

Fig. 4.4 Degradation mechanism by Z-Scheme. In this MoS$_2$ nanosheets acts like support material for CoFe$_2$O$_4$. This format promoted degradation of RhB utilising •O$_2^-$ and h$^+$ radicals. ('Reprinted with permission of Elsevier' from Reference Zeng et al. 2018)

excite electrons in ZnO (Y. Zhang et al. 2012), while visible light will excite electrons in MoS$_2$. In nanocomposites at heterojunction an internal electrostatic field is formed. Fermi energy level (E$_F$) of MoS$_2$ and ZnO are same, which enables them to be better composites, however conduction band and valence band of ZnO is below MoS$_2$ due to different work functions (Choi et al. 2014; Tan et al. 2014).

The electrons of valency band of MoS$_2$ will absorb energy and shift to conduction band and at heterojunction these electrons diffuse to ZnO, due to presence of carrier charge density gradient and good electron acceptors thus avoiding recombination of electron hole pair. Similarly, hole formed at ZnO get diffused to MoS$_2$, forming an internal electrostatic field and band bending at the contact. This configuration calls for distinct separation of electrons and holes at the interface, thus increasing life of charge carriers, hinder recombination of electron hole pairs and finally to required output of enhanced photocatalytic abilities. The equations describing mechanism are as follows:

$$h^+ + H_2O \rightarrow H^+ + OH^\bullet \quad (4.6)$$

$$e^- + O_2 \rightarrow {}^\bullet O^-{}_2 \quad (4.7)$$

$${}^\bullet O^-{}_2 + H_2O \rightarrow HO^\bullet{}_2 + OH^- \quad (4.8)$$

$$HO^\bullet{}_2 + H_2O \rightarrow H_2O_2 + OH^\bullet \quad (4.9)$$

$$H_2O_2 + e^- \rightarrow OH^\bullet + OH^- \quad (4.10)$$

$$Dye + OH^\bullet \rightarrow \text{degradation of dye} \quad (4.11)$$

Z-scheme of degradation pattern is followed for materials which does not convert to activated molecule. Hydrothermal method was utilised to fabricate magnetic Z-scheme $MoS_2/CoFe_2O_4$ (Zeng et al. 2018). In this one side has MoS_2 nanosheets, which acts like support material for $CoFe_2O_4$. This configuration ensures high specific area and prevents aggregation of magnetic $CoFe_2O_4$ nanoparticles. This format promoted degradation of RhB utilising $\bullet O_2^-$ and h^+ radicals. When energy dispenses on the nanocomposite generated electrons and holes of both the elements gets shifted from valance band (VB) to conduction band (CB) (Fig. 4.4) (Zhou et al. 2014). Later the electrons from CB of MoS_2 shift to CB of $CoFe_2O_4$, similarly holes in VB move in opposite direction. The electron in CB of MoS_2 move to surface of nanocomposite and reduce O_2 to $\bullet O_2^-$. However due to difference in potential, OH is not reduced to $\bullet OH$ radical. Thus, complete degradation of RhB is proposed to be conducted by h + and O_2 radical. However, CB of $CoFe_2O_4$ is more positive than O_2.

4.4 Factors Effecting Photocatalysis

4.4.1 Effect of Bandgap

The higher bandgap and thinner MoS_2 nanosheets will promote easy interfacial charge transfer and obviate recombination of charge pairs (X. Zhang et al. 2016). Conventional heterojunction can be classified into four groups on the basis of their Band alignment. Type-I consist of heterojunction in which electrons from the conduction band of semiconductor would transfer into the conduction band of MoS_2 due to higher positive conduction band of MoS_2. In type –II electron from conduction band of semiconductor gets transferred to MoS_2 while holes formed will move in opposite direction; this leads to formation of special suppression of charge carriers with the accumulation of electrons of MoS_2 and holes at semiconductor. In type-III a P-N heterojunction is formed due to inner electric field. Type IV involves a combination in which the band gaps of MoS_2 and semiconductor are not over lapped.

4.4.2 Phases and Morphology

The efficiency of photocatalysis is dependent on number of active sites and edges. MoS_2 nanosheets has more active edge sides than other conventional morphology. MoS_2 with low crystal structure assists in photocatalysis. Nanosheets are plagued with drawbacks like the tendency to stack and aggregate easily. Additionally, morphology of co-catalyst also plays a prominent role in photocatalysis. When elements supporting plasmonic response are used as co-catalyst, efficiency of photocatalysis increases due to efficient interfacial electron transfer and electron injection from anisotropic noble metals to monolayer of MoS_2. Experiments were conducted for a combination of Au-MoS_2 in which Au having a morphology of nanosphere (AuNS), nanorods (AuNR) and nanotriangle (AuNT) were assembled on chemically exfoliated MoS_2 (Zhang et al. 2017). Results confirmed that Au-nanorods displayed highest efficiency of photocatalysis due to surface plasmon response (SPR). Figure 4.5 shows the methodology of assembly of Au on MoS_2 monolayer forming MoS_2/Au nanocomposite. It shows different morphology of Au being assembled on MoS_2 monolayer.

Noble metals like gold and silver when combined with MoS_2 broadens the absorption spectrum, which directly enhances the separation and degradation efficiency (Zhang et al. 2017). Additionally, when these bi-metal nanocomposite are assembled into ternary composite (g-C_3N_4/Ag/MoS_2), it enhances the visible-light absorption of MoS_2 (Akhundi and Habibi-Yangjeh 2016). The inclusion of noble metals like silver, assists in injecting electrons into conduction band due to SPR (Rycenga et al. 2011). Figure 4.6 shows mechanism of the plasmon-induced hot electron generation and subsequent charge separation (CS) over anisotropic Au nanostructures-MoS_2. Rapid interconversion between transverse SPR (TSPR) and

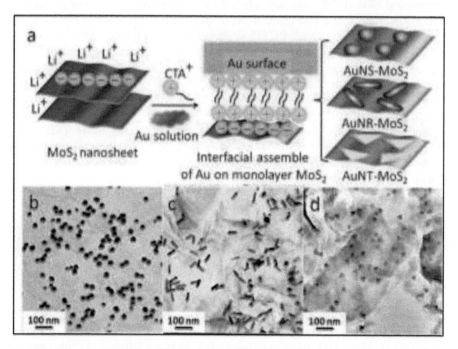

Fig. 4.5 Schematic illustration (**a**) of anisotropic Au nanostructures mono-assembled on monolayer MoS_2. TEM images of (**b**) AuNSMoS_2, (**c**) AuNR-MoS_2, and (**d**) AuNT-MoS_2. ('Reprinted with permission of RSC Pub' from Reference Zhang et al. 2017)

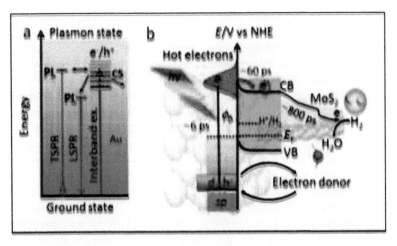

Fig. 4.6 Schematic diagram of (**a**) radiative decays of SPR; and (**b**) interfacial charge transfer from AuNR to MoS_2 for HER. Photoluminescence (PL), transverse SPR (TSPR), longitudinal SPR (LSPR), charge separation (CS), Fermi level EF, and Schottky barrier. ('Reprinted with permission of RSC Pub' from Reference Zhang et al. 2017)

electron-hole occurs due to photoluminiscence (PL). Energy transfer from TSPR to longitudinal SPR (LSPR) happens when the excited electrons lose energy due to nonradiative decay and fast interconvert to the LSPR. This leads to emission of photon leading to a plasmonic PL in low energy level (red line).

It is reported that MoS_2 based heterostructures are influenced by crystal faces formed at the interfaces, as they decide on the efficiencies of charge transmissions. This is validated by degrading MB by MoS_2 having (001) and (101) facet TiO_2. An enhanced photocatalytic behaviour (3.4 times higher) of MoS_2/TiO_2 (001) is attributed to (001) facet of TiO_2 and MoS_2. This proves that conventional physical mixing cannot achieve higher results even though same facets exists in the individual elements (Cao et al. 2015). Similar phenomenon is also reported when MoS_2 of single layer is grown on the surface of Cd rich wurtzite CdS nanocrystals (Chen et al. 2015).

4.4.3 Miscellaneous Factors

Another important factors which influences photocatalytic property of MoS_2 based nanocomposite are calcination temperature and UV irradiation time. These factors decide on crystallisation of MoS_2/TiO_2 with MoS_2 clusters on TiO_2 particles. These reaction follows the first order rate law. Table 4.2 shows the degradation of methylene blue by MoS_2/TiO_2 nanocomposites fabricated with different reaction parameters. The maximum degradation of methylene blue was found at calcination temperature of 573 K and UV irradiation time of one hour. In Table 4.2, 'b' denotes

Table 4.2 Photocatalytic activities of MoS_2 nanocluster sensitized TiO_2 samples under different preparation conditions for methylene blue degradation (Ho et al. 2004)

Samples	UV irradiation time (h)	Calcination temp^b (K)	photocatalytic activity (A/A$_0$, t = 4 h)
MoS_2/TiO_2	1.5	523	0.85
MoS_2/TiO_2	1.5	573	0.68
MoS_2/TiO_2	1.5	623	0.78
MoS_2/TiO_2	0.5	573	0.88
MoS_2/TiO_2	1	573	0.92

calcination in N_2 for 1 h, 'A$_0$' initial absorbance of methylene blue, 'A' is the absorbance of methylene blue after 4 h in visible light irradiation.

Effect of pH is another important factor which effects absorption potential of nanocomposites against contaminants. It influences the surface charge and degree of ionization of adsorbent which decides on absorption of pollutants. The pH of pollutant is directly proportional to absorption/ removal of pollutants. At low pH, free oxygen moieties on absorbent gets protonated and at high pH, heavy metal ions start to precipitate as hydroxides. The contact time and adsorption kinetics decides on adsorption rate and mechanism for degradation (Gusain et al. 2019).

4.5 Degradation of Inorganic Pollutants

MoS_2 based nanocomposites has achieved positive results in degradation cum detoxification of heavy metals and nitric oxide. Removal rate of heavy metals like Cr(VI) by PANI@MoS_2 was reported as 526.3 and 623.2 mg/g at pH 3.0 and 1.5 respectively. This high rate of absorption is attributed to abundant functional groups and flower-like MoS_2 structures encapsulated by Polyaniline (PANI) (Gao et al. 2016). The effect of ionic strength was evaluated using $NaNO_3$, which confirms PANI on the outer surface of MoS_2, changes adsorption mechanism from outer-sphere to inner-surface complexation which is altered by PANI. These results indicate that removal of heavy metals is a function of pH and ionic strength of pollutant and reduction products. $Cr(OH)_3$ a reduction product of Cr(VI) precipitate on active sites of nanocomposites, hindering reusability and effecting its catalytic efficiency.

4.6 Air Purification by MoS_2 Based Nanocomposites

The proposed novel Z-scheme structure $(BiO)_2CO_3/MoS_2$ is reported to have better NO removal efficiency than $(BiO)_2CO_3$ and MoS_2 individually (Xiong et al. 2016). This enhanced degradation is attributed to p-n type semiconductor with 3D

Fig. 4.7 SEM results of $(BiO)_2CO_3/MoS_2$ nanocomposites. ('Reprinted with permission of Elsevier' from Reference Xiong et al. 2016)

Z-scheme photocatalyst format, which preserves redox couple ability. Experimental results proved that h^+ and $\bullet O_2^-$ are major radicals which contribute to NO removal, which is confirmed by electron spin response. The morphology of MoS_2 microflowers existing in microsperes of $(BiO)_2CO_3$ is another contributing factor, for photocatalysis, as many pores are generated as shown in Fig. 4.7.

Figure 4.8 shows mechanism followed for NO removal. $(BiO)_2CO_3$ creates holes and electrons by absorbing energy in visible light irradiation, however h^+ and $\bullet O_2^-$ radical are primary members in reduction of NO. These active holes and electrons then react with $O_2/H_2O/OH^-$ to yield other active species (Krishnan et al. 2019b). Electron spin response of elements show peaks of $\bullet O_2^-$ with peak intensity increasing with time. Nanocomposite of $(BiO)_2CO_3/MoS_2$ with 5% of MoS_2 showed NO removal of 57%, however further increase in MoS_2 decrease removal capacity due to masking effect of light by MoS_2. Similar results of NO removal were obtained with of MoS_2-g-C_3N_4 nanocomposites.

4.7 Degradation of Organic Pollutants

Organic pollutants are major compounds in dyes which has to be degraded as part of water treatment. Bi-functional and ternary nanocomposites have shown enhanced degradation capability due to good conductivity and morphology which assists in formation of layer / wire for transfer of photogenerated electrons. Nanocomposites like $Ag_3PO_4/TiO_2@MoS_2$ with an additional feature of restriction of photo corrosion

4 MoS$_2$ Based Nanocomposites for Treatment of Industrial Effluents

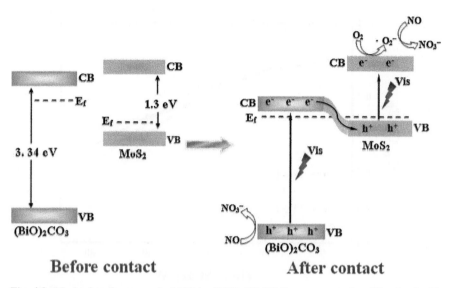

Fig. 4.8 Mechanism for removal of NO by (BiO)$_2$CO$_3$/MoS$_2$ nanocomposite. ('Reprinted with permission of Elsevier' from Reference Xiong et al. 2016)

of Ag$^+$ ensures highly effective degradation of oxytetracycline and reduction in efficiency by only 10% even after 10 repetitions (Shao et al. 2017). Novacron red Hunts- man dye (NRH) and methylene blue (MB) is reported to be degraded by MoS$_2$/ZnO nanocomposite. Figure 4.9 depicts the relation between the degradation efficiency of MB and NRH for different loading of nanocomposite. The steady decrease in intensity of NRH dye confirms the decomposition of dye. 1 g/L loading of MoS$_2$/ZnO nanocomposite showed degradation rate of 81.76% and 80.37% for MB and NRH dye respectively.

Prominent dye like methyl orange (MO) was successfully degraded by TiO$_2$/MoS$_2$@zeolite composite photocatalysts in xenon lamp as irradiation source. Analysis confirms that degradation follows a Langmuir–Hinshelwood kinetic model (pseudo-first order reaction). Figure 4.10 shows the degradation of MO using TiO$_2$/MoS$_2$@zeolite composite which compares the all elements of nanocomposites. It shows the highest reaction rate constant of 2.304 h^{-1}, which is higher than other individual components. The enhancement is attributed to coupled structure of TiO$_2$/MoS$_2$ on the surface of zeolite making a positive synergetic effect. Additionally this distribution makes a large contact area along with short diffusion time (W. Zhang et al. 2015a).

Nitroaromatic chemicals can be rendered non-toxic by removing the nitro group effectively. This would enhance its biodegradability in normal environment. MoS$_2$ and reduced graphene oxide (rGO) are combined to form hybrid nanosheets on which CdS is grown for reducing 4-NP. The nanocomposites are found to be stable and efficiently reduce 4-NP to aminophenol (4-AP), thus assisting in detoxification

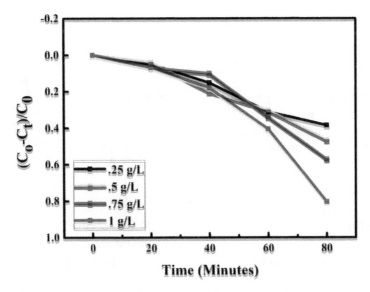

Fig. 4.9 Degradation efficiency of NRH for different loading of MoS_2/ZnO nanocomposite. 1 g/L loading of MoS_2/ZnO nanocomposite showed degradation rate of 80.37% for NRH dye. ('Reprinted with permission of Elsevier' from Reference Krishnan et al. 2019a)

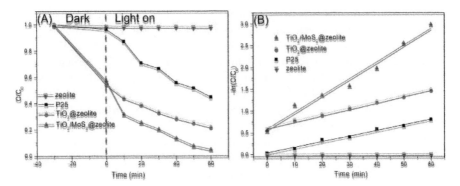

Fig. 4.10 (a) Photocatalytic degradation of MO with TiO_2/MoS_2@zeolite, TiO_2@zeolite, P25 and zeolite; (b) relation curves for the photocatalytic degradation of MO (MO, 20 mg/L; photocatalyst, 500 mg/L) ($-\ln(C/C_0)/t$). ('Reprinted with permission of Elsevier')

and reduction of toxic pollutants to stable form. The weight percentage of 3% of MoS_2/rGO hybrid in composite, provided maximum results of complete conversion of 4-NP to 4-AP in 18 minutes. Further increase of co-catalyst resulted in decrease in results due to shielding effect, which is obstructing the light source on catalyst. The CdS-MoS_2/rGO nanocomposite possess flower-like morphology for hybrid, with rGO sheets meshing adequately close with MoS_2 nanosheets (Peng et al. 2016). Fig. 4.11 displays TEM image in which CdS nanoparticles are supported by MoS_2/

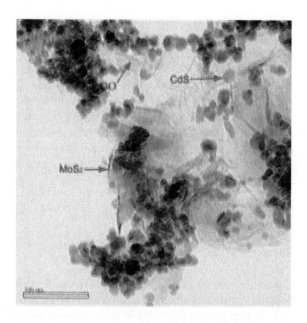

Fig. 4.11 TEM image of CdS-MoS$_2$ /rGO. (Reprinted with permission of Elsevier' from Reference Peng et al. 2016)

rGO hybrid. Heavy metals like Pb(II) and Cd(II) are successfully degraded by MoS$_2$/thiol-functionalized multiwalled carbon nanotube (SH-MWCNT) nanocomposite. Isotherm studies showed that the degradation followed Freundlich adsorption isotherm models and kinetic studies represented that adsorption process followed pseudo-second-order. The degradation attained was 90 mg/g and 66.6 mg/g for Pb(II) and Cd(II) respectively (Gusain et al. 2019). This high efficiency is due to ion-exchange and electrostatic interactions. Additionally, it was noted that Metal-sulfur complex formation was identified as the key contributor for adsorption of heavy-metal ions followed by electrostatic interactions for multilayer adsorption.

4.8 Constraints of Photocatalysis

Photocatalysis has to overcome many drawbacks before commercialisation. Photocatalyst recycling from reaction media is one prominent factor. Retention of costly photocatalyst and its repeatability consists of two steps. First step involves separation and collection of photocatalysis and other step is to maintain degradation efficiency during repeated usage. The later part is dependent on catalyst and usage of methodology for avoiding the aggregation of catalytic element so as to maintain photocatalytic efficiency. Magnetic separation method is proven to possess numerous advantages in comparison to conventional process of filtration and centrifugal separation (Shokouhimehr 2015; Polshettiwar et al. 2011). Due to nanosize of photocatalyst, the conventional methods are not advantageous. Additionally, dissolution of homogenous photocatalyst in the reaction media makes a major hindrance in separating the costly catalyst from resultants. However, heterogeneous photocatalysts are magnetic and can be effectively separated for reuse by magnetic method (Zeng et al. 2018).

The decreased photocatalytic efficiency during cyclic-runs of MoS_2 can be attributed to oxidization of photocatalyst, by OH• and •O_2^- radicals. This affects reusability and efficiency, which can be overcome by addition of co-catalyst materials, which will inhibit oxidation due to active radicals.

MoS_2 and few of its composites are hydrophobic, and have difficulty for complete dispersion in water. However, few nanocomposites of MoS_2 with MMT assists in improving the dispersion of catalyst in water along with its efficiency (Peng et al. 2017).

4.9 Conclusion

Capability of MoS_2 based nanocomposites to degrade the industrial pollutants has exhibited promising results. The emphasis on environment protection and restriction on industries against discharge of toxic elements into atmosphere or water stream has accelerated the research in this sector. Catalytic property of MoS_2 is enhanced by doping and forming heterojunction with other semiconductors. The Z-scheme and active dye degradation mechanisms are extrapolated to many other toxic pollutants in the industry. Usage of noble metals and utilising the SPR has extended the degradation capability of the nanocomposites. The ease of separating the nanocomposite after degradation and maintaining the degradation efficiency even after repetitions will decide its commercialisation. Extended research is being done to produce magnetically separable MoS_2 based nanocomposite. The selection of co-catalyst with good synergistic effect has achieved promising results. These findings ensure the potential capability of MoS_2 based nanocomposites for removal of industrial pollutants in an advantageous manner.

References

Abinaya R, Archana J, Harish S, Navaneethan M, Ponnusamy S, Muthamizhchelvan C, Shimomura M, Hayakawa Y (2018) Ultrathin layered MoS_2 nanosheets with rich active sites for enhanced visible light photocatalytic activity. RSC Adv 8:26664–26675. https://doi.org/10.1039/C8RA02560F

Akhundi A, Habibi-Yangjeh A (2016) Novel G-C_3N_4/Ag_2SO_4 nanocomposites: fast microwave-assisted preparation and enhanced photocatalytic performance towards degradation of organic pollutants under visible light. J Colloid Interface Sci 482:165. https://doi.org/10.1016/j.jcis.2016.08.002

Cao L, Wang R, Wang D, Li X, Jia H (2015) MoS_2-hybridized TiO_2 nanosheets with exposed {001} facets to enhance the visible-light photocatalytic activity. Mater Lett 160:286–290. https://doi.org/10.1016/j.matlet.2015.07.149

Casas N, Blánquez P, Gabarrell X, Vicent T, Caminal G, Sarrà M (2007) Degradation of Orange G by laccase: fungal versus enzymatic process. Environ Technol 28(10):1103–1110. https://doi.org/10.1080/09593332808618874

Chen J, Wu X-J, Yin L, Li B, Hong X, Fan Z, Chen B, Xue C, Zhang H (2015) One-pot synthesis of CdS Nanocrystals hybridized with single-layer transition-metal dichalcogenide nanosheets for efficient photocatalytic hydrogen evolution. Angew Chem Int Ed 54(4):1210–1214. https://doi.org/10.1002/anie.201410172

Choi SH, Shaolin Z, Yang W (2014) Layer-number-dependent work function of MoS_2 nanoflakes. J Korean Phys Soc 64(10):1550–1555. https://doi.org/10.3938/jkps.64.1550

Ding Y, Zhou Y, Nie W, Chen P (2015) MoS_2–GO nanocomposites synthesized via a hydrothermal hydrogel method for solar light photocatalytic degradation of methylene blue. Appl Surf Sci 357:1606–1612. https://doi.org/10.1016/j.apsusc.2015.10.030

Fang, Pengfei (2017) Highly efficient visible-light-induced photoactivity of Z-scheme g-C_3N_4/Ag/MoS_2 ternary photocatalysts for organic pollutant degradation and production of hydrogen. ACS Sustainable Chem. Eng. 5(2):1436–1445. https://doi.org/10.1021/acssuschemeng.6b02010

Gao Y, Chen C, Tan X, Xu H, Zhu K (2016) Polyaniline-modified 3D flower-like molybdenum disulfide composite for efficient adsorption/photocatalytic reduction of Cr(VI). J Colloid Interface Sci 476:62–70. https://doi.org/10.1016/j.jcis.2016.05.022

Guo N, Zeng Y, Li H, Xu X, Yu H (2017) MoS_2 nanosheets encapsulating TiO_2 hollow spheres with enhanced photocatalytic activity for nitrophenol reduction. Mater Lett 209:417–420. https://doi.org/10.1016/j.matlet.2017.08.068

Gusain R, Kumar N, Fosso-kankeu E, Ray SS (2019) Efficient removal of Pb(II) and Cd(II) from industrial mine water by a hierarchical MoS_2/SH-MWCNT nanocomposite. ACS Omega 4:13922–13935. https://doi.org/10.1021/acsomega.9b01603

Ho W, Jimmy CY, Lin J, Jiaguo Y, Li P (2004) Preparation and photocatalytic behavior of MoS_2 and WS_2 nanocluster sensitized TiO_2. Appl Nanosci 25:5865–5869. https://doi.org/10.1021/la049838g

Hu SW, Yang LW, Tian Y, Wei XL, Ding JW, Zhong JX, Chu PK (2014) Non-covalent doping of graphitic carbon nitride with ultrathin graphene oxide and molybdenum disulfide nanosheets: an effective binary heterojunction photocatalyst under visible light irradiation. J Colloid Interface Sci 431:42–49. https://doi.org/10.1016/j.jcis.2014.05.023

Ji P, Zhang J, Chen F, Anpo M (2009) Study of adsorption and degradation of acid Orange 7 on the surface of CeO_2 under visible light irradiation. Appl Catal B Environ 85(3–4):148–154. https://doi.org/10.1016/j.apcatb.2008.07.004

Jo W-K, Lee JY, Selvam NCS (2016) Synthesis of MoS_2 nanosheets loaded ZnO–g-C_3N_4 Nanocomposites for enhanced photocatalytic applications. Chem Eng J 289:306–318. https://doi.org/10.1016/j.cej.2015.12.080

Khan B, Raziq F, Bilal Faheem M, Umar Farooq M, Hussain S, Ali F, Ullah A et al (2019) Electronic and nanostructure engineering of bifunctional MoS_2 towards exceptional visible-light photocatalytic CO_2 reduction and pollutant degradation. J Hazard Mater 381:120972. https://doi.org/10.1016/j.jhazmat.2019.120972

Krishnan U, Kaur M, Kaur G, Singh K, Dogra AR, Kumar M, Kumar A (2019a) MoS_2/ZnO nanocomposites for efficient photocatalytic degradation of industrial pollutants. Mater Res Bull 111:212–221. https://doi.org/10.1016/j.materresbull.2018.11.029

Krishnan U, Kaur M, Singh K, Kaur G, Singh P, Kumar M, Kumar A (2019b) MoS_2/Ag nanocomposites for electrochemical sensing and photocatalytic degradation of textile pollutant. J Mater Sci Mater Electron 30:3711–3721. https://doi.org/10.1007/s10854-018-00653-7

Li B, Cao H (2011) ZnO@graphene composite with enhanced performance for the removal of dye from water. J Mater Chem 21(10):3346–3349. https://doi.org/10.1039/C0JM03253K

Li J, Tang W, Yang H, Dong Z, Huang J, Li S, Wang J, Jin J, Ma J (2014) Enhanced-electrocatalytic activity of $Ni_{1-x}Fe_x$ alloy supported on polyethyleneimine functionalized MoS_2 nanosheets for hydrazine oxidation. RSC Adv 4(4):1988–1995. https://doi.org/10.1039/C3RA42757A

Nidheesh P, Zhou M, Oturan MA, Nidheesh P, Zhou M, Oturan MA (2018) An overview on the removal of synthetic dyes from water by electrochemical advanced oxidation processes. Pergamon Press, Oxford. To Cite This Version: HAL Id: Hal-01721053

ONG SA, TOORISAKA E, HIRATA M, HANO T (2008) Combination of adsorption and biodegradation processes for textile effluent treatment using a granular activated carbon-biofilm configured packed column system. J Environ Sci 20(8):952–956. https://doi.org/10.1016/S1001-0742(08)62192-0

Peng W-c, Chen Y, Li X-y (2016) MoS_2/reduced graphene oxide hybrid with CdS nanoparticles as a visible light-driven photocatalyst for the reduction of 4-nitrophenol. J Hazard Mater 309:173–179. https://doi.org/10.1016/j.jhazmat.2016.02.021

Peng K, Liangjie F, Yang H, Ouyang J, Tang A (2017) Hierarchical MoS_2 intercalated clay hybrid nanosheets. Nano Res 10(2):570–571. https://doi.org/10.1007/s12274-016-1315-3

Polshettiwar V, Luque R, Fihri A, Zhu H, Bouhrara M, Basset J-m, (2011) Magnetically recoverable nanocatalysts. Chem Rev 111:3036–3075. https://doi.org/10.1021/cr100230z

Rycenga M, Cobley CM, Zeng J, Li W, Moran CH, Zhang Q, Qin D, Xia Y (2011) Controlling the synthesis and assembly of silver nanostructures for plasmonic applications. Chem Rev 111(6):3669–3712. https://doi.org/10.1021/cr100275d

Sacco O, Stoller M, Vaiano V, Ciambelli P, Chianese A, Sannino D (2012) Photocatalytic degradation of organic dyes under visible light on N-doped TiO_2 photocatalysts. Int J Photoenergy 2012:626759. https://doi.org/10.1155/2012/626759

Saha N, Sarkar A, Ghosh AB, Dutta AK, Bhadu GR, Paul P, Adhikary B (2015) Highly active spherical amorphous MoS_2: facile synthesis and application in photocatalytic degradation of rose Bengal dye and hydrogenation of nitroarenes. RSC Adv 5(108):88848–88856. https://doi.org/10.1039/c5ra19442c

Shao N, Wang J, Wang D, Corvini P (2017) Preparation of three-dimensional Ag_3PO_4/TiO_2@MoS_2 for enhanced visible-light photocatalytic activity and anti-photocorrosion. Appl Catal B Environ 203:964–978. https://doi.org/10.1016/j.apcatb.2016.11.008

Shokouhimehr M (2015) Magnetically separable and sustainable nanostructured catalysts for heterogeneous reduction of nitroaromatics. Catalysts 5(2):534–560. https://doi.org/10.3390/catal5020534

Tan YH, Yu K, Li JZ, Fu H, Zhu ZQ (2014) MoS_2@ZnO nano-heterojunctions with enhanced photocatalysis and field emission properties. J Appl Phys 116(6):064305. https://doi.org/10.1063/1.4893020

Wen MQ, Xiong T, Zang ZG, Wei W, Tang XS, Dong F (2016) Synthesis of MoS_2/g-C_3N_4 nanocomposites with enhanced visible-light photocatalytic activity for the removal of nitric oxide (NO). Opt Express 24(10):10205–10212. https://doi.org/10.1364/OE.24.010205

Xiong T, Wen M, Dong F, Yu J, Han L, Lei B, Zhang Y, Tang X, Zang Z (2016) Three dimensional Z-scheme $(BiO)_2CO_3$/MoS_2 with enhanced visible light photocatalytic NO removal. Appl Catal B Environ 199:87–95. https://doi.org/10.1016/j.apcatb.2016.06.032

Yan X, Song Z, Wang X, Xu Y, Pu W, Ji H, Xu H, Yuan S, Li H (2019) Construction of 3D hierarchical GO/MoS_2/g-C_3N_4 ternary nanocomposites with enhanced visible-light photocatalytic degradation performance. ChemistrySelect 4:7123–7133. https://doi.org/10.1002/slct.201901472

Zeng X, Niu L, Song L, Wang X, Shi X, Yan J (2015) Effect of polymer addition on the structure and hydrogen evolution reaction property of nanoflower-like molybdenum disulfide. Metals 5(4):1829–1844. https://doi.org/10.3390/met5041829

Zeng Y, Guo N, Song Y, Zhao Y, Li H, Xu X, Qiu J, Yu H (2018) Fabrication of Z-scheme magnetic MoS_2/$CoFe_2O_4$ nanocomposites with highly efficient photocatalytic activity. J Colloid Interface Sci 514:664–674. https://doi.org/10.1016/j.jcis.2017.12.079

Zhang Y, Ram MK, Stefanakos EK, Goswami DY (2012) Synthesis, characterization, and applications of ZnO nanowires. J Nanomater 2012: 624520. https://doi.org/10.1155/2012/624520

Zhang W, Xiao X, Zheng L, Wan C (2015a) Fabrication of TiO_2/MoS_2@zeolite photocatalyst and its photocatalytic activity for degradation of methyl orange under visible light. Appl Surf Sci 358:468–478. https://doi.org/10.1016/j.apsusc.2015.08.054

Zhang W, Xiao X, Zheng L, Wan C (2015b) Fabrication of TiO_2/MoS_2 composite photocatalyst and its photocatalytic mechanism for degradation of methyl orange under visible light. Can J Chem Eng 93(9):1594–1602. https://doi.org/10.1002/cjce.22245

Zhang X, Lai Z, Tan C, Zhang H (2016) Solution-processed two-dimensional MoS_2 nanosheets: preparation, hybridization, and applications. Angew Chem Int Ed 55(31):8816–8838. https://doi.org/10.1002/anie.201509933

Zhang P, Fujitsuka M, Majima T (2017) Hot electron-driven hydrogen evolution using anisotropic gold nanostructure assembled monolayer MoS_2. Nanoscale 9(4):1520–1526. https://doi.org/10.1039/C6NR07740D

Zhou P, Yu J, Jaroniec M (2014) All-solid-state Z-scheme Photocatalytic systems. Adv Mater 26(29):4920–4935. https://doi.org/10.1002/adma.201400288

Chapter 5
Removal of Priority Water Pollutants Using Adsorption and Oxidation Process Combined with Sustainable Energy Production

Sheen Mers Sathianesan Vimala, Omar Francisco González-Vázquez,
Ma. del Rosario Moreno-Virgen, Sathish-Kumar Kamaraj,
Sheem Mers Sathianesan Vimala, Virginia Hernández-Montoya, and
Rigoberto Tovar-Gómez

Abstract Water pollution and sustainable energy production are the two major concerns in the current scenario. The total freshwater content of the earth is less than 1% part of which is polluted with organic and inorganic debris from industries. Energy production is another issue which arises due to the depletion of fossil fuels and hence the need for finding an alternate source of sustainable energy is in demand. Both of these issues can be resolved concurrently using the integrated adsorption or oxidation based microbial fuel cells. Such integrated systems are highly advantageous especially when recalcitrant pollutants are present in the water ecosystem. In a nutshell, this chapter deals with the list of priority organic and inorganic pollutants in water ecosystem, the available adsorption and oxidation techniques for the disintegration of organic water pollutants, specific systems formed by combined oxidation and adsorption techniques, integration of pollutant removal with sustainable energy-producing systems and the advances in integrated

---

S. M. Sathianesan Vimala
Department of Chemical Engineering, Indian Institute of Technology, Chennai, Tamil Nadu, India

O. F. González-Vázquez · M. d. R. Moreno-Virgen (✉) · V. Hernández-Montoya
R. Tovar-Gómez
Instituto Tecnológico de Aguascalientes (ITA), Tecnológico Nacional de México (TecNM), Aguascalientes, Mexico

S.-K. Kamaraj (✉)
Instituto Tecnológico El Llano Aguascalientes (ITEL), Tecnológico Nacional de México (TecNM), Aguascalientes, Mexico
e-mail: sathish.k@llano.tecnm.mx; sathish.bot@gmail.com

S. M. Sathianesan Vimala
Department of Basic Engineering, Government Polytechnic College, Nagercoil, Tamil Nadu, India

© The Author(s), under exclusive license to Springer Nature Switzerland AG 2021
S. Rajendran et al. (eds.), *Metal, Metal-Oxides and Metal-Organic Frameworks for Environmental Remediation*, Environmental Chemistry for a Sustainable World 64, https://doi.org/10.1007/978-3-030-68976-6_5

prototypes for environmental remediation. One of the advanced integrated oxidation based microbial fuel cell system is the bioelectro-Fenton system where Fenton reaction is carried out for the degradation of pollutants through OH radical formation with the simultaneous formation of electrons.

Keywords Water purification · Adsorption · Oxidation

5.1 Introduction

In the current scenario, the two major problems that need to be concerned are scarcity of clean water and renewable source of energy. Out of the total water content in the earth, 97% is composed of saltwater and the freshwater availability is less than 1% (Brillas et al. 2009). Despite this limited availability, most of the water bodies are polluted with organic wastes from industries, and the predominant among them are dyes, toxic metals, oil effluents etc. that exist in dissolved form and make it inappropriate for common use. The outstanding techniques for addressing these issues are adsorption and oxidation based integrated systems which degrade the organic pollutants and in some cases with the simultaneous generation of electrons as in the case of a microbial fuel cell (Cha et al. 2010). The general approaches for the removal of contaminants are through the direct removal of organic debris by physicochemical method (like adsorption based on π-π^* stacking, H-bonding and electrostatic interactions) or through the conversion of organic pollutants to eco-friendly products (Sharma and Das 2013; Zhao et al. 2017). In this chapter, we highlight the priority water pollutants, available adsorption and oxidation based remedies for water purification and the integrated systems for the simultaneous production of electrical energy during pollutant removal.

5.2 Priority Pollutants in Water

The contamination of both surface and underground water resources is a problem with very broad connotations, as it not only poses a risk to the environment but also represents a danger to public health and the economic growth of a region and its surroundings. The origin of the pollutants that affect the water is very varied; however, they can be classified into two large groups: those of natural origin that are related to biogeochemical cycles, food chain, geological phenomena and the same water dynamism (Li et al. 2017); and on the other hand, those of anthropogenic origin, caused by the various products and service activities (Palma et al. 2010). The impact of pollutants of anthropogenic origin is much more significant than those of natural origin. Within the pollutants of anthropogenic origin, we find two different types of sources: direct or indirect sources. Direct sources are those that discharge pollutants in the form of effluents, such as industries, homes and treatment plants. On the other hand, indirect sources are those in which the pollutant comes into contact with water through leaching, draining and dilutions. It seems that the impact of a pollutant directly affects the place where it originated; however, several studies indicate that there may be a contact of a pollutant that originated thousands of kilometres from the body of water due to the transport of goods within the commercial and production supply chains (Mekonnen and Hoekstra 2015).

Pollutants can be classified according to their nature in physical, biological or chemical, being the chemical type pollutants one of the most common agents presents in contaminated bodies of water. Chemical contaminants represent a serious problem within water bodies, since most of these are usually outside any biogeochemical and metabolic cycle, so finding techniques for their removal is of the most importance. Normally, chemical contaminants are dissolved or suspended (Goel 2006), and therefore, the techniques for their elimination must be very specific taking into account the physical and chemical characteristics of the polluting agent. Both the impact and the level of contamination of chemical agents will depend on the characteristics of the contaminant as well as the type of body of water, its location and the types of beneficial uses it supports (Schweitzer and Noblet 2018).

Among the pollutants of a chemical nature, there are some that due to their persistence and their serious harmful effect on human beings and the environment are classified as "priority pollutants". The damages that these could cause are very diverse, in the environment they can cause from eutrophication to sterility of waters (SA'AT SKBM 2006); while the effects on humans range from carcinogenicity, mutagenicity, teratogenicity or acute toxicity (Sciortino et al. 1999). These pollutants were first classified by the Environmental Protection Agency of the United States in 1976, having a list of 129 compounds of different chemical nature, which can be subdivided into nine large groups (Keith and Telliard 1979): metals, asbestos, total cyanides, pesticides, compounds extracted under acidic conditions, compounds extracted under alkaline conditions, neutrally removable compounds, total phenols and purgeable compounds.

For its part, the European Union also has a similar list which was updated in 2013, with 45 substances, some grouped by their chemical characteristics (Ue et al. 2013). Both the American and European lists have to be continually updated due to the presence of "emerging contaminants", which are organic substances that are the subject of great concern among the scientific community due to their frequent detection in aquatic environments. Highlights pharmaceutical or personal care products that are highly consumed by modern society (Bueno et al. 2012). In this way, the spectrum of priority chemical pollutants present in the water is defined; however, it increases as the presence of new compounds is observed and the damages that these could cause are not completely known.

5.2.1 *Priority Organic and Inorganic Pollutants*

The growing development of agricultural and industrial activities has resulted in the synthesis of various organic and inorganic substances that lack studies on their impact on the environment and public health. Products such as hydroxybenzenes or phenols and their analogues are recognized as priority pollutants (Sarkar et al. 2003), enter the environment through numerous chemical industries, as well as the use of herbicides and pesticides in agriculture (Hijosa-Valsero et al. 2013). These compounds are mainly carcinogenic, and their impact on the environment is important, causing damage to the human being and the biota that comes in contact with them (Azizullah et al. 2011). Due to its chemical nature, most organic pollutants are hydrophobic which makes them insoluble in water and accumulates on the surface of water bodies. It is known that these compounds can agglutinate with other substances, forming sediments that are ingested by aquatic animals, creating bioaccumulation within their tissues, resulting in danger for the trophic chains and therefore, for the human being, since they could ingest intoxicated animals (Vrana et al. 2006).

Among the recognized priority organic pollutants are three groups: chlorinated organic pesticides, polychlorinated biphenyls and polycyclic aromatic hydrocarbons.

Organochlorine pesticides (OCP) are substances that began to be used in the '70s in the United States and part of Europe. These compounds tend to bioaccumulate in adipose tissues, and there is evidence that there is a close relationship between these substances and the development of neuronal diseases such as Parkinson's disease (Fleming et al. 1994). Although these pollutants have already been banned, they are still found in the environment, since they have a period of degradation close to 20 years. Likewise, in developing countries, some compounds such as dichloro diphenyl trichloroethane DDT continue to be used to combat disease transmission vectors such as malaria and dengue (Harner et al. 1999), what ultimately keeps them current as contaminants.

On the other hand, the polychlorinated biphenyls (PCB) are part of the group of halogenated aromatic compounds. The PCBs were produced by the chlorination of biphenyl, and the resulting products were marketed according to their chlorine content percentage (Safe 1985). The chemical properties that are mainly responsible for many of the industrial applications of PCBs, that is, their flammability, chemical stability and miscibility with organic compounds (lipophilicity), are also the same properties that have contributed to their environmental problems. Once introduced into the environment, stable PCBs degrade relatively slowly and undergo transport cycles within the various components of the global ecosystem (Safe 1994).

Likewise, polycyclic aromatic hydrocarbons (PAH) are the product of the incomplete combustion of fossil fuels and have a high presence in the environment (Haritash and Kaushik 2009). Some of these chemicals are worrisome to the environment due to their genotoxic and carcinogenic potential and their persistence in the environment (Cerniglia 1993). There are multiple forms of degradation of these compounds; however, compounds with rings of 5 or 6 carbons are usually difficult to break down by biological and physicochemical means.

Concerning the inorganic pollutants, it is necessary to mention first that an inorganic compound can be considered as a compound that does not contain a carbon-hydrogen bond besides since many inorganic compounds contain some metal, which tends to be able to conduct electricity. The principal inorganic water pollutants include *(a)* acidity, especially sulfur dioxide from coal-burning or crude-oil-burning power plants, *(b)* ammonia, *(c)* chemical waste, like fertilizers, *(d)* heavy metals, and *(e)* silt or sediment (Speight 2017a; Speight 2017b). In general, inorganic pollutant has harmful effects on the human body, as they tend to accumulate in living organisms and cause various diseases, including cancer, damage to the nervous system, reduced growth and development, and in extreme cases, death.

In summary, the priority organic pollutants are substances that share great chemical stability, which prevents easily, degrade these molecules. Their danger is serious since they are usually mutagenic and accumulate easily in the adipose tissues of living beings. The inclusion of halogenated within the organic chains generate greater stability to physical conditions of heat, which limits the manoeuvrability in the physicochemical techniques of elimination of said compounds. On the other hand, the priority inorganic pollutants are considered a priority due to their toxicity as well as their mutagenic and recalcitrant properties; both have negative effects on human health and environment.

5.3 Adsorption Processes for the Removal Contaminants Present in the Water

5.3.1 *Adsorption of Organic Pollutants*

One of the techniques for the removal of pollutants in aqueous phase most used for its high efficiency is adsorption. Adsorption is a phenomenon in which impurities (adsorbates) of fluid on the surface of a material (adsorbent) including pores or internal surface are retained by adhering. The adsorption is governed by electrostatic and non-mechanical forces; even so, it is considered as a very fine type of filtration (De Boer 1956). It can be caused by "physical" forces, comparable to those that are responsible for the liquefaction of gases, or by "chemical" forces, similar to those that act in the formation of normal chemical compounds. It is usual, therefore, to distinguish between physical adsorption or physisorption (also called Van der Waals adsorption, due to the nature of these physical forces of cohesion) and chemical adsorption, or chemisorption.

Without a doubt, one of the key pieces in the successful removal of any contaminant through the adsorption process is the choice of good adsorbent material. This must be chosen based on the characteristics of the pollutant. Information such as the shape and size of the molecule, polarity, chemical nature, hydration, concentration, among others, are data that will make it possible to design an effective adsorption system. The variety of pollutants classified as the priority, it is intuited how diverse the adsorption process of each compound can be.

Specifically, the adsorption of organic compounds is usually very simple, since the size, molecular weight and configuration of the organic molecules allows a greater fixation in the porous solids both in gas and liquid phase (Mok and Kim

Table 5.1 Removal of priority organic pollutants by the adsorption method

Organic pollutant removed	Adsorbent Material	Adsorption capacity (mg/g)-(µg/g)*	Reference
Ametryn, Aldicarb, Dinoseb, Diuron	Carbon clothes	354.61, 421.58, 301.84, 213.06	Ayranci and Hoda (2005)
Lindane, Heptaclhor, Aldrin, Dieldrin	Pine bark	2.8*, 2.7*, 4.76*, 2.96*	Brás et al. (1999)
Heptaclorobifenyl	Fly ashes	0.149	Nollet et al. (2003)
PCB3, PCB4, PCB5, PCB6	Modified Montmorillonite	4.74*, 7.04*, 11.24*, 12.39*	Barreca et al. (2014)
PCB3, PCB4, PCB5	Corn straw coal	22.5, 10.1, 39.5	Wang et al. (2016)
PCB28, PCB52	β-Cycledextrine iron oxide cover	39.91 mol/g, 30.26 mol/g	Wang et al. (2015)
Dioxines, dibenzofurans	Activated carbon	2.981×10^{-6}, 9.682×10^{-6}	Zhou et al. (2016)

2011). The most commonly used materials for the removal of organic pollutants are granular activated carbon and carbon fibers (Dąbrowski et al. 2005), synthetic resins (Abburi 2003), natural zeolites (Canli et al. 2013; Damjanović et al. 2010), among other.

In adsorption in a liquid phase, the adsorption capacity of materials for organic compounds depends on several factors (Haghseresht et al. 2002): *(1)* physical nature of the adsorbent: pore structure, ash content, functional groups, *(2)* chemical nature of the adsorbate: its pKa, functional groups present, polarity, molecular weight and size, and *(3)* conditions of the solution: the pH, the ionic strength and the concentration of adsorbate.

In this sense, several studies have proven the effectiveness in the removal of organic compounds by the adsorption method (Table 5.1):

Although various materials are known for the removal of organic contaminants, there is still the problem of disposing of the adsorbent material saturated with the contaminant, that is, the contaminant only passed from one medium to another, which is why techniques are preferred help to degrade these pollutants, being the oxidation methods one of the most used.

5.3.2 Inorganic Pollutants

On the other hand, the adsorption of the main inorganic pollutants depends on the initial concentration of the pollutant, its pH and operating temperature, as well as the physical and chemical characteristics of the adsorbent. Several studies have focused on the removal of inorganic contaminants such as heavy metals, sulfur dioxide and ammonia; some of these studies are shown below in Table 5.2.

Table 5.2 Removal of priority inorganic pollutants by the adsorption method

Inorganic pollutant	Adsorbent Material	Adsorption capacity	Reference
Co^{2+}, Cd^{2+}, Li^{2+}	Nanoparticle of wild herbs	40.82, 52.91, 181.82 mg/g	Ghadah and Foziah (2018)
Pb^{2+}, Cd^{2+}	ZnO TiO$_2$@ZnO	Pb^{2+}: 790 mg/g, Cd^{2+}:643 mg/g Pb^{2+}: 978 mg/g, Cd^{2+}:786 mg/g	Manisha et al. Manisha et al. (2019)
Pb^{2+}, Cd^{2+}	Biochar	0.75, 0.55 mmol/g	Bing et al. (2019)
Ag^+, Cu^{2+}, Hg^{2+}, Cr^{3+}, Cr^{6+}	Magnetic chitosan beads	117.7, 147, 338, 63.5, 89.6 mg/g	Chunzhen et al. (2018)
Sulfur dioxide	Natural zeolitic tuff	230 µmol/g	Al-Harahsheh et al. (2014)
Sulfur dioxide	Carbon silica composites	0.2 mol/kg	Furtado et al. (2013)
Ammonia	Carbon silica composites	0.1 mol/kg	Furtado et al. (2013)
Ammonia	Bentonite	5.85 mg/g	Houming et al. (2019)
Ammonia	Silica gel	99.808 mg/g	Shaojuan et al. (2019)
Ammonia	Nanoporous carbon	10 mmol/g	Qajar et al. (2015)

5.4 Oxidative Processes for the Degradation of Priority Organic Pollutants

Most techniques for the remediation of contaminated water only retain the contaminant by physical or physico-chemical systems. However, these methods do not degrade pollutants and the problem does not end there, because they must have the means of retention saturated with contaminants, in this sense, the trend in research is aimed at changing the chemical nature of pollutants by degrading them even simpler elements or compounds that do not represent a danger to the environment or health. Processes such as oxidation are considered a highly competitive water treatment technology for the elimination of those contaminants that can't be treated with conventional techniques due to their high chemical stability and/or low biodegradability. Oxidative processes can be classified generally into two types: chemical and biological. Chemical oxidation has the characteristics that it can have a slow or moderate rate of degradation, besides being able to be selective; it may even be quick but not selective, which depends on the reactors and reactants used and translates into high costs. On the other hand, aerobic biological oxidation is usually of low cost, but operationally it has enough limitations since the fragility of the used strains is usually high, and the degradation is limited when the food is resistant to biodegradation, inhibitory or toxic for the bioculture (Scott and Ollis 1995).

5.4.1 Chemical Oxidative Processes

Chemical oxidative processes can be defined as treatment processes for water purification, carried out at room temperature and normal pressure, based on the in-situ generation of a powerful oxidizing agent, such as hydroxyl radicals (•OH), in a concentration enough to decontaminate the water effectively (Glaze et al. 2008).

The chemical oxidative processes constitute a series of promising, efficient and environmentally friendly methods to eliminate persistent organic pollutants from the waters. The formation of oxidant radicals is produced by various chemical, photochemical, sonochemical or electrochemical reactions (Gogate and Pandit 2001; Parsons 2004; Tarr 2003).

5.4.1.1 Fenton Type Oxidation

The oldest physicochemical oxidative process used is the Fenton method, in which a mixture of a soluble iron (II) salt and H_2O_2, known as Fenton's reagent, is applied to degrade and destroy organic compounds (Andreozzi et al. 1999). Fenton's chemistry began at the end of the nineteenth century, when Fenton published, in pioneering work, a detailed study on the use of a mixture of H_2O_2 and Fe^{2+} for the oxidation and destruction of tartaric acid. The catalytic decomposition of H_2O_2 by iron salts obeys a radical complex and chain mechanism. More recent studies have shown that the Fenton process was initiated by the formation of hydroxyl radicals, and could be applied to the degradation of several organic pollutants (Gallard et al. 1998; Sudoh et al. 1986), according to the following reaction (Oturan and Aaron 2014):

$$Fe^{2+} + H_2O_2 \rightarrow Fe^{3+} + \bullet OH + OH^- \quad (5.1)$$

If the reaction is carried out in an acid medium, the reaction can be rewritten as follows:

$$Fe^{2+} + H_2O + H^+ \rightarrow Fe^{3+} + H_2O + \bullet OH \quad (5.2)$$

The Fenton process can be applied efficiently when the optimum pH value of the contaminated aqueous medium is approximately 2.8–3.0.

In fact, under these conditions, the Fenton reaction can be propagated by the catalytic behaviour of the Fe^{3+}/Fe^{2+} pair. It is worth noting that only a small catalytic amount of Fe^{2+} is required since this ion is regenerated from the so-called Fenton type reaction between Fe^{3+} and H_2O_2:

$$Fe^{3+} + H_2O_2 \rightarrow Fe^{2+} + HO\bullet_2 + H^+ \quad (5.3)$$

Fu and group have studied the role of Fe^{2+} ion concentration in the degradation of amaranth dye using the Fenton process and observed that after a particular concentration of Fe^{2+} ions the degradation process is known to be depleted. The explanations given for such lowering of efficiency is due to two reasons: *(1)* the •OH radicals are increased to a maximum extent, and it favours the reaction among Fe^{2+} ions and H_2O_2 and *(2)* there is a possibility that •OH radicals interact with H_2O_2 to form •HO_2 radicals. Because of these two processes, the availability of OH* for dye degradation becomes less. However, at low concentration, since the kinetics of the reaction: RH + •OH→R• + H_2O is faster than the interaction of •OH with Fe^{2+} ions, and hence the Fe^{2+} regeneration or •HO_2 formation will not occur (Fu et al. 2010).

$$Fe^{2+} + \bullet OH \rightarrow Fe^{3+} + OH^- \quad k = 3.2 \times 10^8 M^{-1} s^{-1} \tag{5.4}$$

$$RH + \bullet OH \rightarrow R\bullet + H_2O \quad k = 10^9 - 10^{10} M^{-1} s^{-1} \tag{5.5}$$

$$\bullet OH + H_2O_2 \rightarrow HO^{2-} + \bullet HO_2 + H_2O \quad k = 2.7 \times 10^7 M^{-1} s^{-1} \tag{5.6}$$

In comparison with •OH, the HO_2 radical formed is characterized by a lower oxidation power and, therefore, is significantly less reactive towards organic compounds; however, a faster reaction can be generated. Therefore, it can be established that the combined reaction could be a good option for the oxidation of organic matter.

Likewise, variants of Fenton oxidation are known that make them more effective and easier to apply, such as photolytic methods (Foto-Fenton), which take advantage of irradiation with UV light from artificial or natural sources such as the sun, which allows the formation of oxidizing agents with greater ease, sometimes they are aided by photocatalytic substances in heterogeneous phases such as the use of TiO particles (Herrmann et al. 1999; Konstantinou and Albanis 2003), or the use of other oxidizing agents other than H_2O_2, such as O_3 aided by UV irradiation (Rosenfeldt et al. 2006). Likewise, there are electrolytic variants of the Fenton method (Electro-Fenton) in which an electric current is added having the advantage that they use a clean reagent, the electron, avoiding or reducing considerably the use of chemical reagents. The •OH is generated directly by the oxidation of the water in an anode overvoltage of high evolution of O_2 (anodic oxidation (AO)) (Panizza and Cerisola 2009), or indirectly in a bulk solution using the Fenton reagent generated electrochemically from electrode reactions (Brillas et al. 2008).

The Fenton process has several important advantages for water (Bautista et al. 2008):

1. A simple and flexible operation that allows easy implementation in existing plants;
2. Chemical products that are easy to handle and relatively inexpensive, there is no need for energy input.

However, the following drawbacks have also been observed:

1. Rather high costs and risks due to the storage and transport of H_2O_2, need for significant amounts of chemicals to acidify effluents at pH 2–4 before decontamination and/or to neutralize treated solutions before disposal;

2. Accumulation of iron sludge that must be removed at the end of the treatment;
3. Impossibility of the general mineralization due to the formation of Fe (III) carboxylic acid complexes, which can't be destroyed efficiently with the volume •OH.

5.4.1.2 Photochemical oxidation

Photochemical technologies have the advantages of being simple, clean, relatively economical and, in general, more efficient than simple chemical oxidative processes. In addition, they can disinfect water and destroy contaminants. Consequently, UV radiation has coupled with powerful oxidants such as O_3 and H_2O_2, which include, in some cases, catalysis with Fe^{3+} or TiO_2, which results in several types of important photochemical oxidative processes. These photochemical processes can degrade and/or destroy contaminants employing three possible reactions, including photodecomposition, based on UV radiation, excitation and degradation of contaminating molecules, oxidation by the direct action of O_3 and H_2O_2, and oxidation by photocatalysis (with Fe^{3+} or TiO_2), inducing the formation of radicals •OH (Oturan and Aaron 2014).

As an example, there are photochemical oxidation techniques that have shown good results for the degradation of organic compounds in the aqueous phase, such as:

1. Photolysis of H_2O_2 (H_2O_2/UV) (Antonaraki et al. 2002; Hernandez et al. 2002)
2. Photolysis of O_3 (O_3/UV) (Parsons 2004; Zaviska et al. 2009)
3. Heterogeneous photocatalysis (TiO_2/UV) (Fujishima et al. 2000; Mills and Le Hunte 1997)
4. Photo-Fenton (H_2O_2/Fe^{2+}/UV) (Oturan and Aaron 2014; Pera-Titus et al. 2004)

5.4.1.3 Sonochemical Oxidation

Ultrasounds in aqueous media constitute a particular technology that can proceed through two different types of actions, be it a chemical (indirect) or physical (direct) mechanism. In the indirect action, generally performed at high frequency, the water and dioxygen molecules undergo homolytic fragmentation and yield the radicals • OH, HO_2• and •O (Trabelsi et al. 1996).

Direct action, called sonication, involves the formation of ultrasonic cavitation bubbles that grow, then collapse, creating powerful breaking forces at extremely high temperatures (2000–5000 K) and pressures (approximately 6×10^4 kPa). In these extreme conditions, a sonolysis of water molecules occurs, which produces very reactive radicals capable of reacting with organic chemical species present in the aqueous medium, and / or degradation by pyrolysis of organic compounds (Zaviska et al. 2009).

5.4.1.4 Electrochemical Oxidation

Electrochemical oxidation is a technique based on the transfer of electrons, which makes it particularly interesting from the environmental point of view since it is a clean and effective way to produce hydroxyl radicals in situ (•OH) that can destroy a large variety of organic pollutants. These OH radicals can be produced electrochemically on the cathode side either directly or indirectly through the Fenton reagent (Electro-Fenton process) in acidic medium. In the cathode compartment, apart from the Fenton process, there is another possible reaction that is the production of H_2O_2 through oxygen reduction reaction which is also strongly influenced by pH change. Typically, in an integrated electro-Fenton system, the impact of pH on cathodic reaction can be explained using the Nernst equation [Eq. (5.3)]:

$$E = E^o + \frac{RT}{nF} \ln \frac{[Ox]}{[Red]} \tag{5.7}$$

where E is the electrode potential at a known temperature, E^o is the standard electrode potential, R is the gas constant (R = 8.314 J K^{-1} mol^{-1}), T is the temperature in kelvin, F is the Faraday constant (1 F = 96,486 C mol^{-1}), n is the number of moles of electrons transferred in the cathodic reaction.

At equilibrium, this Eq. (5.3) gets reduced to

$$E^o = -\frac{RT}{nF} \ln \frac{[Ox]}{[Red]} \tag{5.8}$$

And at 298 K Eq. (5.4) becomes

$$E^o = -\frac{25.7 \, mV}{n} \ln \frac{[H_2O_2]}{[H^+]} \tag{5.9}$$

$$E^o = -\frac{59.2 \, mV}{n} p^H \tag{5.10}$$

Equation (5.6) implies that the pH of the catholyte is directly proportional to the electrode potential (E) and at equilibrium, the cell potential increases by a factor of $\frac{52.7}{n}$ for a unit change in pH (Birjandi et al. 2016).

The effectiveness of the process can be further increased by combining both electrochemical processes, as indicated by several studies of the application of anodic oxidation (Brillas et al. 2008).

Both the electrochemical oxidation processes that use direct electrochemical or indirect electrochemical oxidations have several advantages in their use for the decontamination of water with organic agents (Nidheesh and Gandhimathi 2012):

1. Allow rapid degradation of organic pollutants while preventing the formation of new toxic species
2. It leads to the total mineralization of organic pollutants.
3. Use few or no chemical reagents.
4. Have energy costs as low as possible.

The electrochemical oxidation has variants in its application, which have advantages to each other; all have been successfully tested for the degradation of organic compounds:

1. Anodic oxidation (Martínez-Huitle et al. 2008; Wu et al. 2012).
2. Electro-Fenton (Ammar et al. 2007; Balci et al. 2009).

5.4.2 Biological Oxidative Processes

Domestic and some industrial wastewaters contain large amounts of biodegradable organic compounds in addition to relatively small concentrations of recalcitrant compounds. These last persistent compounds can lead to failure to discharge effluents if they are not specifically eliminated. The biological treatment for the degradation of organic compounds mineralizes a large portion that is biodegradable, effectively reducing the residual chemical oxygen demand of the water (Manilal et al. 1992), so this method is widely used in the treatment of wastewater that has a significant load of organic matter.

The biological degradation of a chemical compound refers to the elimination of the contaminant by the metabolic activity of living organisms, generally microorganisms and in particular bacteria and fungi that live in water and soil naturally (Oller et al. 2011). In this context, conventional biological processes do not always provide satisfactory results, especially for the treatment of industrial wastewater, since many of the organic substances produced by the chemical industry are toxic or resistant to biological treatment (Lapertot et al. 2006; Muñoz and Guieysse 2006). Industrial waste streams often contain biodegradable compounds that have a certain degree of toxicity or activity inhibition for biocultivation. These compounds can be degraded by comet activity or by the presence of specific degrading species in the microbial population. The bioculture that treats a toxic or inhibitory effluent is often less robust and more susceptible to a system disorder (Scott and Ollis 1995), In this sense, we have opted for the combination of previous treatments such as chemical pre-oxidation, which helps the microbiological culture to metabolize more easily and efficiently the sub compounds degraded by the chemical oxidation process.

5.4.2.1 Choice of Biological Agents for the Oxidation of Organic Pollutants

The appropriate biological scheme depends on the characteristics of the wastewater and the purpose of the treatment. Wastewater with multiple biodegradable organic compounds can be treated more efficiently with a robust community of activated sludge. Conversely, specialized bacterial species or highly acclimated strains may be appropriate for individual, bio-resistant contaminants.

You can use combinations of sludge strains that have been acclimated to a type of water similar to the one to be treated; however, the difficulty for degradation will be evident, which suggests that it is preferable to use fresh sludge that acclimates faster (Miller et al. 1988).On the other hand, it is possible to choose to use specialized strains mixed with common sludge, and it has been shown that the mineralization of the elements is significantly increased (Kearney et al. 1988). However, it has also been shown that if a consortium of selected strains is used and its efficiency is compared against a common mud, no increase in the rate of degradation or the percentage of mineralization of the compounds is detected (Scott and Ollis 1995). In conclusion, it is very difficult to have a suitable strain from the beginning of the operation; it is preferable to make combinations of strains and wait until they acclimate to the effluent.

Comparison Between Acclimated and Non-acclimated Strains

Acclimation of microorganisms to the substrate of interest is widely used to maximize the elimination efficiency of specific compounds that are difficult to biodegrade. This is achieved by increasing the feed concentration of the compound of interest over a while to allow the growth of microorganisms that can use the compound as a carbon source or nutrients. However, the substrate degraded by the bioculture may not be the original compound to which it was acclimatized, but one or more reaction intermediates.

The degradation of these intermediates will depend on whether the specific microorganisms or others present in the microbial community have the necessary enzymes and the ability to classify them successfully. In some cases, acclimated crops may have an advantage in the degradation of intermediates due to their specialized metabolic pathways. Studies have been reported in which a diversified mullet was contacted and acclimated to the byproducts of the degradation reactions of the original organic pollutant is more efficient than one that is only acclimated to the original organic contaminants (Sud et al. 2008). Likewise, it has been reported that non-acclimated strains are more resistant to by-products since they are easier to degrade, however, a decrease of some of the species acclimated to the original compound was also noted (Hu and Yu 1994; Jones et al. 1985), so it can be concluded that the set of strains will undergo some change throughout the process since some strains will die while others will become stronger.

Pure Culture

Pure microbial cultures have been cultivated and genetically engineered to degrade chemical compounds that are generally resistant to biodegradation by conventional treatment. This approach can be effective for highly defined wastes or that are highly toxic to other microorganisms. Treatment with pure cultures can be effective only if the objective is to reduce the concentration of the target compound. Complete mineralization may not be achieved, resulting in the accumulation of final products and requires the addition of a second or more robust microbial community (Scott and Ollis 1995).

5.5 Combined Adsorption and Oxidation Processes for the Removal of Priority Pollutants

Adsorption is a simple, effective and relatively cost-effective process used for the removal of various contaminants. However, surface phenomena that involve the accumulation of unwanted contaminants on the solid surface end in large amounts of "spent adsorbent" and/or "regeneration solutions", which must be handled as "hazardous waste" (Ince and Apikyan 2000).In this way, the adsorption fulfils its objective of removing organic and inorganic substances from an aqueous medium; however, it does not eliminate the said compounds, but only change the medium. In this sense, oxidative processes have been adapted for the elimination of pollutants, since it offers an effective response to the degradation and total elimination of the contaminants present, so to think about the combination of both processes is a reasonable idea.

Normally the combination of adsorption-oxidation processes can be considered as synergistic processes in which they could be taken as heterogeneous reactors (Kurniawan and Lo 2009). The process could be considered as a catalytic reaction in which an oxidizing agent is already mounted on the surface of the adsorbent material (Shukla et al. 2010; Zhang et al. 2012), in other cases the adsorbent material functions as only support to concentrate the contaminating organic substances and then use chemical oxidizing agents and/or assisted by some physicochemical process (Kurniawan and Lo 2009; Mok and Kim 2011). In such a way that they could be subdivided as oxidative catalytic adsorption and adsorption assisted by an oxidative method.

5.5.1 Oxidative Catalytic Adsorption

Most of the studies presented on oxidative catalytic adsorption consist of an adsorbent material that functions as a matrix to load some catalytic agent that triggers the oxidative reaction. The study presented by Crittrnden-Suri shows how a material with known characteristics as adsorbent material (Silica gel), a photoactive catalyst (Pt-TiO_2) is loaded in its active sites, it was demonstrated that under controlled laboratory tests it was possible to eliminate 96% of the organic compounds present in the water, as long as the irradiation of UV light was effective (Crittenden et al. 1997). In this case, the oxidation is carried out by a highly catalytic agent such as TiO_2, which allows the dissociation of water, forming highly oxidizing agents.

On the other hand, the efficiency of Fenton-type oxidative processes is known, so it is not uncommon to use Fenton reactions catalyzed on an adsorbent material. You can find studies in which the metallic agent such as Fe is loaded onto the surface of adsorbent material, such as the case studied by Shukla-Wang, in which Fe catalysts were loaded employing the impregnation technique on mesoporous silica. In this study, it was found that the amount of Fe impregnated in the surface of the silica to diminish the capacity of adsorption of it is not determinant. However, both the concentration of H_2O_2 and the water in which the pollutant is dissolved, reduce the amount of Fe available on the surface of the material, preventing the degradation efficiencies from being as expected (Shukla et al. 2010), despite this, at specific pH levels, the total mineralization of 60% of the degraded organic pollutant was determined. A similar study showed that the effectiveness of degradation of an organic compound by Fenton reaction could benefit if a small amount of Fe is impregnated on the surface of a zeolite contrasted with a high concentration of Fe dissolved in the aqueous bed. The heterogeneous catalytic reaction is benefited since the contact of Fe with hydrogen peroxide is stimulated by the transport forces related to the phenomenon of adsorption (Gonzalez-Olmos et al. 2013), In this sense, the combination of the processes simultaneously benefits the degradation of organic pollutants.

The support of Fe can be given in multiple materials, such as carbonaceous matrices both granular and fibers, it was demonstrated that these materials have a high capacity of adsorption and allow a very high degradation of organic compounds, besides, it has been demonstrated that a high efficiency at low concentrations of H_2O_2 (Wang et al. 2014). The synthesis of materials that have favorable properties as a catalyst has been the subject of study. We have studied the formation of nanomaterials composed of iron oxide as the FeOOH, supported on a carbonic matrix, which acts as a reducing agent within the Fenton reactions, it was found that this type of materials allows using a wider range of pH what facilitates the operation of degradation of organic compounds. Likewise, the assistance of an electric current helps achieve this objective, transforming the process into an electrocatalyzed adsorption, through Electro-Fenton reaction. The liberation of electrons in the anode, in turn, allows the regeneration of reduced Fe, which allows reaching the total degradation of the organic pollutant since there is no limitation of the reducing agent (Zhang et al. 2012). Likewise, Photocatalytic adsorption systems have been studied in which a carbonaceous photoreceptor material radicalizes H_2O_2, generating decompositions of 92.5% (Ince and Apikyan 2000). Systems like these allow reaching high efficiencies due to the conjunction of a suitable adsorbent material, a catalytic element capable of regeneration and a correct application of photolytic and electrolytic elements.

There are also non-ferrous catalysts for use in Fenton reactions, the application of Cu supported in a mesoporous matrix of MnO giving good results in degradation with efficiencies of between 56–89% and can be used at practically neutral pH

(Zhang et al. 2016). Not only have oxidation of organic compounds been achieved, but inorganic compounds have also been oxidized (Asaoka et al. 2012; Buatier de Mongeot et al. 1998); although this does not enter into Fenton reactions, it shows that it is a good option to use the combination of adsorption and oxidative processes simultaneously.

5.5.2 Adsorption Assisted by an Oxidative Method

Unlike oxidative catalytic adsorption, the adsorption processes assisted by an oxidative method do not carry out the adsorption-oxidation process simultaneously; rather, these are a process train, that is, it passes from one process to another, the most common order being oxidation followed by adsorption. The reason for this arrangement is due to the adsorption of degraded elements that present a lower risk than the original contaminants before the oxidation process (Wang et al. 2010).

The degradation of contaminants by these assisted methods is usually greater than that of oxidative catalytic adsorption since it is easier to control both processes separately you can get efficiencies of up to 98% mineralization (Akrout et al. 2015), although the operating time is significantly longer. Furthermore, it must be dealt with that the adsorption process can be inhibited if it is not selective since part of the reactants and byproducts of the reactions compete to occupy active sites, resulting in a not very efficient removal (Kurniawan and Lo 2009). See Fig. 5.1.

5.6 Sustainable Energy from an Integrated Microbial Fuel Cells-Electro-Fenton System

Since the electro-Fenton process is highly effective in the removal of refractory effluents, the energy output from such systems is not always very much concentrated. A few reports on bioelectro-Fenton systems that can efficiently function as effluent removal and the corresponding energy outputs are tabulated in the table: 3. It can be seen that mostly the power output is of the order of 2 mWm^{-2}. Even though the energy output is low in bioelectroFenton system, being an integrated technique, it has the advantage of the dual application. If we consider the design of microbial fuel cells, the electrodes are mostly made of low-cost graphite/carbon-based materials and wastewater is used as an electrolyte. This is a very cheaper prototype for the production of electric power (Yu et al. 2018; Zhao and Kong 2018). But for the practical implementation of bioelectroFenton in day-to-day life, the usage of power from microbial fuel cells (MFCs) is very difficult since the energy production is not constant and it will not be enough for operating devices. Alternatively, MFCs can be used for powering some specific applications like MFC integrated sensors, metal ion removal from wastewater through electrochemical reduction, dye degradation and

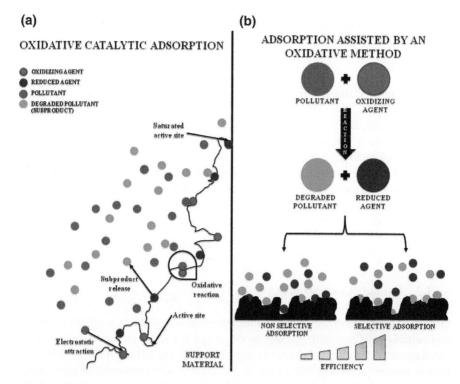

Fig. 5.1 (a) The type of interaction of the adsorbent at different sites of the reactant and (b) schematic illustration of adsorption process assisted by an oxidative method

for enhancing the efficiency of chemical oxygen demand removal through advanced oxidation processes (Fallaha and Rahimnejada 2019; Fu et al. 2018; Mathuriya and Yakhmi 2014; Nordin et al. 2017). BioelectroFenton system is one of the oxidation processes (Fenton process) integrated MFC system that is found to be efficient in removing water pollutants and also for the production of sustainable energy.

5.6.1 Configuration of a bioelectroFenton Cell

The experimental set up of bioelectroFenton system is similar to that of normal MFC cell and the difference is the use of Fenton's reagents in an electrolyte and the type of electrode material used. Similar to MFCs, the integrated cell can be single stacked or multiple stacked. In normal MFCs, carbon-based materials like graphite felt, carbon fiber etc. are used as electrodes whereas, in MFC-electro-Fenton cell, the cathode is modified in order to carry out two types of reactions viz. Oxygen reduction reaction to form H_2O_2 or in few cases, it is modified with an iron-based catalyst for the Fenton process to occur (Cha et al. 2010; Ieropoulos et al. 2010; Li et al. 2009; Tokumura et al. 2009). The components of anodic and cathodic compartments in MFC-electro-Fenton cell are described below:

5.6.1.1 Anode Compartment

The anode compartment consists of an anode which is usually made up of graphite as base material on which either enzymes or biofilms are formed for the conversion of organic residues to CO_2 and water (Santiago et al. 2016). Degradable organic effluents like wastewater or added molecules like glucose, lactate and acetate form the electrolyte. The structure of the anode compartment is schematically represented in Fig. 5.2. More specific details of the bioelectroFenton cell such as the type of electrodes used, cell design, nature of electrolyte and the efficiency of operation are depicted in Table 5.3.

5.6.1.2 Cathode Compartment

In cathode compartment, even though the base material of anode is made of carbon, it is suitably modified with suitable metal oxides for H_2O_2 production or with Fe based materials for electro-Fenton reaction (Li and Zhang 2019; Yu 2018). Since the Fenton process is mostly carried out in the cathode compartment, the electrolyte should include H_2O_2 and Fe^{2+} (incase if Fe based cathode is not used). Further, there is a possibility that during Fenton reaction, Fe^{2+} ions are converted to Fe^{3+} ions and

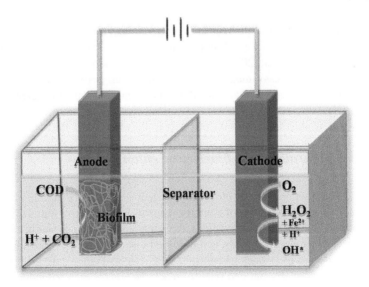

Fig. 5.2 Structure of a bioelectroFenton system showing the biofilm formed anode, separator and the production of H_2O_2 in cathode for Fenton reaction

Table 5.3 Comparison of the output from microbial fuel cell integrated ElectroFenton Cell

Anode	Anolyte	Cathode	Catholyte	Pollutant	Pollutant removal (%)	Output	References
Au@CF	Glucose solution	ZIF-8/CF	AO7, FeSO$_4$, Na$_2$SO$_4$	AO7	90% in 10 h	170 mW m^{-2}	Le et al. (2016)
Shewanella decolorationis S12 on CF	Lactate solution	(CNT)/γ-FeOOH	Orange II	Orange II	100% within 14 h	230 mW m^{-2}	Feng et al. (2010)
Carbon plate	Reactive black 5	Iron sheet	Reactive black 5	Reactive black 5	86.4% in 8 h	11.39 mW cm^{-2}	Nordin et al. (2017)
Pt sheet	RhB, Na$_2$SO$_4$	Fe@Fe$_2$O$_3$/ACF	RhB, Na$_2$SO$_4$	RhB	~80% in 2 h	–	Li et al. (2009)
Fe anode	Orange II containing effluent	Carbon plate	Orange II containing effluent	Orange II	84.2%	16.5 Wh kg^{-1}	Tokumura et al. (2009)
Carbon brush	Acetate, CBZ	Gas diffusion cathode	Na$_2$SO$_4$ in CBZ solution	CBZ	90% in 24 h	112 ± 11 mW m^{-2}	Wang et al. (2018)
S. Oneidensis MR-1/ graphite felt	Sodium lactate	Fe@Fe$_2$O$_3$/graphite felt	Air cathode	TPTC	78.32 ± 2.07% in 100 h	57.25 mW m^{-2}	Yong et al. (2017)

(continued)

Table 5.3 (continued)

Anode	Anolyte	Cathode	Catholyte	Pollutant	Pollutant removal (%)	Output	References
Graphite brushes	WAS	(AQS/PPy-Fe$_2$(MoO$_4$)$_3$/rGO)	Sludge	Organic waste	–	–	Yu et al. (2018)
CF	Activated sludge, pyraclostrobin	CF	K$_3$[Fe(CN)$_6$] in KCl	Pyraclostrobin	83% in 16 h	789 mA m^{-2}	Zhao and Kong (2018)
Graphite	Marine sediment	Graphite	Iron alignate beads	LG	98.2% in 1 h	–	Dios et al. (2014a, b)
				CV	96.2% in 1 h	–	
				IC	97.2% in 1 h	–	
				RB5	88.2% in 1 h	–	
				Poly R-478	19.1% in 1 h	–	

CF Carbon Felt, *ZIF* Zeolitic imidazolate frameworks, *AO7* Acid Orange II sodium salt, *ACF* Active Carbon Fiber, *RhB* Rhodamine B, *CBZ* carbamazepine, *TPTC* triphenyltin chloride, *WAS* Waste Activated Sludge, *LG* Lissamine Green B, *CV* Crystal Violet, *IC* Indigo Carmine, *RB5* Reactive Black 5 the reduced Fe^{2+} ions are reformed by accepting the electrons formed during the conversion of chemical oxygen demand to CO_2 in the anode side (Gao et al. 2015).

5.6.2 Novel Bioelectro-Fenton System for Environmental Remediation

Zhu and Ni first designed the bioelectro-Fenton system in 2009. (Li et al. 2018; Zhu and Ni 2009) The combination of microbial fuel cell (MFC) set up with Fenton reaction has the advantage of powering Fenton reaction and lowering the cost of operation since no external power source is used, production H_2O_2 within the system and the formation of Fe^{2+} ions simultaneously from Fe electrodes. Moreover, instead of the using sewage sludge as anolyte, marine sediments are used which significantly improved the power efficiency of MFC (Dios et al. 2014a; Dios et al. 2014b). In simple words, the essential characteristics that a cathodic material should inherit for a bioelectroFenton cell are good conductivity, Fe^{2+} source, 2e oxygen reduction catalyst and are resistant to various pH values. (Li et al. 2018).

While associating Fenton system with microbial fuel cell (MFC), the efficiency of the overall cell can be improved through *(1)* enhancing the current generation from MFC and *(2)* choosing the appropriate electrode material for cathode reaction. Distinct cathodic electrode materials have been used for enhancing the efficiency of Fenton processes. For example, Feng and his group used (CNT)/γ-FeOOH for the degradation of an azo dye Orange II at neutral pH. Here CNT highly favours the two-electron oxygen reduction reaction whereas γ-FeOOH acts as the Fe^{2+} source. (Feng et al. 2010) Xu et al. used $FeVO_4$/CF cathode for the treatment of coal gasification wastewater. Here both Fe^{3+} and V^{5+} will act as a catalyst for the production of OH• radicals which has considerably increased the overall efficiency for the removal of pollutant. (Xu et al. 2018) Le et al. used carbon Felt@Au (CF@Au) and porous carbon deposited CF as the cathode materials. Here ferrous sulphate solution is taken as catholyte which provides the Fe^{2+} source for carrying out Fenton reaction. (Huongle et al. 2016) Zhuang et al. used Fe@Fe_2O_3/carbon felt for the controlled release of Fenton's reagent and has used expanded polytetrafluoroethylene laminated cloth as separator. The separator enhanced the efficiency of MFC by four times. The proposed mechanism is shown in Fig. 5.3 (Zhuang et al. 2010) Dios and group used iron-containing zeolite for the degradation of Black 5 dye and phenanthrene. The dye was decolourized nearer to completion after 90 min, and 78% of phenanthrene degradation was obtained after 30 h. Since zeolite is a naturally existing adsorbent iron can be easily loaded on the highly surfaced zeolites. Furthermore, preliminary reusability tests of the developed catalyst showed high degradation levels for successive cycles (Dios et al. 2014a; Dios et al. 2014b).

In general, the factors that affect the working of bioelectroFenton system are pH, temperature, the substrate used and the nature of the biocatalyst. (Kahoush et al.

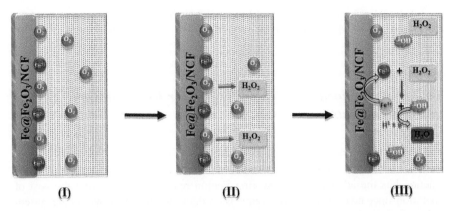

Fig. 5.3 Mechanism of Electro-fenton reaction.: (**a**) Diffusion of O_2 towards the electrode surface (**b**) conversion of O_2 to H_2O_2 and (**c**) Reaction of Fe^{2+} with H_2O_2. (Image reproduced from reference Zhuang et al. 2010)

2018) Zhang et al. have come up with serially connected four MFCs for powering the cathode reaction. This prototype is found to show better activity, yet the removal of excess H_2O_2 becomes difficult. (Zhang et al. 2015) The usage of noble metal platinum as a cathode material has the drawback of tearing nafion membrane easily. One of the reasons is that platinum gets break off from the electrode surface and gets deposited over the nafion membrane where it acts as a Fenton like reagent to produce •OH radicals. Since •OH is highly reactive it will affect the side groups in nafion membrane and cause defective sites. (Yu et al. 2011) Another interesting work was carried out by Wang et al. in which the graphene-poly(vinyl alcohol) on indium tin oxide coated glass substrate in BEF-MFC system was used as cathode and decolourised reactive black 5 dye by 60.25%. Moreover, at open circuit potential (0.42 V), the optimal power density obtained using this method is 74.1 mW/m^2. (Wang and Wang 2017).

5.7 Conclusions and Perspectives

In conclusions, integrated oxidation and adsorption based systems are proven to be highly effective in the removal of priority organic and inorganic pollutants and for the production of sustainable energy. Combination of adsorption or oxidation process with sonochemical, electrochemical and photochemical methods demonstrates the improvement in the production of •OH radical production in oxidation processes and adsorption efficiency. Further, the combination of microbial fuel cell with oxidation processes especially electro-Fenton reaction is an economically feasible system since the reactants of Fenton process like Fe^{2+} and H_2O_2 can be produced within the reaction system with the simultaneous generation of electric power as output.

References

Abburi K (2003) Adsorption of phenol and p-chlorophenol from their single and bisolute aqueous solutions on Amberlite XAD-16 resin. J Hazard Mater 105:143–156. https://doi.org/10.1016/j.jhazmat.2003.08.004

Akrout H, Jellali S, Bousselmi L (2015) Enhancement of methylene blue removal by anodic oxidation using BDD electrode combined with adsorption onto sawdust. Comptes Rendus Chim 18:110–120. https://doi.org/10.1016/j.crci.2014.09.006

Al-Harahsheh M, Shawabkeh R, Batiha M, Al-Harahsheh A, Al-Zboon K (2014) Sulfur dioxide removal using natural zeolitic tuff. Fuel Process Technol 126:249–258. https://doi.org/10.1016/j.fuproc.2014.04.025

Ammar S, Oturan N, Oturan MA (2007) Electrochemical oxidation of 2-nitrophenol in aqueous medium by using electro-Fenton technology. J Environ Eng Manag 17:89–96

Andreozzi R, Caprio V, Insola A, Marotta R (1999) Advanced oxidation processes (AOP) for water purification and recovery. Catal Today 53:51–59. https://doi.org/10.1016/S0920-5861(99)00102-9

Antonaraki S, Androulaki E, Dimotikali D et al (2002) Photolytic degradation of all chlorophenols with polyoxometallates and H2O2. J Photochem Photobiol A Chem 148:191–197. https://doi.org/10.1016/S1010-6030(02)00042-4

Asaoka S, Hayakawa S, Kim K et al (2012) Combined adsorption and oxidation mechanisms of hydrogen sulfide on granulated coal ash. J Colloid Interface Sci 377:284–290. https://doi.org/10.1016/j.jcis.2012.03.023

Ayranci E, Hoda N (2005) Adsorption kinetics and isotherms of pesticides onto activated carbon-cloth. Chemosphere 60:1600–1607. https://doi.org/10.1016/j.chemosphere.2005.02.040

Azizullah A, Khattak MNK, Richter P, Häder D-P (2011) Water pollution in Pakistan and its impact on public health — a review. Environ Int 37:479–497. https://doi.org/10.1016/j.envint.2010.10.007

Balci B, Oturan MA, Oturan N, Sirés I (2009) Decontamination of aqueous glyphosate, (Aminomethyl)phosphonic acid, and glufosinate solutions by electro-Fenton-like process with Mn2+ as the catalyst. J Agric Food Chem 57:4888–4894. https://doi.org/10.1021/jf900876x

Barreca S, Orecchio S, Pace A (2014) The effect of montmorillonite clay in alginate gel beads for polychlorinated biphenyl adsorption: isothermal and kinetic studies. Appl Clay Sci 99:220–228. https://doi.org/10.1016/j.clay.2014.06.037

Bautista P, Mohedano AF, Casas JA et al (2008) An overview of the application of Fenton oxidation to industrial wastewaters treatment. J Chem Technol Biotechnol 83:1323–1338. https://doi.org/10.1002/jctb.1988

Bing Jie N, Qi-Su H, Chen W, Tian-Yi N, Jing S, Wei W (2019) Competitive adsorption of heavy metals in aqueous solution onto biochar derived from anaerobically digested sludge. Chemosphere 219:351–357. https://doi.org/10.1016/j.chemosphere.2018.12.053

Birjandi N, Younesi Y, Ghoreyshi AA, Rahimnejad M (2016) Electricity generation through degradation of organic matters in medicinal herbs wastewater using bio-electro-Fenton system. J Environ Manag 180:390–400. https://doi.org/10.1016/j.jenvman.2016.05.073

Brás IP, Santos L, Alves A (1999) Organochlorine pesticides removal by pinus bark sorption. Environ Sci Technol 33:631–634. https://doi.org/10.1021/es980402v

Brillas E, Garrido JA, Rodríguez RM et al (2008) Oxidation processes using a BDD anode and electrogenerated H_2O_2 with Fe(II) and UVA light as catalysts. Port Electrochim Acta 26:15–46. https://doi.org/10.4152/pea.200801015

Brillas E, Sirés I, Oturan MA (2009) Electro-Fenton process and related electrochemical technologies based on Fenton's reaction chemistry. Chem Rev 109:6570–6631. https://doi.org/10.1021/cr900136g

Buatier de Mongeot F, Scherer M, Gleich B et al (1998) CO adsorption and oxidation on bimetallic Pt/Ru(0001) surfaces – a combined STM and TPD/TPR study. Surf Sci 411:249–262. https://doi.org/10.1016/S0039-6028(98)00286-6

Bueno MJM, Gomez MJ, Herrera S et al (2012) Occurrence and persistence of organic emerging contaminants and priority pollutants in five sewage treatment plants of Spain: two years pilot survey monitoring. Environ Pollut 164:267–273. https://doi.org/10.1016/j.envpol.2012.01.038

Canli M, Abali Y, Bayca SU (2013) Removal of methylene blue by natural and ca and k-exchanged zeolite treated with hydrogen peroxide. Physicochem Probl Miner Process 49:481–496. https://doi.org/10.5277/ppmp130210

Cerniglia CE (1993) Biodegradation of polycyclic aromatic hydrocarbons. Curr Opin Biotechnol 4:331–338. https://doi.org/10.1016/0958-1669(93)90104-5

Cha J, Choi S, Yu H, Kim H, Kim C (2010) Directly applicable microbial fuel cells in aeration tank for wastewater treatment. Bioelectrochemistry 78:72–79. https://doi.org/10.1016/j.bioelechem.2009.07.009

Chunzhen F, Kan L, Yi H, Yaling W, Xufang Q, Jinping J (2018) Evaluation of magnetic chitosan beads for adsorption of heavy metal ions. Sci. Total Environ 627:1396–1403. https://doi.org/10.1016/j.scitotenv.2018.02.033

Crittenden JC, Suri RPS, Perram DL, Hand DW (1997) Decontamination of water using adsorption and photocatalysis. Water Res 31:411–418. https://doi.org/10.1016/S0043-1354(96)00258-8

Dąbrowski A, Podkościelny P, Hubicki Z, Barczak M (2005) Adsorption of phenolic compounds by activated carbon – a critical review. Chemosphere 58:1049–1070. https://doi.org/10.1016/j.chemosphere.2004.09.067

Damjanović L, Rakić V, Rac V et al (2010) The investigation of phenol removal from aqueous solutions by zeolites as solid adsorbents. J Hazard Mater 184:477–484. https://doi.org/10.1016/j.jhazmat.2010.08.059

De Boer JH (1956) Adsorption phenomena. In: Chemical Sensors and Biosensors, pp 17–161

Fallaha M, Rahimnejada M (2019) MFC-based biosensors known as an integrated and self-powered system. Biomed J Sci Tech Res 14:10608–10609. https://doi.org/10.26717/BJSTR.2019.14.002538

Feng C, Li F, Mai H, Li X (2010) Bio-electro-Fenton process driven by microbial fuel cell for wastewater treatment. Environ Sci Technol 44:1875–1880. https://doi.org/10.1021/es9032925

Dios MAF, Iglesias O, Bocos E, Pazos M, Sanromán MA (2014a) Application of benthonic microbial fuel cells and electro-Fenton 3 process to dye decolourisation. J Ind Eng Chem 20:3754–3760. https://doi.org/10.1016/j.jiec.2013.12.075

Dios M, Iglesias O, Pazos M, Sanromán MA (2014b) Application of electro-Fenton technology to remediation of polluted effluents by self-sustaining process. Sci World J 2014:801870. https://doi.org/10.1155/2014/801870

Fleming L, Mann JB, Bean J et al (1994) Parkinson's disease and brain levels of organochlorine pesticides. Ann Neurol 36:100–103. https://doi.org/10.1002/ana.410360119

Fu B, Shen C, Ren J, Chen J, Zhao L (2018) Advanced oxidation of biorefractory organics in aqueous solution together with bioelectricity generation by microbial fuel cells with composite FO/GPEs. IOP Conf Series: Earth Environ Sci 127:012015. https://doi.org/10.1088/1755-1315/127/1/012015

Fu L, You S, Zhang G, Yang F, Fang X (2010) Degradation of azo dyes using in-situ Fenton reaction incorporated into H2O2 producing microbial fuel cell. Chem Eng J 160:164–169. https://doi.org/10.1016/j.cej.2010.03.032

Fujishima A, Rao TN, Tryk DA (2000) Titanium dioxide photocatalysis. J Photochem Photobiol C: Photochem Rev 1:1–21. https://doi.org/10.1016/S1389-5567(00)00002-2

Furtado AMB, Wang Y, Levan MD (2013) Carbon silica composites for sulfur dioxide and ammonia adsorption. Microporous Mesoporous Mater 165:48–54. https://doi.org/10.1016/j.micromeso.2012.07.032

Gallard H, de Laat J, Legube B (1998) Influence du pH sur la vitesse d'oxydation de compose's organiques par FeII/H2O2. Me'canismes re'actionnels et mode'lisation. New J Chem 22:263–268. https://doi.org/10.1039/A708335A

Gao G, Zhang Q, Hao Z, Vecitis CD (2015) Carbon nanotube membrane stack for flow-through sequential regenerative electro-Fenton. Environ Sci Technol 49:2375–2383. https://doi.org/10.1021/es505679e

Ghadah MA, Foziah FA (2018) Adsorption study of heavy metal ions from aqueous solution by nanoparticle of wild herbs. Egypt J Aquat Res 44:187–194. https://doi.org/10.1016/j.ejar.2018.07.006

Glaze WH, Kang J, Douglas H (2008) Ozone : science & engineering the chemistry of water treatment processes involving ozone, hydrogen peroxide and ultraviolet radiation. Ozone Sci Eng:335–352

Goel PK (2006) Water pollution: causes, effects and control. New Age International

Gogate PR, Pandit AB (2001) Hydrodynamic cavitacion reactors: a satete of the art review. Rev Chem Eng 17:1–85. https://doi.org/10.1515/REVCE.2001.17.1.1

Gonzalez-Olmos R, Kopinke F, Mackenzie K, Georgi A (2013) Hydrophobic Fe-zeolites for removal of MTBE from water by combination of adsorption and oxidation. Environ Sci Technol 47:2353–2360. https://doi.org/10.1021/es303885y

Haghseresht F, Nouri S, Finnerty JJ, Lu GQ (2002) Effects of surface chemistry on aromatic compound adsorption from dilute aqueous solutions by activated carbon. J Phys Chem B 106:10935–10943. https://doi.org/10.1021/jp025522a

Haritash AK, Kaushik CP (2009) Biodegradation aspects of polycyclic aromatic hydrocarbons (PAHs): a review. J Hazard Mater 169:1–15. https://doi.org/10.1016/j.jhazmat.2009.03.137

Harner T, Wideman JL, Jantunen LMM et al (1999) Residues of organochlorine pesticides in Alabama soils. Environ Pollut 106:323–332. https://doi.org/10.1016/S0269-7491(99)00110-4

Hernandez R, Zappi M, Colucci J, Jones R (2002) Comparing the performance of various advanced oxidation processes for treatment of acetone contaminated water. J Hazard Mater 92:33–50. https://doi.org/10.1016/S0304-3894(01)00371-5

Herrmann JM, Guillard C, Arguello M et al (1999) Photocatalytic degradation of pesticide pirimiphos-methyl determination of the reaction pathway and identification of intermediate products by various analytical methods. Catal Today 54:353–367. https://doi.org/10.1016/S0920-5861(99)00196-0

Hijosa-Valsero M, Molina R, Schikora H et al (2013) Removal of priority pollutants from water by means of dielectric barrier discharge atmospheric plasma. J Hazard Mater 262:664–673. https://doi.org/10.1016/j.jhazmat.2013.09.022

Houming C, Qi Z, Zipeng X (2019) Adsorption of ammonia nitrogen in low temperature domestic wastewater by modification bentonite. J Clean Prod 233:720–730. https://doi.org/10.1016/j.clepro.2019.06.079

Hu S-T, Yu Y-H (1994) Preozonation of chlorophenolic wastewater for subsequent biological treatment. Ozone Sci Eng 16:13–28. https://doi.org/10.1080/01919519408552377

Ieropoulos I, Greenman J, Melhuish C (2010) Improved energy output levels from small-scale microbial fuel cells. Bioelectrochemistry 78:44–50. https://doi.org/10.1016/j.bioelechem.2009.05.009

Ince NH, Apikyan IG (2000) Combination of activated carbon adsorption with light-enhanced chemical oxidation via hydrogen peroxide. Water Res 34:4169–4176. https://doi.org/10.1016/S0043-1354(00)00194-9

Jones BM, Sakaji RH, Daughton CG (1985) Effects of ozonation and ultraviolet irradiation on biodegradability of oil shale wastewater organic solutes. Water Res 19:1421–1428. https://doi.org/10.1016/0043-1354(85)90309-4

Kahoush M, Behary N, Cayla A, Nierstrasz V (2018) Bio-Fenton and bio-electro-Fenton as sustainable methods for degrading organic pollutants in wastewater. Process Biochem 64:237–247. https://doi.org/10.1016/j.procbio.2017.10.003

Kearney PC, Muldoon MT, Somich CJ et al (1988) Biodegradation of ozonated atrazine as a wastewater disposal system. J Agric Food Chem 36:1301–1306. https://doi.org/10.1021/jf00084a044

Keith L, Telliard W (1979) ES&T Special Report: priority pollutants: I-a perspective view. Environ Sci Technol 13:416–423. https://doi.org/10.1021/es60152a601

Konstantinou IK, Albanis TA (2003) Photocatalytic transformation of pesticides in aqueous titanium dioxide suspensions using artificial and solar light: intermediates and degradation pathways. Appl Catal B Environ 42:319–335. https://doi.org/10.1016/S0926-3373(02)00266-7

Kurniawan TA, Lo W (2009) Removal of refractory compounds from stabilized landfill leachate using an integrated H_2O_2 oxidation and granular activated carbon (GAC) adsorption treatment. Water Res 43:4079–4091. https://doi.org/10.1016/j.watres.2009.06.060

Lapertot M, Pulgarín C, Fernández-Ibáñez P et al (2006) Enhancing biodegradability of priority substances (pesticides) by solar photo-Fenton. Water Res 40:1086–1094. https://doi.org/10.1016/j.watres.2006.01.002

Le TXH, Esmilaire R, Drobek M, Bechelany M, Vallicari C, Nguyen DL, Julbe A, Tingrya S, Cretin M (2016) Design of a novel fuel cell-Fenton system: a smartapproach to zero energy depollution. J Mater Chem A 4:17686–17693

Li G, Zhang Y (2019) Highly selective two-electron oxygen reduction to generate hydrogen peroxide using graphite felt modified with N-doped graphene in an electro-Fenton system. New J Chem:134605. https://doi.org/10.1039/C9NJ02601K

Li H, Yang Z, Liu G et al (2017) Analyzing virtual water pollution transfer embodied in economic activities based on gray water footprint: a case study. J Clean Prod 161:1064–1073. https://doi.org/10.1016/j.jclepro.2017.05.155

Li J, Ai Z, Zhang L (2009) Design of a neutral electro-Fenton system with $Fe@Fe_2O_3$/ACF composite cathode for wastewater treatment. J Hazard Mater 164:18–25. https://doi.org/10.1016/j.jhazmat.2008.07.109

Li X, Chen S, Angelidaki I, Zhang Y (2018) Bio-electro-Fenton processes for wastewater treatment: advances and prospects. Chem Eng J 354:492–506. https://doi.org/10.1016/j.cej.2018.08.052

Manilal VB, Haridas A, Alexander R, Surender GD (1992) Photocatalytic treatment of toxic organics in wastewater: toxicity of photodegradation products. Water Res 26:1035–1038. https://doi.org/10.1016/0043-1354(92)90138-T

Manisha S, Jasminder S, Satyajit H, Soumen B (2019) Adsorption of heavy metal ions by mesoporous ZnO and TiO_2@ZnO monoliths: adsorption and kinetic studies. Microchem J 145:105–112. https://doi.org/10.1016/j.microc.2018.10.026

Martínez-Huitle CA, De Battisti A, Ferro S et al (2008) Removal of the pesticide Methamidophos from aqueous solutions by electrooxidation using Pb/PbO2, Ti/SnO2, and Si/BDD electrodes. Environ Sci Technol 42:6929–6935. https://doi.org/10.1021/es8008419

Mathuriya AS, Yakhmi JV (2014) Microbial fuel cells to recover heavy metals. Environ Chem Lett 12:483–494. https://doi.org/10.1007/s10311-014-0474-2

Mekonnen MM, Hoekstra AY (2015) Global Gray water footprint and water pollution levels related to anthropogenic nitrogen loads to fresh water. Environ Sci Technol 49:12860–12868. https://doi.org/10.1021/acs.est.5b03191

Miller RM, Singer GM, Rosen JD, Bartha R (1988) Sequential degradation of chlorophenols by photolytic and microbial treatment. Environ Sci Technol 22:1215–1219. https://doi.org/10.1021/es00175a015

Mills A, Le Hunte S (1997) An overview of semiconductor photocatalysis. J Photochem Photobiol A Chem 108:1–35. https://doi.org/10.1016/S1010-6030(97)00118-4

Mok YS, Kim DH (2011) Treatment of toluene by using adsorption and nonthermal plasma oxidation process. Curr Appl Phys 11:S58–S62. https://doi.org/10.1016/j.cap.2011.05.023

Muñoz R, Guieysse B (2006) Algal–bacterial processes for the treatment of hazardous contaminants: a review. Water Res 40:2799–2815. https://doi.org/10.1016/j.watres.2006.06.011

Nidheesh PV, Gandhimathi R (2012) Trends in electro-Fenton process for water and wastewater treatment: an overview. Desalination 299:1–15. https://doi.org/10.1016/j.desal.2012.05.011

Nollet H, Roels M, Lutgen P et al (2003) Removal of PCBs from wastewater using fly ash. Chemosphere 53:655–665. https://doi.org/10.1016/S0045-6535(03)00517-4

Nordin N, Ho L, Ong S, Ibrahim AH, Wong Y, Lee S, Oon Y, Oon Y (2017) Hybrid system of photocatalytic fuel cell and Fenton process for electricity generation and degradation of reactive black 5. Sep Purif Technol 177:135–141. https://doi.org/10.1016/j.seppur.2016.12.030

Oller I, Malato S, Sánchez-Pérez JA (2011) Combination of advanced oxidation processes and biological treatments for wastewater decontamination-a review. Sci Total Environ 409:4141–4166. https://doi.org/10.1016/j.scitotenv.2010.08.061

Oturan MA, Aaron JJ (2014) Advanced oxidation processes in water/wastewater treatment: principles and applications. A review. Crit Rev Environ Sci Technol 44:2577–2641. https://doi.org/10.1080/10643389.2013.829765

Palma P, Alvarenga P, Palma VL et al (2010) Assessment of anthropogenic sources of water pollution using multivariate statistical techniques: a case study of the Alqueva's reservoir, Portugal. Environ Monit Assess 165:539–552. https://doi.org/10.1007/s10661-009-0965-y

Panizza M, Cerisola G (2009) Electro-Fenton degradation of synthetic dyes. Water Res 43:339–344. https://doi.org/10.1016/j.watres.2008.10.028

Parsons S (2004) Advanced oxidation processes for water and wastewater treatment. IWA Publishing

Pera-Titus M, García-Molina V, Baños MA et al (2004) Degradation of chlorophenols by means of advanced oxidation processes: a general review. Appl Catal B Environ 47:219–256. https://doi.org/10.1016/j.apcatb.2003.09.010

Qajar A, Peer M, Reza Andalibi M, Rajagopalan R, Foley HC (2015) Enhanced ammonia adsorption on functionalized nanoporous carbons. Microporous Mesoporous Mater 218:15–23. https://doi.org/10.1016/j.micromeso.2015.06.030

Rosenfeldt EJ, Linden KG, Canonica S, von Gunten U (2006) Comparison of the efficiency of {radical dot}OH radical formation during ozonation and the advanced oxidation processes O_3/H_2O_2 and UV/H_2O_2. Water Res 40:3695–3704. https://doi.org/10.1016/j.watres.2006.09.008

SA'AT SKBM (2006) Subsurface flow and free water surface flow constructed wetland with magnetic field for leachate treatment. Universiti Teknologi Malaysia

Safe S (1985) Polychlorinated biphenyls (PCBs) and polybrominated biphenyls (PBBs): biochemistry, toxicology, and mechanism of action. Crit Rev Toxicol 13:319–395. https://doi.org/10.3109/10408448409023762

Safe SH (1994) Polychlorinated biphenyls (PCBs): environmental impact, biochemical and toxic responses, and implications for risk assessment. Crit Rev Toxicol 24:87–149. https://doi.org/10.3109/10408449409049308

Sarkar M, Acharya PK, Bhattacharya B (2003) Modeling the adsorption kinetics of some priority organic pollutants in water from diffusion and activation energy parameters. J Colloid Interface Sci 266:28–32. https://doi.org/10.1016/S0021-9797(03)00551-4

Schweitzer L, Noblet J (2018) Chapter 3.6: Water contamination and pollution. In: Green chemistry, pp 261–290

Sciortino JA, Ravikumar R, Programme. B of B (1999) Fishery harbour manual on the prevention of pollution

Scott JP, Ollis DF (1995) Integration of chemical and biological oxidation processes for water treatment: review and recommendations. Environ Prog 14:88–103. https://doi.org/10.1002/ep.670140212

Shaojuan Z, Junli W, Pengfei L, Haifeng D, Hui W, Xiaochun Z, Xiangping Z (2019) Efficient adsorption of ammonia by incorporation of metal ionic liquids into silica gels as mesoporous composites. Chem Eng J 370:81–88. https://doi.org/10.1016/j.cej.2019.03.180

Sharma P, Das MR (2013) Removal of a cationic dye from aqueous solution using graphene oxide nanosheets: investigation of adsorption parameters. J Chem Eng Data 81:151–158. https://doi.org/10.1021/je301020n

Shukla P, Wang S, Sun H et al (2010) Adsorption and heterogeneous advanced oxidation of phenolic contaminants using Fe loaded mesoporous SBA-15 and H_2O_2. Chem Eng J 164:255–260. https://doi.org/10.1016/j.cej.2010.08.061

Speight JG (2017a) Properties of inorganic compounds. Environ Inorg Chem Eng:171–229. https://doi.org/10.1016/b978-0-12-849891-0.00004-7

Speight JG (2017b) Sources and types of inorganic pollutants. Environ Inorg Chem Eng:231–282. https://doi.org/10.1016/b978-0-12-849891-0.00005-9

Sud D, Mahajan G, Kaur MP (2008) Agricultural waste material as potential adsorbent for sequestering heavy metal ions from aqueous solutions - a review. Bioresour Technol 99:6017–6027. https://doi.org/10.1016/j.biortech.2007.11.064

Sudoh M, Kodera T, Sakai K, Zhang JQ, Koide K (1986) Oxidative degradation of aqueous phenol effluent with electrogenerated fenton's reagent. J Chem Eng Jpn 19:513–518. https://doi.org/10.1252/jcej.19.513

Tarr MA (2003) Chemical degradation methods for wastes and pollutants: environmental and industrial applications. CRC Press

Tokumura M, Morito R, Shimizu A, Kawase Y (2009) Innovative water treatment system coupled with energy production using photo-Fenton reaction. Water Sci Technol 60:2589–2597. https://doi.org/10.2166/wst.2009.582

Trabelsi F, Aït-Lyazidi H, Ratsimba B et al (1996) Oxidation of phenol in wastewater by sonoelectrochemistry. Chem Eng Sci 51:1857–1865. https://doi.org/10.1016/0009-2509(96)00043-7

Yu Q, Jin X, Zhang Y (2018) Sequential pretreatment for cell disintegration of municipal sludge in a neutral bioelectro-Fenton system. Water Res 135:44–56. https://doi.org/10.1016/j.watres.2018.02.012

Ue D, Parlamento DEL, Del EY (2013) Directiva 2013/39/Ue Del Parlamento Europeo Y Del Consejo. 2013:1–17

Vrana B, Mills GA, Dominiak E, Greenwood R (2006) Calibration of the Chemcatcher passive sampler for the monitoring of priority organic pollutants in water. Environ Pollut 142:333–343. https://doi.org/10.1016/j.envpol.2005.10.033

Wang F, Ren X, Sun H et al (2016) Sorption of polychlorinated biphenyls onto biochars derived from corn straw and the effect of propranolol. Bioresour Technol 219:458–465. https://doi.org/10.1016/j.biortech.2016.08.006

Wang K, Wei M, Peng T et al (2010) Treatment and toxicity evaluation of methylene blue using electrochemical oxidation, fl y ash adsorption and combined electrochemical oxidation- fl y ash adsorption. J Environ Manag 91:1778–1784. https://doi.org/10.1016/j.jenvman.2010.03.022

Wang L, Yao Y, Zhang Z et al (2014) Activated carbon fibers as an excellent partner of Fenton catalyst for dyes decolorization by combination of adsorption and oxidation. Chem Eng J 251:348–354. https://doi.org/10.1016/j.cej.2014.04.088

Wang M, Liu P, Wang Y et al (2015) Core-shell superparamagnetic Fe_3O_4@β-CD composites for host-guest adsorption of polychlorinated biphenyls (PCBs). J Colloid Interface Sci 447:1–7. https://doi.org/10.1016/j.jcis.2015.01.061

Wang W, Lu Y, Luo H, Liu G, Zhang R, Jin S (2018) A microbial electro-Fenton cell for removing carbamazepine in wastewater with electricity output. Water Res 139:58–65. https://doi.org/10.1016/j.watres.2018.03.066

Wang Y, Wang Y (2017) Investigation of the lamination of electrospun graphene-poly (vinylalcohol) composite onto an electrode of bio-electro-Fenton microbial fuel cell. Nanosci Nanotechnol 7:1–12. https://doi.org/10.1177/1847980417727427

Wu J, Zhang H, Oturan N et al (2012) Application of response surface methodology to the removal of the antibiotic tetracycline by electrochemical process using carbon-felt cathode and DSA (Ti/RuO_2 –IrO_2) anode. Chemosphere 87:614–620. https://doi.org/10.1016/j.chemosphere.2012.01.036

Xu P, Xu H, Shi Z (2018) A novel bio-electro-Fenton process with $FeVO_4$/CF cathode on advanced treatment of coal gasification wastewater. Sep Purif Technol 194:457–461. https://doi.org/10.1016/j.seppur.2017.11.073

Yong X, Gu D, Wua YD, Yan Z, Zhou J, Wu X, Wei P, Jia H, Zheng T, Yonge Y (2017) Bio-Electron-Fenton (BEF) process driven by microbial fuel cells fortriphenyltin chloride (TPTC) degradation. J Hazard Mater 324:178–183. https://doi.org/10.1016/j.jhazmat.2016.10.047

Yu T, Sha Y, Liu W, Merinov BV, Shirvanian P, Goddard WA (2011) Mechanism for degradation of Nafion in PEM fuel cells from quantum mechanics calculations. J Am Chem Soc 133:19857–19863. https://doi.org/10.1021/ja2074642

Zaviska F, Drogui P, Mercier G, Blais J-F (2009) Procédés d'oxydation avancée dans le traitement des eaux et des effluents industriels: Application à la dégradation des polluants réfractaires. Rev des Sci l'eau / J Water Sci 22:535–564. https://doi.org/10.7202/038330ar

Zhang G, Wang S, Yang F (2012) Efficient adsorption and combined heterogeneous/homogeneous Fenton oxidation of Amaranth using supported Nano-FeOOH as cathodic catalysts. J Phys Chem C 116:3623–3634. https://doi.org/10.1021/jp210167b

Zhang Y, Liu C, Xu B et al (2016) Degradation of benzotriazole by a novel Fenton-like reaction with mesoporous Cu/MnO 2 : Combination of adsorption and catalysis oxidation. Appl Catal B Environ 199:447–457. https://doi.org/10.1016/j.apcatb.2016.06.003

Zhang Y, Wang Y, Angelidaki I (2015) Alternate switching between microbial fuel cell and microbial electrolysis cell operation as a new method to control H_2O_2 level in bioelectro-Fenton system. J Power Sources 291:108–116. https://doi.org/10.1016/j.jpowsour.2015.05.020

Zhao H, Kong C (2018) Elimination of pyraclostrobin by simultaneous microbial degradation coupled with the Fenton process in microbial fuel cells and the microbial community. Bioresour Technol 258:227–233. https://doi.org/10.1016/j.biortech.2018.03.012

Zhao L, Yang S, Feng S, Ma Q, Peng X, Wu D (2017) Preparation and application of carboxylated graphene oxide sponge in dye removal. Int J Environ Res Public Health 14:1301. https://doi.org/10.3390/ijerph14111301

Zhou XJ, Buekens A, Li XD et al (2016) Adsorption of polychlorinated dibenzo-p-dioxins/dibenzofurans on activated carbon from hexane. Chemosphere 144:1264–1269. https://doi.org/10.1016/j.chemosphere.2015.10.003

Zhu X, Ni J (2009) Simultaneous processes of electricity generation and p-nitrophenol degradation in a microbial fuel cell. Electrochem Commun 11:274–277. https://doi.org/10.1016/j.elecom.2008.11.023

Zhuang L, Zhou S, Li Y, Liu T, Huang D (2010) In situ Fenton-enhanced cathodic reaction for sustainable increased electricity generation in microbial fuel cells. J Power Sources 195:1379–1382. https://doi.org/10.1016/j.jpowsour.2009.09.011

Chapter 6
Metal Oxides for Removal of Arsenic Contaminants from Water

Tamil Selvan Sakthivel, Ananthakumar Soosaimanickam, Samuel Paul David, Anandhi Sivaramalingam, and Balaji Sambandham

Abstract Arsenic (As), one of the highest harmful pollutants found in drinking/groundwater, is owing to have unfavourable impacts, for example, skin disease, on human health. The new Environmental Protection Agency (EPA) assigned the maximum contaminations of arsenic in groundwater is 10μg/L, and several drinking water plants are needing extra treatment to accomplish this standard. In recent years, several researchers have been attempting to discover practical and expendable adsorbents for some water filtration systems that are utilized in many arsenic endemic territories. Metal oxide-based adsorbents had been proved to be the best strategies for arsenic expulsion/removal. This chapter reviews the removal of both arsenite (As III) and arsenate (As V) species from drinking/groundwater. Also, we give an overview of traditionally applied strategies to expel both arsenic species to incorporate coagulation-flocculation, oxidation, and membrane techniques. More focus has been given to adsorption methods, type of adsorption and factors affecting adsorption. Moreover, brief summary has been given for an advancement on the

---

T. S. Sakthivel (✉)
Department of Materials Science and Engineering (MSE), Advanced Materials Processing and Analysis Center (AMPAC), University of Central Florida, Orlando, FL, USA
e-mail: Tamilselvan.Sakthivel@ucf.edu

A. Soosaimanickam
Institute of Materials (ICMUV), University of Valencia, Valencia, Spain

S. Paul David
HiLASE Centre, Institute of Physics of the Czech Academy of Science,
Dolni Brezany, Czech Republic

Department of Physics, Kalasalingam Academy of Research and Education,
Krishnankoil, Tamil Nadu, India

A. Sivaramalingam
Department of Physics, Sathyabama Institute of Science and Technology, Chennai, Tamil Nadu, India

B. Sambandham
Department of Materials Science and Engineering, Chonnam National University,
Gwangju, South Korea

efficacy of different nanomaterials and composites for the polluted water treatment. A basic examination of the most generally explored nanomaterials is highlighted.

Keywords Arsenic · Removal technology · Metal oxide · Adsorption Process · Drinking water

6.1 Introduction

Water is by and large acquired from two leading normal sources: surface water, for example, freshwater lakes, streams, waterways, and groundwater, for example, well water (McMurry and Fay 2004). Water is essential for all types of plants and creature lifecycles (VanLoon and Duffy 2017) and it has one of a kind material property because of its extremity and hydrogen securities which mean it can break up, retain, adsorb or suspend various mixes (Organization 2007). The main source of drinking water is considered as under groundwater in the world due to the accessibility and consistent quality. Under groundwater is likewise the favoured drinking water source for the provincial territories, especially in creating nations, since no or minimal treatment is needed, and it is frequently situated close to consumers. Nonetheless, in nature, water isn't unadulterated because it secures toxins from its environment and those emerging from people and creatures similar to other organic exercises (Mendie 2005). This part of the review provides information on the water quality, the issues identified with groundwater contamination, and the multiple strategies utilized in the investigation and expulsion of heavy metal ions from groundwater. This section likewise gives the extent of the hypothesis.

6.1.1 Ground Water Quality and Availability

Groundwater is located underneath the earth surface that occupies the area between the cleft and grains or splits in rocks. Mostly the groundwater collected through the dirt by downpour and permeation down. Groundwater has various fundamental favorable circumstances when contrasted with surface water because of higher quality, well shielded from conceivable contamination, a reduced amount of subject to occasional and lasting changes, and consistently spread all over places than surface water. It can be accessible in dry places even if surface water is no available. Additionally, collecting groundwater through well fields into activity is less expensive in contrast with collecting the surface water which regularly involves extensive capital ventures. Because of these favorable circumstances combined with decreased groundwater powerlessness to contamination especially have brought about widespread use of groundwater for drinking water supply. As of now, 97% of the earth's liquefied freshwater is put away in springs. Numerous nations on the planet subsequently depend to an enormous degree on groundwater for the fundamental use of

Table 6.1 Drinking water collection from groundwater by territory

Territory	Use of Drinking Water from Ground Water (%)	People Used (million)
World		2000 (2.0 billion)
Australia	15	3
United States	51	135
Latin America	29	150
Europe	75	200–500
Asia and Pacific	32	1000–2000

Table 6.2 Countries (selected) uses groundwater to produce drinking water

Continent	Country	Percentage (%)
Europe	United Kingdom	27
	Belgium	83
	Netherlands	75
	Slovakia	82
	Germany	75
	Denmark	100
Asia	India (rural)	80
	Philippines	60
	Thailand	50
	Nepal	60
Africa	Tunisia	95
	Morocco	75
	Ghana	45
America	United States (rural)	96

drinking water. Table 6.1 demonstrates that several billion individuals mainly depend on groundwater to use for drinking water.

Numerous nations in the world uses groundwater for the purpose of drinking water as a main source (Table 6.2). Sadly, it was identified that Africa is utilizing the little quantity of groundwater for drinking purpose. Table 6.2 demonstrates that groundwater is broadly utilized for drinking water in Europe, particularly in Denmark which generates 100% of the drinking water from groundwater. In provincial regions of India and the United States, groundwater additionally signifies to the essential source of consumable water (80% and 96%, respectively). In Tunisia, groundwater signifies 95% of the nation's water requirements, 83% in Belgium, and 75% in the Germany, Netherlands, and Morocco. In most European nations (Belgium, Austria, Hungary, Denmark, Switzerland, and Romania) groundwater usage surpasses 70% of the complete water utilization (Vrba and van der Gun 2004). In numerous countries, the greater part of the well-drawn groundwater is for household water supplies which gives 25 to 40 % of the earth's drinking water (Lambrechts et al. 2003).

Different human exercises can bring about huge deviations in the states of groundwater assets arrangement which causes exhaustion and contamination. Groundwater contamination, as a rule, is an immediate aftereffect of ecological contamination. Groundwater is contaminated basically by nitrogen mixes (nitrate, alkali, and ammonium), oil-based commodities, phenols, iron mixes, and substantial metals (zinc, copper cadmium, mercury, lead, etc.) (Vrba and van der Gun 2004).

Groundwater is firmly interconnected with different parts of the earth. In the least progressions in ecological rainfall can cause variations in the earth's water body or groundwater system, assets, and quality. Siphoning could concentrate the chemical compounds in groundwater which may not suitable for drinking in profound springs and could attract saline seawater in beach front zones. These conditions ought to be viewed when pumping groundwater in use. If the groundwater is contaminated, cleaning up the polluted water can be a generally long haul (several years), in fact challenging and overpriced (Urba 1985).

Groundwater frameworks are recharged by rainfall and surface water. Generally, groundwater flow is not exactly distinctive, thus far what is stored underneath the ground's surface is the biggest savings of waterbody. Its complete volume denotes to 96% of total world's freshwater (Shiklomanov and Rodde 2011). Around 60% of the under groundwater is collected and utilized for cultivation in numerous nations wherever dry and semi-dry environments prevails; balance 40% of the groundwater is similarly separated between the industrial and household segments (Shiklomanov and Rodde 2011). Table 6.3 demonstrates the 15 nations that utilize to a great extent of groundwater for horticulture, household and industry use (Margat and Van der Gun 2013).

Table 6.3 Top 15 countries with the major estimated yearly groundwater extractions (2010)

	Groundwater Extraction			
	Estimated Groundwater Extraction (km^3/yr)	Breakdown by Sector		
Country		Horticulture (%)	Household Use (%)	Industry (%)
Italy	10.40	67	23	10
Thailand	10.74	14	60	26
Japan	10.94	23	29	48
Syria	11.29	90	5	5
Russia	11.62	3	79	18
Turkey	13.22	60	32	8
Indonesia	14.93	2	93	5
Saudi Arabia	24.24	92	5	3
Mexico	29.45	72	22	6
Bangladesh	30.21	86	13	1
Iran	63.40	87	11	2
Pakistan	64.82	94	6	0
United States	111.70	71	23	6
China	111.95	54	20	26
India	251.00	89	9	2

6.2 Heavy Metals Present in Ground Water

Groundwater polluting is the greatest significant ecological issues today in numerous nations (Sharma and Al-Busaidi 2001). Between the several variety of pollutants influencing water body, heavy metallic ions or elements get specific concern as their solid poisonousness at very low focuses/concentrations (Marcovecchio et al. 2007). The non-degradable metal ions are naturally existing and active in the soil and dangerous to existing creatures. Along these lines, the exclusion of substantial heavy metal ions from groundwater is essential to ensure good national health.

It was reported earlier that ventures, for example, plating, earthenware production, mining, lead crystal, and battery assembling industries are treated as the primary causes of substantial heavy metal ion present in water body because of their chemical wastes mix in the groundwater (Momodu and Anyakora 2010). Moreover, landfill leachates are the other potential source of metallic contamination found in groundwater (Marcovecchio et al. 2007). The action of landfill frameworks considered for waste transfer is the principle strategy utilized for strong waste transfer in most creating nations. In any case, a "landfill" in a building up nation's setting is generally shallow and unprotected (frequently not more profound than 20 inches). This is as a rule followed for a long way from standard suggestions (Adewole 2009) and adds to the pollution of groundwater, surface and air, and soil, which is a hazard to living creatures in the earth. Additionally, Fred & Jones in 2005, demonstrated that certified landfills can also be deficient in the counteractive action of water body pollution (Fred Lee and Jones-Lee 2005).

A specific gravity of metallic ions are much higher than water, so it can be in the form of particulate, or colloidal, or disintegrated stages in groundwater (Adepoju-Bello et al. 2009). Several substantial metal ions are highly essential to living creatures such as copper, cobalt, manganese, iron, zinc, and molybdenum that are required at lower concentration as an stimulus for protein exercises (Adepoju-Bello et al. 2009). In any situation, abundance introduction of metal ions could be poisonousness to living creatures. The nature of water body is influenced by the available channels which it passes through to the groundwater region (Adeyemi et al. 2007), in this manner, the metallic substances released by ventures, city wastes, farming wastes and incidental oil spillages from transporters can bring about an unfaltering ascent in pollution of groundwater (Igwilo et al. 2006).

6.3 Toxicity and Chemistry of Heavy Metals

Contingent upon the characteristic property and amount of metal consumed, substantial metals can originate serious medical issues (Adepoju-Bello and Alabi 2005). Dangerousness of heavy metal is identified with the arrangement of structures with nutrients, in which amine ($- NH_2$), thiol ($- SH$) groups and carboxylic corrosive ($- COOH$) are included. At the point when metals tie to these buildings, significant catalyst and nutrients are influenced. The most hazardous metals to humans are aluminum, cadmium, mercury, arsenic, and lead. Aluminum is related to Parkinson's and Alzheimer's ailment, presenile dementia and infirmity. The arsenic can cause different types of cancers such as skin, lungs, and bladder cancer and heart diseases. Cadmium introduction can produce hypertension and kidney damage. Lead is also a dangerous toxic substance and a conceivable human cancer-causing agent (Bakare-Odunola 2005). The dangerous mercury brings about mental aggravation and hindrance of discourse, hearing, and vision issues (Warner et al. 2013). Moreover, mercury and lead might cause the advancement of auto-invulnerability where an individual's insusceptible framework attacks its very own cells, that can produce joint sicknesses and the kidney illness, cardiac disease, and neuron infections. At much higher concentrations, mercury and lead can originate permanent cerebrum harm.

The toxicity, source, and chemistry of few substantial metallic elements are explained below.

(a) Arsenic (As)

The proximity of arsenic in the earth is ever-present on the planet because of characteristic and anthropogenic bases. It happens in the world's exterior and it is activated via regular enduring responses, organic action, geochemical responses, and volcanic outflows (Mohan and Pittman 2007). The highest concentration of arsenic was found to be 5 ppm (mg/L) in arsenic-rich groundwater zones, and geothermal impacts could expand arsenic concentrations, upto 50 ppm. Herbicide, purifying, and removal procedures are the best instances for arsenic contamination generated from anthropogenic sources (Smedley and Kinniburgh 2002).

Arsenic threat relies upon its oxidation states, and the characteristic feature of arsenic species in groundwater/drinking water are in the form of inorganic structures (Tuutijärvi 2013). Consuming the arsenic contaminated water is dangerous to lung, skin, kidney, and gives bladder disease, skin thickening (hyperkeratosis) neurological disarranges, solid shortcoming, loss of hunger, and queasiness (Yuan et al. 2002; Basu et al. 2014). This varies from intense harming, which commonly causes stomach cancer, and bleeding with watery stools. High groupings of arsenic in water could likewise bring about an expansion in death of a fetus and unconstrained premature births (Tuutijärvi 2013).

Arsenic in groundwater is an overall issue. Drinking an arsenic contaminated water from groundwater represents a serious health issues in a few emerging regions (Mohan and Pittman 2007; Ranjan et al. 2009a, b). On the other hand, the WHO temporary regulation of 10 ppb (µg/L) of arsenic in water is currently perceived as an overall issue in numerous nations, particularly in Southeast Asia, including China, Bangladesh and India (Guo et al. 2007). The biggest section of populace right now is in danger in eastern part of India and the Bengal Bowl region of Bangladesh (Pandey et al. 2009; Prasad et al. 2013); nevertheless, both India and Bangladesh have held WHO's previous standard regulation, i.e., 50 ppb arsenic concentration, in dr/inking water (Pennesi et al. 2012; Kamsonlian et al. 2012). Around 70 million individuals are experiencing arsenic issue alone in these areas; this is maybe the biggest harming on the planet's history (Nigam et al. 2013a, b). As of 2011, in West Bengal, the polluted arsenic concentration in groundwater, and in drinking water was accounted for in the range from 50 to 3600 ppb in 111 squares of 12 regions of the state (Mondal et al. 2011); influencing around 1 million individuals (Basu et al. 2014). Therefore, the expulsion of arsenic contamination from water has gotten huge consideration and significant worry to many water utilities and administrative organizations

Arsenic exists in As^{-3}, As^0, As^{+3}, and As^{+5} oxidation states, however in groundwater it is typically identified in inorganic structure as oxyanions: trivalent arsenite-As(III) species and pentavalent arsenate -As(V) species (Kumari et al. 2005; Chiban et al. 2012; Basu et al. 2014). In eastern part of India, the arsenic species in drinking water or groundwater are seen as As (V) and As(III) in 1:1 proportion (Roy et al. 2013b; De Anil 2003). Both As (V) and As (III) are sensitive to the assembly at the pH esteems typically identified in groundwater (pH 6.5–8.5). Figure 6.1 demonstrates the significant components that control arsenic oxidations: redox potential (Eh) and pH (Tuutijärvi 2013; Mohan and Pittman 2007; Chiban et al. 2012).

Arsenite prevails in modest reducing anaerobic conditions, for example, groundwater which can be seen in Fig. 6.2; uncharged arsenite H_3AsO_3 overwhelms in reducing circumstances at pH < ~9.2. On the other hand, in oxidizing situations, distinctive separated types of arsenate are imposing. Speciation of arsenate and

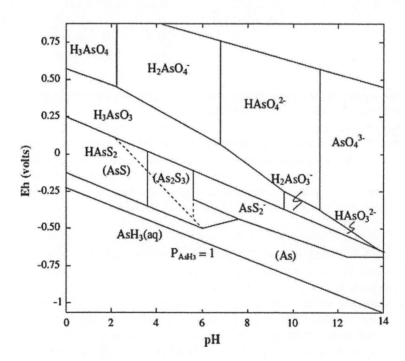

Fig. 6.1 The Eh vs pH diagram of arsenic at room temperature and 101.3 kPa. (Reprinted with permission from (Wang and Mulligan 2006))

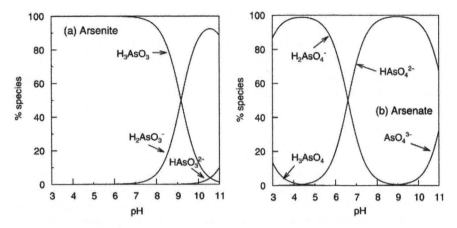

Fig. 6.2 classification of As (II) and As (V) as a function of pH (Reprinted with permission from (Smedley and Kinniburgh 2002))

Table 6.4 Separation of arsenite-As (III) and arsenate-As (V)

Specification	Dissociation Reactions	pKa
Arsenate [As(V)]	$H_3AsO_4 \leftrightarrow H^+ + H_2AsO_4^-$	2.24
	$H_2AsO_4^- \leftrightarrow H^+ + HAsO_4^{2-}$	6.69
	$HAsO_4^{2-} \leftrightarrow H^+ + AsO_4^{3-}$	11.5
Arsenite [As(III)]	$H_3AsO_3 \leftrightarrow H^+ + H_2AsO_3^-$	9.2
	$H_2AsO_3^- \leftrightarrow H^+ + HAsO_3^{2-}$	12.1
	$H_2AsO_3^- \leftrightarrow H^+ + HAsO_3^{3-}$	13.4

arsenite separations in various pKa qualities are exhibited in Table 6.4 (Chiban et al. 2012; Mohan and Pittman 2007; Tuutijärvi 2013).

Figure 6.2 demonstrates the circulation of As (III) and As (V), as a component of pH. As(III) occurs in non–separated state at nonpartisan and marginally at higher pH (pH > 8) extensive measure of anionic species are identified. Then again As (V), is totally separated and present as monovalent, divalent and trivalent anions. In general, examination reports in speciation information did not consider about the level of protonation (Chowdhury et al. 2013; Chiban et al. 2012).

(b) Copper (Cu)

In the liquid state, copper (Cu) exists in 3 classes: colloidal, dissolvable and particulates. The broke up stage might contain free particles just as copper material and inorganic ligands. Copper structures can be easily connecting with the base compounds like sulfate, carbonate, nitrate, chloride, smelling salts, humic compounds and hydroxide. The arrangement of unsolvable malachite ($Cu_2(OH)_2CO_3$) is a central point in regulating the degree of unrestricted copper (II) particles in liquid states. An oxidation state of copper (+1), (+2) and (+3) are well known in spite of the fact that Cu(+2) is highly usual. Cu(+1) is a run of the mill delicate corrosive. In ground water, Cu(+2) is significantly available at acidic pH (up to 6); liquefied

CuCO$_3$ is common at basic pH (6.0–9.3) and at higher pH (9.3–10.0), the fluid {Cu(CO$_3$)$_2$}$^{2-}$ particle dominates (Morgan and Stumm 1970). The predominant copper-containing species of free copper particles can be identified at a pH below 6.5. Alkalinity profoundly affects the free copper particle fixation (Ohlinger et al. 1998).

Copper is quite compelling a result of its danger and it is across the board nearness in the machine-driven applications like paint industries, metal completing and electroplating manufacturing. The occurrence of copper in groundwater could cause dermatitis, tingling, keratinization in the feet and hands. Hence, the copper concentration in water should be decreased to the point that fulfil the ecological guideline for different waterways (Uwamariya 2013). Broken down copper bestows shading and undesirable, metallic, unpleasant taste to the water. Recolouring of clothing and pipes connections happen where the copper level in water exceeds 1.0 ppm (mg/L). Regurgitating, loose bowels, queasiness, and some intense side effects probably because of surrounding disturbance by consumed copper (II) particles that have been depicted in a few incidents. The most extreme satisfactory convergence of copper in water is 2 ppm (mg/L) depends on wellbeing contemplations (Edition 2011).

(c) Lead (Pb)

Lead pollution could happen in water because of mining and refining exercises, battery discharge and waste reprocessing centers, car fumes outflows, lead contained fuel spills, municipal solid waste incinerator disposal leachates. Mostly, the actions of lead in groundwater is a mixer of inorganic complexing materials and usual ligands. In higher pH (above 5), the precise amount of lead becomes unfavorably charged, yet is reduced to some degree by reacting with solvents, substances and metal chelates (Moore and Ramamoorthy 2012). In surface and groundwater, the prevalent type of lead is present as a particles. However, their concentration is based on the pH and the redox potential. At higher pH (above pH 8), the lead concentration could be 10μg/L, while at neutral pH, the dissolvability could approach and exceed 100μg/L of concentration.

Lead is a commonly available material in the world, and it is known for a considerable length of time to be a collective metabolic toxic substance. Lead is a profoundly toxic metal (regardless of whether breathed in or gulped), influencing pretty much every organ and framework in the human body. Creating focal sensory systems of kids might be influenced, prompting hyperactivity, crabbiness, migraines, and learning and focus troubles. The lead concentration should be around 0.01 mg/L in water (Edition 2011).

(d) Cadmium (Cd)

Cadmium is usually found related to zinc (Zn) in sulphide and carbonate metals. In addition, it is found as a result of processing different metals. Wet chemistry of cadmium is, generally, commanded by means of cadmium carbonate-CdCO$_3$(s) (otavite), Cd^{2+} oxidation, and cadmium hydroxide-Cd(OH)$_2$(s) (Faust and Aly 2018). An acidity condition of water generally affects the solubility of cadmium in water. At higher pH (greater than 10), the solubility of cadmium carbonate is usually around 300 ppb (µg/L), cadmium hydroxide is from 44 to 225 ppb (µg/L) for the matured structures. Different structures of cadmium are available in the pH scope of groundwater, however at lower pH (less than 6), the previously mentioned cadmium structures are absent significantly (Benjamin 2014). At pH level between 5 and 9, the solubility conditions of cadmium species ($CdOH^+$, Cd^{2+}, and $Cd(OH)_2$) are quite higher, as revealed earlier (Lai et al. 2002).

Cadmium compounds are available in water from a varies assortment of sources in nature and from the industry wastes. One of the important sources is ingestion of food products, particularly green vegetable and grain, which promptly ingest cadmium compound from the dirt. The cadmium compounds may happen in surface/ground water normally or a contaminants from sewage muck, composts, mining wastes or contaminated groundwater (Fauci et al. 1998).

Cadmium compounds have no basic organic capacity and is amazingly dangerous for the humans. In ceaseless introduction, cadmium additionally amasses in the human body, especially in the liver and the kidneys. Intense harming can happen from the inhalation of cadmium chloride gas and absorption of chloride salts which has been resulted to death (Baldwin and Marshall 1999). It was identified with an examination on humans that cadmium can cause numerous infections and highly dangerous whenever inhaled at higher dosages. In view of the conceivable poisonous quality of cadmium, the WHO wellbeing-based rule an incentive for drinking water is 3µg/l (Edition 2011).

(e) Chromium (Cr)

The valance state of chromium present in groundwater are Cr (VI) and Cr (III). Numerous chromium, particularly Cr (III), complexes are moderately insoluble in water. Cr (III) hydroxide and oxide are the main components soluble in water. Cr (VI) mixes are steady under high-impact conditions yet are reduced to Cr (III) mixes under reducing environment. another probability in an oxidizing situation is the invert procedure. The formation of Cr (III) species in water is depends on pH, particularly $Cr(OH)^{2+}$ is the usual chromium species present in groundwater with a pH somewhere in the range of 6–8 (Calder 1988). Cr (VI) in water system occurs solely as oxyanions (CrO_4^{2-}, $Cr_2O_7^{2-}$). In weaken arrangements (<1 ppm (mg/L)), the transcendent structure is CrO_4^{2-}; negatively charged component which doesn't perplex through an anionic particulate issue. On the other hand, Cr (VI) anions are mostly adsorbed by the positively charged surfaces, for example, the hydroxides and oxides of Al, Fe, and Mn. However, Cr (VI) adsorption on these positively charged adsorbents is generally restricted and reduces with increasing pH (Benders 2012). Subsequently, Cr (VI) is more transferrable than Cr (III). Because Cr (III) species are heavy in acidic pH (pH under 3), and, in higher pH (above 3.5), hydrolysis occurs in Cr (III) and forms trivalent hydroxy species such as $Cr(OH)_2^+$, $Cr(OH)_3^0$, $Cr(OH)_4^-$, $Cr(OH)^{2+}$.

Chromium is necessity as a nutritional for various living beings. This nutritional behaviour just applies to Cr (III). Cr (VI) is dangerous to plants and vegetable. The standard acceptable concentration of chromium in drinking water is 50 ppb (µg/L) (Edition 2011). This standard rule is temporary because of vulnerabilities identified with the good impact, therefore, the acceptable concentration would be lowering in near future. Complete chromium has been determined as a result of troubles in investigating the hexavalent structure.

6.4 Conventional Techniques to Remove Arsenic Metal Ions from Water

The chemistry and structure of arsenic-exposed groundwater are the central point of deciding the expulsion of arsenic (Singh et al. 2015). A large portion of the accessible expulsion innovations is progressively proficient for As (V) given that As (III) is generally neutral at pH lower than 9.2 (Johnston et al. 2001). Because of the non-charge surfcae, As (III) species are less accessible for precipitation, adsorption, and/or particle trade. Appropriately, advanced methods are acceptable for complete removal of arsenite by utilizing a two-advance methodology which oxidize the As (III) to As (V) pursued by a system for the expulsion of arsenate (Pous et al. 2015).

6.4.1 Oxidation Method

Oxidation includes the transformation of solvent arsenite to arsenate. This by itself doesn't expel arsenic from the arrangement, accordingly, an expulsion system, for example, adsorption, coagulation, or particle trade, must pursue (Johnston et al. 2001). For redox state of groundwater, oxidation is a significant advancement since As (III) is the common type of arsenic at close neutral pH (Singh et al. 2015). Beside environmental oxygen, numerous synthetic substances, just as microbes, have just been utilized to legitimately oxidize As (III) to As (V) in groundwater and these findings are identified in Table 6.5.

In many different nations, oxygen, permanganate and hypochlorite are the commonly utilized oxidants. As (III) oxidation with oxygen is an extremely moderate procedure, which can take hours or weeks to finish (Ahmed 2001). Then again, synthetic chemicals, for example, ozone, chlorine, and permanganate could quickly oxidize arsenite to arsenate as exhibited in Table 6.5. On the other hand, in spite of this improved oxidation, interfering elements present in the contaminated water should be measured in choosing the correct oxidant material as interfering materials can significantly influence and direct the energy of As(III) oxidation (Singh et al. 2015). For example, the oxidation pace of As (III) by ozone could be significantly diminished when S^{2-} is present in the water (Dodd et al. 2006). Additionally, in other examination, it was demonstrated that influencing of other anions and natural issue in water significantly influence the utilization of UV/titanium dioxide (TiO_2) material in As (III) oxidation (Guan et al. 2012). Besides, this includes an

Table 6.5 Various oxidants utilized to oxidize As (III) to As (V), and their properties, working conditions, and efficiencies

Oxidants	Solution pH	Initial Concentration (μg/L)	Type of Water	Comments	References
Ozone and oxygen	7.5–8.5	45–62	Ground water	Ozone method is very quicker than unadulterated oxygen/air for oxidizing As (III). For example, ozone technique takes 20 min for the complete oxidation of As (III) while unadulterated oxygen and air oxidizes only 57% and 54%, respectively, after 5 days	Kim and Nriagu (2000)
Type of Chlorine	8.3	300	DI water	The complete oxidation of As(III) to As(V) occurs by the addition of active chlorine material when its underlying fixation was more prominent than 300 ppb (μg/L). The estimated adsorption capacity was 0.99 mg of Cl_2/mg of As (III).	Hu et al. (2012)
Chlorine dioxide	8.12	50	Ground water	86% of oxidation accomplished within 1hr. This high value is for the most part because of the nearness of certain metals present in water which could act as a catalyst and help in oxidizing As (III) to As (V)	Sorlini and Gialdini (2010)
Monochloramine	8.12	50	Ground water	Longer time is needed to acquire complete As (III) oxidation. Generally, it could oxidize only 60% of As (III) even after 18 h of contact time.	Sorlini and Gialdini (2010)

(continued)

Table 6.5 (continued)

Oxidants	Solution pH	Initial Concentration (μg/L)	Type of Water	Comments	References
Hypochlorite	7	500	Ground water	Complete oxidation of As (III) to As (V) was occurred with higher loading of hypochlorite, typically, 500μg/L	Viet et al. (2003)
Hydrogen peroxide	7.3–10.5	50	Sea water and fresh water	Highly efficient As (III) oxidation occurs when the solution pH lever was from 7.3 to 10.5	Pettine et al. (1999)
Potassium permanganate	8.12	50	Ground water	Complete oxidation occurs within a minute, very quick process.	Sorlini and Gialdini (2010)
Photocatalytic oxidation (UV/H2O2)	8	100	Ground water	Hydrogen peroxide with UV irradiation works effectively to oxidize As (III) to As (V). oxidation can be expanded by increasing the power of UV irradiation. Typically, 85% of oxidation was occurred by applying UV portion of 2000 mJ/cm^2	Sorlini et al. (2014)
Biological oxidation	N/A	N/A	N/A	Chemoautotrophic arsenite-oxidizing microscopic organisms can include in the process of As (III) oxidation by means of the oxygen as electron acceptors through the absorption process with inorganic carbon materials into cell	Katsoyiannis and Zouboulis (2004)
In situ oxidation	N/A	N/A	Ground water	oxygenated or extra oxygen loaded water is siphoned with the contaminated water to decrease the concentration of As less than 10μg/L	Gupta et al. (2009)

unpredictable treatment, which delivers an As-bearing build-up that is hard to discard. Subsequently, to productively expel arsenic from a reaction by oxidation, oxidants ought to be chosen cautiously. In addition, all referred to impediments of oxidation alone is a less skilful technique for arsenic expulsion.

6.4.2 Coagulation and Flocculation

These methods are the most utilized and archived strategies for expulsion of arsenic from contaminated water (Choong et al. 2007). In coagulation, positively charged coagulant material such as ferric chloride ($FeCl_3$), aluminum sulfate ($Al_2(SO_4)_3$), etc., reduces the negatively charged colloids, in this way causing the coagulant material to impact and get higher efficiency. Flocculation, then again, includes the expansion of negatively charged or an anionic flocculant which can cause spanning or charge balance between the enclosed materials that motivates the arrangement of flocculation. During these procedures, contaminated arsenic species is altered by the synthetics material into an insoluble strong complex material that precipitates later (Mondal et al. 2013). On the other hand, solvent arsenic species could be consolidated in the form of metal hydroxide (Johnston et al. 2001). In any case, solids can be evacuated a short time later through sedimentation as well as filtration.

Arsenic expulsion proficiency of various coagulants fluctuates based on solution pH. Lower than pH 7.5, $FeCl_3$ and $Al_2(SO_4)_3$ compounds are similarly powerful in expelling arsenic from contaminated water (Garelick et al. 2005). Between the arsenic and arsenate, specialists recommended that As (V) is the more effectively contrasted with As (III) and reports indicates that ferric chloride is a superior coagulant material than aluminum sulfate, particularly at higher pH (greater than 7.6) (Hering et al. 1996; Cheng et al. 1994). With the influence of other metal ions present in the contaminated water, ferric chloride coagulants worked out very well to reduce the arsenic contaminations less than the Maximum Concentration Level (MCL) (10μg/L) in contrast with aluminum-based coagulants (Hu et al. 2012). Table 6.6 demonstrates a summary of the coagulants utilized in the arsenic evacuation, organized with their efficiencies, working circumstances, and properties.

The real downside of coagulation-flocculation method is the creation of high measures of arsenic-concentrated discharge (Singh et al. 2015). The administration of this muck is vital in order to avert the outcome of optional contamination of nature. Also, action of slime is exorbitant. These impediments cause this procedure lower the possibility, particularly in the real conditions (Mondal et al. 2013).

6.4.3 Membrane Technologies

Membrane filtration is a system which can be utilized in drinking water production aimed at the expulsion of arsenic species as well as any additional contaminants

Table 6.6 Various coagulants utilized for arsenic removal, and their properties, working conditions, and efficiencies.

Coagulant Material	pH Condition	Initial Concentration of Arsenic	Type of Water	Comments	References
Iron components	7	2 mg/L	Deionized	As (V) and As (III) was effectively removed around 75% and 45%, respectively, with FeCl3 loading of 30 mg/L. With the higher loading of iron FeCl3, the arsenic removal rate was increased, however, lingering iron content was increased in drinking water after coagulation.	Hesami et al. (2013)
Aluminium components	7	20μg/L	River	With the $Al_2(SO_4)_3 \cdot 18H_2O$ utilization of 40 mg/L, As(V) was removed effectively upto 90%. However, As (III) elimination with aluminium material was irrelevant even after increasing the aluminium dosages.	Hering et al. (1997)
$ZrCl_4$	5.5	50μg/L	Deionized	With 2 mg/L addition of ZrCl4, around 55% of As (V) was effectively removed. In addition, the As (V) removal rate was expanded at pH 6.5 and reduced at pH8.5. on the other hand, As (III) removal rate was roughly 8% irrespective of pH.	Lakshmanan et al. (2008)
$TiCl_3$	7.5	50μg/L	Deionized	With 2 mg/L addition of TiCl3, As (V) and As (III) expulsion efficiency was 75% and 32%, respectively. Both arsenic species evacuation was exceptionally pH subordinate.	Lakshmanan et al. (2008)

(continued)

Table 6.6 (continued)

Coagulant Material	pH Condition	Initial Concentration of Arsenic	Type of Water	Comments	References
$TiCl_4$	7.5	50μg/L	Deionized	With 2 mg/L of TiCl4 dosage, around 55% of As (V) and 26% of As (III) was removed. As(V) expulsion was profoundly pH depended, Although As(III) evacuation was free of pH.	Lakshmanan et al. (2008)
$TiOCl_2$	7.5	50μg/L	Deionized	With 2 mg/L of TiOCl2 dosage, around 37% of As (V) and 20% of As (III) was removed. Both As (V) and As (III) expulsion were profoundly pH depended.	Lakshmanan et al. (2008)
$ZrOCl_2$	7.5	50μg/L	Deionized	With 2 mg/L of ZrOCl2 dosage, around 59% of As (V) and 8% of As (III) was removed. As (V) evacuation was profoundly pH subordinate, while As(III) expulsion was free of pH.	Lakshmanan et al. (2008)
$Fe_2(SO_4)_3$	7	1 mg/L	Double Deionized	With 25 mg/L of Fe2 (SO4)3 dosage, around 80% of As (III) was removed.	Sun et al. (2013a)
$Ti(SO_4)_2$	7	1 mg/L	Double Deionized	around 90% of As (III) was removed using 25 mg/L of Ti (SO4)2 dosage,	Sun et al. (2013a)

present in water. Generally, membranes are manufactured compounds contains several billions of micro pores connected through particular boundaries, which don't stop the water to go through (Shih 2005). A main impetus, for example, weight contrast between the penetrate and feed sides, is expected to pass through the water by the layer (Van der Bruggen et al. 2003). For the most part, two types of weight driven membrane filtrations are available: (1) low-weight film forms, for example, ultrafiltration (UF) and microfiltration (MF); and (2) pressure membrane forms, for example, nanofiltration (NF) and Reverse Osmosis (RO) (Mondal et al.

Table 6.7 Outline of pressure driven membrane forms and their attributes

Parameters	Microfiltration	Ultrafiltration	Reverse Osmosis	Nanofiltration
Pressure (bar)	0.1–2	0.1–5	3–20	5–120
Permeability (1/h.m^2.bar)	>1000	10–1000	1.5–0	0.05–1.5
Pore size (nm)	100–10,000	2–100	0.5–2	<0.5
Multivalent ions	–	–/+	+	+
Organic compounds	–	–	–/+	+
Macromolecules	–	+	+	+
Particles	+	+	+	+
Separation mechanism	Filtering	Filtering	Filtering Charge effects	Solution-Diffusion
Applications	Purification, pre-treatment, Clarification	Exclusion of Bacteria, macromolecules, viruses	Exclusion of dissolved salts, and organic compound	Exclusion of dissolved salts

2013; Shih 2005). The attributes of these four procedures are abridged in Table 6.7 (Bottino et al. 2009; Van der Bruggen et al. 2003).

Utilizing films with pore measures somewhere in the range of 0.1μm to 10μm, MF- (microfiltration) alone can't be utilized to expel arsenic species from polluted water. Accordingly, the molecule size of arsenic-bearing species should be expanded preceding MF; the flocculation and coagulation is the most prevalent procedures for this presence (Singh et al. 2015). Another investigation (Han et al. 2002), utilized the combination of flocculation and MF in which ferric sulfate (Fe$_2$(SO$_4$)$_3$) and ferric chloride (FeCl$_3$) was used as flocculants to eliminate arsenic contaminations especially from drinking water. Results confirmed that flocculation by ferric materials before MF prompts successful arsenic adsorption onto the flocculants and complete expulsion of arsenic in the penetrate. In any case, the water pH level and the resemblance of different particles are central point influencing the productivity of arsenic species immobilization which is a weakness of this system. Particularly, when managing As (III) evacuation as it has an independent charge in the scope of pH 4–10 (Shih 2005). Since As (V) is contrarily charged in pH scope of 4–10, it can attach on the surface to form a complex which would have a productive As (V) evacuation. Along these lines, to have a powerful system, it should have finalized oxidation of As (III) to As (V).

Similarly, like MF, UF alone isn't a viable system for the arsenic-polluted water treatment because of enormous membrane pores (Velizarov et al. 2004). To utilize this strategy in arsenic expulsion, surfactant-based detachment procedures, for example, micellar-enhanced ultrafiltration (MEUF) can be used (Beolchini et al. 2007; Gecol et al. 2004) such as, cationic surfactant can be added to arsenic polluted water in higher level than the critical micelle concentration (CMC) of drinking water would prompt the arrangement of micelles that can be bonded to the negatively charged component of arsenic. As a result, arsenic expulsion will occur in high level as the surfactants are sufficiently enormous to go through the layer pores. There are many reports already available in removing arsenic species using MFUF. One of the studies utilized a cationic surfactant as a source to investigate the efficiency of arsenic removal (Iqbal et al. 2007). The highest removal efficiency (96%) was observed on hexadecyl pyridinium chloride (CPC) surfactant. However, it was reported that the removal efficiency decreases while decreasing pH of the solution. Besides, in spite of the viable expulsion of arsenic, additional treatment is required with powdered activated carbon (PAC) to remove the excess concentration of surfactant in the solution.

If a molecular weight of the dissolved compound is over 300 g/mol, then NF and RO methods are appropriate for the water filtration (Van der Bruggen et al. 2003). These filtration systems can reduce the contaminant arsenic from water significantly, however there should not be any suspended solids in the feed and arsenic should be in the form of arsenate (Figoli et al. 2010). It was demonstrated that the As (V) removal efficacy surpassed 85% for all type of NF membrane involved in the investigation (Sato et al. 2002), whereas the As(III) removal efficacy was extremely low. This study was sustained by the discoveries of Uddin et al. (2007), who showed that without oxidation of arsenite to arsenate, NF can't fulfil to the maximum concentration level of arsenic present in water. Similar situation for RO as well, as appeared in many investigations (Brandhuber and Amy 1998).

Diatomaceous earth (DE) is an another filtration system works similar like other membranes, however, technically it does not come under membrane filtration system (Bhardwaj and Mirliss 2005). DE filtration system contains pasty sedimentary material which has microscopic water plants (fossil-like skeletons) that is called as diatoms (Logsdon et al. 2002). Diatoms are made up of porous structure with a small opening of around 50 nm in radius and total size of the diatoms in the range of 5–100µm. The consolidated impact of high porosity and small pore sizes present in DE makes the best channels used to evacuate tiny particles in the water filter system (Bhardwaj and Mirliss 2005; Logsdon et al. 2002). Moreover, this kind of channel is tasteless, odourless, and chemically inactive which makes it safe use in drinking water filtration. There was a method developed by Misra and Lenze utilized the combination of mixed hydroxides and DE filter system for the effective removal of arsenic and other heavy metals present in the drinking water (Misra and Lenz 2008). This development was performed in the lab scale with a reagent loading of 1000 mg/L and 100 ppb (µg/L) as an initial concentration of arsenic and showed 90% of removal efficiency. However, the pH alteration, a long conditioning time and reagents are required which are considered as a drawbacks (Misra and Lenz 2008).

6.4.4 Ion Exchange and Adsorption

In a process where solid materials are used as a medium to remove any substances from liquids are called "Adsorption process" (Singh et al. 2015). Basically, in the adsorption process, the substances are isolated from one phase and collected on the surface of another phase. This adsorption procedure mainly occurs by either electrostatic force and/or van der Waals force between the adsorbent surface ions and adsorbate molecules. Therefore, it is highly imperative to find the surface properties of adsorbent first (e.g., polarity and surface area) before to consider in adsorption process (Choong et al. 2007).

There are many different adsorbents has been utilized for the adsorption process as appeared in Table 6.8. The absorbents are coal, activated carbon, fly ash, kaolinite, red mud, montmorillonite, zeolites, goethite, iron hydroxide, chitosan, zero-valent iron, titanium dioxide, cation-exchange resins and activated alumina. Among these absorbents, the Table 6.8 outlines that iron-based adsorbents in adsorption process is an emergent strategy for the removal process of arsenic-polluted water. The iron based adsorbents works very well because of the higher affinity towards inorganic arsenic (Gupta et al. 2012). Iron could expel arsenic from water either by behaving as a reductant, acting as a absorbent, contaminant-immobilizing or co-precipitant (Mondal et al. 2013).

There are many reports available in adsorption process and considered as the extensively used method for the removal of arsenic from contaminated water because of its multiple advantages such as high efficiency in removing arsenic (Singh and Pant 2004; Mohan and Pittman 2007), easy handling and operation (Jang et al. 2008), cost-viability (Anjum et al. 2011), and no sludge generation (Singh et al. 2015). However, the maximum adsorption efficiency relies upon the concentration of system and operating pH. At lower pH, As (V) adsorption is good, while, for As (III), most extreme adsorption can be acquired between pH 4 and 9 (Lenoble et al. 2002). In addition, the competing ions, for example, silicate, and phosphate, also present in the arsenic contaminated water which should be consider for the adsorption locations (Giles et al. 2011). Besides the conditions of system, the viability of arsenic adsorption can likewise be delayed by the kind of adsorbent itself. In Table 6.8, various adsorbents have been reported for the expulsion of arsenic with different adsorbent loading. However, traditional adsorbents shows intermittent pore structures and low explicit surface areas, prompting lower adsorption limits. Absence of selectivity, feeble connections with metallic ions and recovery challenges can likewise limit the capacity of these sorbents in bringing arsenic concentration down to levels lower than MCL (Habuda-Stanić and Nujić 2015; Samiey et al. 2014).

Table 6.9 gives brief depictions of arsenic removal technology, where the correlation among those ordinary arsenic removal innovations along with their removal viability and operational expenses are outlined in Table 6.10 (Mohan and Pittman 2007; Tuutijärvi 2013).

Among the previously mentioned innovations talked about up until now, adsorption is considered as the best for the treatment of water and wastewater as far as accommodation, benefit, plan and critical arsenic expulsion effectiveness. Also, it is most reasonable because of the accessibility of a wide scope of adsorbents.

Table 6.8 Similar assessment of various adsorptive media recently utilized for arsenic removal

Adsorbents	pH*	Adsorbent Dose Used (g/L)	Arsenic Concentration Range (mg/L)	Adsorption Capacity (mg/g)		Source
				As(III)	As (V)	
Fly ash	4	5	1–100	–	2.40	Lorenzen et al. (1995)
Red mud	7.25	20	0.1–50	0.884	0.941	Altundoğan et al. (2000)
Kaolinite	5	100	0–200	–	0.86	Mohapatra et al. (2007)
Goethite	6–8	1.6	0.1–200	28	7	Lenoble et al. (2002)
Alumina	7.6	1–13	1–100	0.18	–	Singh and Pant (2004)
TiO_2	7	1	1–200	32.4	41.4	Bang et al. (2005)
Fe_3O_4-GO	7	0.1	0–550	85	38	Yoon et al. (2016)
Fe_3O_4– reduced GO	7	0.1	0–550	57	12	Yoon et al. (2016)
NZVI-Reduced GO	7	0.4	1–15	35.83	29.04	Wang et al. (2014)
GNP-Hydrous Cerium oxide	4	0.1	1–80	–	62.33	Yu et al. (2015)
CeO_2-GO	–	0.5	0.1–200	185	212	Sakthivel et al. (2017)
Fe_3O_4-MWNT based electrodes	–	0.08	200–400	39	53	Mishra and Ramaprabhu (2010)
NZVI/AC	6.5	0.5–6	2	11	15	Zhu et al. (2009)
Mg-Al double hydroxide/GO	5	0.5	0.1–150	–	180.26	Wen et al. (2013)
GN-α-FeOOH Aerogel	8–9	0.05	1–16	13.42	81.3	Andjelkovic et al. (2015)
rGO-Fe_3O_4-TiO_2	7	0.2	3–10	147.05	–	Benjwal et al. (2015)
GO-$MnFe_2O_4$	1–2	0.2	10–50	–	240.3	Huong et al. (2016) and Luo et al. (2013)
β-FeOOH@GO-COOH	6.5	1	1–200	77.5	45.7	Chen et al. (2015)
$FeMnO_x$/RGO		0.2	0.2–7	22.17	22.05	Zhu et al. (2015)
GO–ZrO(OH)$_2$	7	0.5	2–80	95.15	84.89	Luo et al. (2013)
Eu-MGO/Au@MWCNT	7	3	0–65	320	298	Roy et al. (2016)

Table 6.9 Narratives of arsenic removal technologies

Technology	Description
Oxidation	Arsenite oxidized to arsenate by mixed oxidizers through air/unadulterated oxygen, microbiological or photochemical. Regularly applied specialists are chemical oxidizers, for example, chlorine, ozone, chlorine dioxide, sodium hypochlorite, hydrogen peroxide, Fenton's reagent, and potassium permanganate.
Coagulation-flocculation	Utilizations of synthetic chemical to separate arsenic addicted to an unsolvable solid that is precipitated or to adsorb arsenic on additional unsolvable solid that is encouraged (coprecipitation). The arsenic adsorbed solid is then expelled from the fluid stage by separation. The pH of the media in this procedure exceptionally impacts the effectiveness of expulsion. Usually utilized synthetic compounds are manganese sulfate, ferric salts, and alum.
Membrane filtration	Isolates contaminants from water passing through semi-porous hindrance or layer. The layer enables a few elements to pass, while stops others. Categories of layer filtration incorporate ultrafiltration (UF), microfiltration (MF), invert assimilation (RO) nanofiltration (NF).
Ion exchange and adsorption	Exchange ions controlled electrostatically on the outside of a solid surface with ions of a comparative charge in the liquid medium. The ion exchanged solution is generally stuffed into a section and polluted water is pass through the section to remove the contaminants.
	Pollutants interact with the surface of an adsorbent, subsequently diminishing their concentration in the liquid. The adsorption media is typically packed in a section and polluted water pass through the segment to adsorb the contaminants. Conventionally utilized adsorbents are commercial and engineered activate carbons or cost-effective adsorbents.

6.5 Adsorption Mechanism

Various investigations are described in the literature on the adsorption of metal ions by utilizing various adsorbents. The component of adsorption differs as indicated by the metal species and kind of adsorbent utilized up until this point. Late examinations uncover that the instruments in procedures of metal ion adsorption are of six sorts. The component of adsorption is abbreviated as pursues (Chowdhury et al. 2013; Kwok 2009; Wang et al. 2010):

a) *Transport over the cell membrane*: This concept is related with cell digestion by living adsorbent. The procedure is normally intervened by a similar system as utilized by living organs in the digestion of basic particles, for example, magnesium, potassium, and sodium. However, adsorption might be associated by the presence of heavy metal ions of a similar charge and ionic radius which ought to be considered cautiously.

b) *Complexion*: The metal ion adsorption from contaminated solution may happen however complex development on the cell surface after the cooperation between the metal and active sites. Metal ions can tie with a monovalent ligand or through chelation with polyvalent ligands. The cell surface complexion is on

Table 6.10 Advantage and disadvantages of arsenic removal technologies

Technology	Advantage	Disadvantage	Removal %	Relative cost
Oxidation method	Easy method; in situ arsenic evacuation; oxidizes other inorganic and natural constituents present in water.	The slow procedure, For the most part, evacuates arsenic(V) and quicken the oxidation procedure	–	low
	Oxidizes different pollutions and kills organisms; generally straightforward and quick procedure; least residual mass	Productive PH control and oxidation step is required		
Coagulation/ electrocoagulation/ coprecipitation:	Solid powder synthetic substances are accessible; straight forward in activity; successful over a more extensive scope of pH	Produces dangerous slimes; low expulsion of arsenic; pre-oxidation might be required	20–90	Low
	Basic synthetic substances are accessible; more effective than alum coagulation on a weight premise	Medium expulsion of arsenic(III); sedimentation and filtration required		
	Synthetic substances are accessible economically with ease	Sulfate particles impact productivity; slime development; rearrangement of pH; auxiliary treatment is regularly required		
Membrane filtration	Well–characterized and high evacuation proficiency	Preconditioning, high water dismissal, high capital expense	≥ 90	High
	No harmful strong waste is created	Cutting edge activity and upkeep, produce a high measure of rejected water which is dangerous		
Ion-exchange	Well-characterized medium and limit; pH autonomous; selective particle explicit gum to evacuate arsenic	Cutting edge activity and support; recovery makes muck transfer issue; As(III) is hard to evacuate; the life of resins is low	≥ 90	High
Absorption	Effectively accessible; straightforward in activity; powerful for a household treatment plant	Needs substitution after 4–5 recovery	≥ 90	Low

the idea of surface charge created from the amphoteric surface locales, which are fit for the response with adsorbing cationic or anionic species to frame surface complex.

c) *Coordination*: The binding of metals to the ligands is depends on the arrangement of a coordination compound. For this situation, the metal goes about as lewis acid, i.e., end to gain enough electrons to arrive at a latent state, and the ligand goes about as a lewis base, i.e., has electron pair that can be imparted to the metal. Coordination at that point is a lewis corrosive – lewis base balance process

d) *Ion Exchange*: This one plays a significant job in adsorption and modeled the binding of substantial metal ions and protons as an element of metal concentration and ideal pH. The light metal particles presence in cell divider and film, for example, potassium, sodium, calcium, and magnesium can likewise be exchanged with the reasonable metal cations.

e) *Chelation*: Chelation happens when ligand structures facilitate bonds with metal ions through more than one sets of shared electrons, in this manner framing a ring structure. Contingent upon the necessity for electrons of the metal and the development of the ligand, there can be a sharing of up to eight electron combines between a single metal ion and ligand.

f) *Microprecipitation*: Microprecipitation might be either reliant on the cell digestion or independent of it. In the previous case, the metal adsorption from the solution is frequently with an active resistance arrangement of microorganism. They respond within the sight of a harmful metal, delivering mixes which support the precipitation procedure. For the situation where microprecipitation isn't subject to cell digestion, it might be a result of the compound cooperation among metal and cell surface.

6.6 Adsorption Theory

Adsorption phenomena are renowned for an extremely lengthy timespan. Sorption maybe used as a treatment procedure to expel exceptionally unwanted mixes from feedwater. It includes the detachment of unwanted mixes from the fluid stage, and compounds bind to a surface of the solid material. The binding to the surface of materials is mostly weaker and its changeable. The two extensive characterizations of sorption are Chemisorption and physisorption. Chemisorption is the development of solid bonds among adsorbate particles and explicit surface areas, otherwise called active destinations. Accordingly, it is fundamentally utilized to quantify the surface-active destinations, that take part in advancing (catalyse) compound responses. Electrostatic and Van der Waals forces between adsorbate particles and also the molecules that make the adsorbent surface outcome in physisorption. Along these lines, surface properties, for example, surface region and extremity assume a critical job in adsorbent analysis. The issue of recognizing chemisorption and physisorption is fundamentally equivalent to that of recognizing physical and chemical interaction in normal. A totally sharp differentiation is uncommon, and transitional case exist.

The main considerations influencing sorption are the adsorbate's concentration and its nature, the solution pH and temperature, the existence of competing solutes, and adsorbent's properties, for example, size, pore size and surface area. A porous strong adsorbent is exceptionally vital within the sorption method. Adsorbents are often grouped as porous and nonporous material. Comparatively, nonporous adsorbents have lower active exterior adsorptive surfaces; such materials are embodying glass, mud and steel dabs. Other hand, permeable adsorbents generally have enormous inside adsorptive surfaces. Especially, a portion of the significant adsorbent attributes influencing isotherms are active surface areas, pore volume and pore diameter. In case of nonporous materials, adsorption is relative to the available active surfaces in the adsorbents. In any case, the porosity of adsorbents isn't the central impact on sorption limit (Rice et al. 2012; Lin and Wu 2001).

There are three specific adsorption mechanisms available: equilibrium, kinetic and steric mechanisms (Do 1998; Lee et al. 2004). Generally, thermodynamic is related to the steric mechanism, because it relates to the isosteric heat produced in an adsorption process for a particular amount of adsorbents. In fluid stage sorption frameworks, sorption of solute particles is commonly combined via the water desorption, and accordingly, moderately small measures of steric heat are developed. For sorption equilibria that pursue a Langmuir design, the isosteric heat of adsorption is steady as a result of the suggested vigorous homogeneity of the adsorbing surface. In this case, kinetic mechanism and the equilibria isotherm are increasingly significant for fluid stage adsorption frameworks.

The principle characterization of specific adsorption procedure are adsorption isotherms and kinetics (for example adsorption equilibria and the rate of adsorption). These two adsorption principles are discussed further in detail.

6.6.1 Adsorption Isotherms

The sorption equilibrium for a specific adsorbate-adsorbent framework is so-called an adsorption isotherm since it is the conveyance of a solute among the fluid stage and the adsorbed stage at a predefined temperature.

The sorption process occurs based on any one of the following adsorption isotherms, i.e., Brunauer Emmet and Teller (BET), Freundlich, Polanyi, Dubinin and Raduskevich (D-R), and Langmuir, The proper isotherm model for a specific part relies upon the attributes of the framework. The active heterogeneity or the adsorptive surface uniformity is a significant factor in determining an appropriate isotherm model for a specific adsorbate.

In case of single-solute sorption, the Langmuir and the Freundlich are the typical isotherm models (Pontius and Association 1990; Ruthven 2006). The accompanying surely understood experimental and applied Freundlich equation clarifies adsorption information sensibly well:

6 Metal Oxides for Removal of Arsenic Contaminants from Water

$$q_e = KC_e^{1/n} \tag{6.1}$$

The linear form of equation is given below:

$$\log q_e = \log k + \frac{1}{n} \log C_e \tag{6.2}$$

where qe equilibrium surface (unit mass of adsorbate/mass of adsorbent), Ce solution concentration, 1/n and K are the specific constants for a given framework. the unit for K is dictated by the units of C_e and q_e. but 1/n is unitless. K expresses the adsorption capacity for the adsorbate and 1/n denotes an adsorption quality. For fixed estimations of 1/n and Ce, higher K would estimate higher qe. On the other hand, with fixed estimations of Ce and K, smaller 1/n would estimate the stronger adsorption bond. If 1/n turns out to be extremely smaller, the limit will be independent of Ce, and the isotherm model plot moves toward the flat level; the estimation of qe is then fundamentally steady, and the isotherm is named irreversible. If the estimation of 1/n is enormous, the adsorption bond is weak, and the estimation of the qe varies with little changes in Ce.

Furthermore, the Freundlich isotherm model depends on the speculation that the adsorbent has a heterogeneous surface made of various classes of adsorption site. The linearized Langmuir condition is given below.

$$q_e = \frac{q_{max} b C_e}{1 + b C_e} \text{ or } \frac{1}{q_e} = \frac{1}{q_{max} b C_e} + \frac{1}{q_{max}} \tag{6.3}$$

where q_{max} and b are constants. q_{max} denotes the most extreme estimation of q_e that could be accomplished as C_e is expanded. The consistent q_{max} relates to the surface concentration. The consistent b is identified with the adsorption energy and increments by the expansion in adsorption bond quality. The fundamental presumption of the Langmuir isotherm is that solutes adsorption happens at clear homogeneous places and forms a monolayer structure on the surface.

6.6.2 Kinetic Mechanisms

The sorption kinetics is the most significant factors in assessing the proficiency of sorption and in deciding the size of water treatment unit developments. To measure the adsorption kinetics and recognize the adsorptive performance, the pseudo first-order and second order equations are generally utilized (Ho and McKay 2000). A straightforward dynamic examination of adsorption is the pseudo first-order equation and it is given below.

$$\frac{dq_1}{dt} = k_1(q_e - q_t) \tag{6.4}$$

where k_1 (1/min) is the rate constant of pseudo first-order adsorption, qc is the measure of metal ion adsorbed at equilibrium and the unit is mg/g. q_t is the measure of metal ion on the outside of the sorbent at given time t (min) and the measurement unit is mg/g. By applying the limit qt = 0 at t= 0, above equation converts to

$$\log(q_e - q_t) = \log q_e - k_1 t \tag{6.5}$$

on the other hand, a pseudo second-order equation was described lately to describe the adsorption kinetics and the equation is given below

$$\frac{t}{q_1} = \frac{1}{k_2 q_e^2} + \frac{1}{q_e} t \tag{6.6}$$

where k_2 is the rate constant of sorption and the measure unit is g/mg min and h is the initial adsorption rate and measure unit is mg/g min. When time becomes zero (t – 0), h can be mentioned as

$$h = k_2 q_e^2 \tag{6.7}$$

The initial adsorption rate (h), the equilibrium adsorption limit (q_e), and the pseudo second-order rate constant (k_2) could be measured experimentally from the intercept and slop of the plot of t/q vs t.

To find the diffusion state of sorbate on sorbent, the rate constant for intraparticle diffusion (k_{id}) is provided elsewhere (Peak and Sparks 2002; Namasivayam and Ranganathan 1995). The equation is given below:

$$q = k_{id} t^{1/2} \tag{6.8}$$

The sharp linear portions usually denote intraparticle diffusion within the sorbent, while the plateaus are attributed to the equilibrium.

6.6.3 Type of Adsorption

Column and batch activities are basically applied to decide the execution of adsorbents in adsorption frameworks. Batch activities are typically performed to assess the capacity of a material to adsorb and the adsorption limit of the adsorbent (Cui et al. 2012). The information got from batch activities are, constrained to a research scale and hence try not to give information which can be precisely applied in household and industrial frameworks. Column tasks, then again, give information

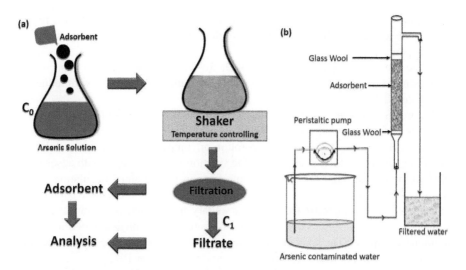

Fig. 6.3 Schematic diagram of (**a**) batch test, and (**b**) column study

which could be productive used in household and Industrial areas (Mohan and Pittman 2007). The batch and column tasks are schematically exhibited in Fig. 6.3a and b, respectively.

6.6.4 Factors Affecting Adsorption

Various components influence the adsorption procedure, such as, specific qualities of adsorbent, adsorbate nature, ionic focus, natural issue, temperature and pH. The impact of various elements is examined below.

One of the important qualities is surface are which is influencing the adsorption limit since the adsorbent's adsorption limit is relative to the specific surface area, for example, the sorption of a specific component enhances by an expansion of specific surface area (Faust and Aly 2018). The specific surface area of non-permeable adsorbent enhances with the reduction in size of a particle (Sharma 2001). Subsequently, the adsorptive limit of adsorbent raises with a decrease in particle size.

An adsorbent composition may be valued basic data in the starting point when evaluating the appropriateness of adsorbent for heavy metal evacuation. The surface hydroxides in the case of iron oxide can adsorb both anions and cations present in any water body (Sharma et al. 2002). The degree of adsorption relies upon the density and type of the adsorption and the adsorbing material's nature.

Adsorption is affected by physical and chemical properties of the substance that needs to be removed, i.e., adsorbate. These are surface charge, dissolvability, adsorbate ion size, ionic radii, and atomic weight. Dissolvability/solubility is the efficient property influencing the sorption limit. The higher dissolvability shows a more robust solvent-solute interaction of liking and the degree of sorption is relied upon to be less due to the need of breaking the solvent-solute links before sorption can happen.

There are few things being interrelated, for example the sorption capacity of transition metal improves with increasing atomic number and reduces with reducing ionic size (Sharma et al. 2002).

Natural groundwater comprises a blend of numerous interfering ions and organic mixtures instead of a solitary one. These interfering ions and compounds may act independently or may increase the adsorption or compete with each other. Adsorption of one individual ion is delicate to the solution's ionic strength with adsorption limit expanding through diminishing ionic fixation (Melia and Coagulation 1972; Salomons et al. 2012; Faust and Aly 2018). Common restraint could be anticipated to happen if sorption is restricted towards a solitary and a couple of atomic layers, the solute's adsorption affinities don't change and there is no interaction between solutes to improve sorption. The level of joint restraint is identified with sizes, fixations, and sorption affinities of interfering ions (Melia and Coagulation 1972).

Ligands, generally anions, could influence the sorption of heavy metal ions onto surface of oxide materials in multiple different ways which are given below, (Sharma et al. 2002).

- complex formation of metal-ligand in solution and adsorb very little
- The ionic species might interact obliquely on the surface of the oxide material and change the electrical properties
- enhanced metal-ion sorption may so happened due to the formation of strong metal-ligand complex
- The complex formation may have no impact on metal-ion sorption

Subsequently, the metal-ion sorption on transition metal oxide surface may increase/decrease/unalter contingent upon the anion's nature present in the solution and complex formed with ligands,

The natural Organic matter (OM) derived directly from plants or microbial build-ups. In their unique or artificially adjusted structure, the debris of OM delivered ashore are accessible to be moved from the dirt to the hydrosphere. The movement often happens because of precipitation that keeps running off or permeates through the dirt column conveying particulate organic matter (POM) and Dissolved Organic Matter (DOM) to groundwater, lakes, seas, and streams. Humic substances (HS) which are a type of natural OM, other than considering as a proton acceptor for charge balance in watery frameworks, additionally respond with metal ions through ionic or covalent bond formation. The arrangement of some of this relationship among metal ions and humic corrosive relies upon the underlying condition of the metal ions and the HS and their fixations (Sharma et al. 2002).

The porosity and size of OM additionally influence the adsorption. Particularly, Organic compound's solubility in water body diminishes with expanding chain length and expanding molecule's size (Faust and Aly 2018).

Since the majority of the organic mixes in groundwater are typically liquified, their essence extensively influences the sorption of metal ions present in the water body. Generally, when many absorbable substances are available, and sorption destinations are restricted, focused adsorption happens and the adsorption of certain ions might be constrained via the absence of energetic sites (Sharma et al. 2002; Salomons et al. 2012).

Commonly, there is a sensational increment or decline in anion and cation sorption due to pH increments. It has been seen that, for guaranteed adsorbent/adsorbate proportion, with limited pH extend, in which the anion and cation sorption on hydrous oxides increments to 100%, yielding regular pH vs rate adsorption curve identified as adsorption pH edges. When the adsorbent/adsorbate proportion is expanded, the partial adsorption at a given pH is decreased and thus the cation "pH edge" movements to one side (Abdus-Salam and Adekola 2005).

6.7 Analysis and Modelling of Column Study

The arsenic to be expelled from water in a fixed–bed column holding adsorbent can be determined by breakthrough curves (Ranjan et al. 2009a; Carabante 2012; Bansal and Goyal 2005). The ideal opportunity for breakthrough presence and state of the breakthrough curve are significant attributes for deciding the activity and the dynamic reaction of a sorption column (Fig. 6.4). Consequently, breakthrough curve that is the proportion of conclusive arsenic concentration; C_e and introductory arsenic fixation; C_0 i.e., C_t/C_0 as a function of time (t) vs time was plotted. The breakthrough time (t_b, the time at which the arsenic fixation in the emanating came to less than 50μg/L) (Ranjan et al. 2009b; Guo and Chen 2005; Pennesi et al. 2012) was utilized to assess the breakthrough curves.

6.8 Metal Oxide/Composites for Removal of Arsenic from Water

In the last decades, progresses in nanoscience and nanotechnology have made ready for the advancement of different nanomaterials for the remediation of polluted water (Mondal et al. 2013). Because of their higher reactivity, specific surface area with porosity, and higher specificity, nanomaterials have been extensively considered as an excellent adsorbents of pollutants, for example, lead, arsenic, cadmium, and chromium, from drinking water (Hristovski et al. 2007). Titanium-based nano adsorbents, Carbon contains nanocomposites, iron-based nanomaterials, and other metal- metal oxide nanoparticles are the most broadly utilized and explored nanomaterials for the remediation of arsenic-polluted water (Qu et al. 2013; Hua et al. 2012; Hristovski et al. 2007). Table 6.11 exhibits a short analysis of the similar assessment of few nano-based adsorbents utilized for arsenic expulsion.

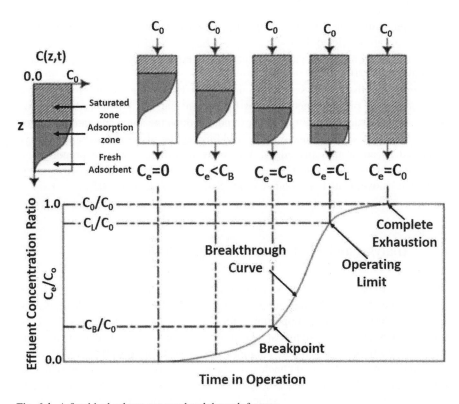

Fig. 6.4 A fixed-bed column process breakthrough features

6.8.1 Carbon Based Composite Materials

Carbon based nanocomposite materials have been described as an effective adsorbent to remove different organic chemicals as well as heavy metals from water (Pan and Xing 2008; Li et al. 2003; Lu and Liu 2006). Roy et al. (2013a), reported that 34.22% of As(III) contamination was removed from the initial concentration of 542μg/L by utilizing 1 g/L concentration of carbon composite material. Although, the carbon-based composite materials able to reduce the arsenic concentration below the EPA level, but it works only in low initial concentration of arsenic in water.

Additionally, Carbon-based material can be functionalized to improve their removal efficacy towards heavy metals present in water. Velickovic et al. (2013) prepared the functionalized carbon-based materials with polyethylene glycol (PEG) to remove arsenate and other contaminated metal ions from drinking water. However, it was determined that the removal efficacy of metal ions on PEG-Carbon based material is mainly depends on the solution pH. In addition, with 10 mg/L of initial centralization at pH 4, the most extreme adsorption limit of arsenate on this functionalized carbon material was shown as 13.0 mg/g.

Table 6.11 A similar assessment of different nano-adsorbents for arsenic expulsion

Adsorbent	Adsorbent Dosage (g/L)	Initial As Conc./ Range (mg/L)	Max. As (V) Adsorption Capacity (mg/g)	Max. As (III) Adsorption Capacity (mg/g)	References
Ceria-GO Composite	0.5	200–0.01	212	185	Sakthivel et al. (2017)
Iron doped TiO$_2$	4	100–10	20.4	–	Liu et al. (2012)
Ceria nanoparticles	5	100–1	18.02	18.	Pena et al. (2005)
Magnetic graphene oxide	0.4	60–10	59.6	–	Sheng et al. (2012)
GNP-HCO composite	0.1	80–1	62.33	–	Yu et al. (2015)
Fe(III)/La(III)-chitosan	–	1–0.05	–	109	Shinde et al. (2013)
Manganese-modified activated carbon fiber	0.80	80–1.4	23.77	–	Sun et al. (2013b)
Hydrous Cerium Oxide (HCO)	0.1	100–1	107	170	Li et al. (2012)
ZrPACM-43	13	100–10	–	41.48	Mandal et al. (2013)
Fe$_3$O$_4$-MWNTs based electrodes	0.08	400–200	53	39	Mishra and Ramaprabhu (2010)
TiO$_2$ nanoparticles	1	90–5	46.08	31.35	Deedar and Aslam (2009)
Zirconium dioxide impregnated GAC	0.03	0.12	8.95	–	Sandoval et al. (2011)
CuO nanoparticles	2	100–0.1	22.6	26.9	Martinson and Reddy (2009)
Fe$_3$O$_4$–graphene–MnO$_2$ composites	0.50	10–0.01	12.22	14.04	Luo et al. (2012)
Magnetite-reduced graphitic oxide	0.2	7–3	5.83	13.10	Chandra et al. (2010)
TiO$_2$-coated CNT filter	0.31	10–0.1	14.1	13.2	Liu et al. (2014a)
Cu$_2$O-reduced graphene oxide	0.09	2–0.25	4.80	–	Dubey et al. (2015)
Fe$_3$O$_4$-graphene composite	2	1–0.1	–	0.313	Guo et al. (2015)

But, in over-all, carbon-based material might not be a superior option for surface activated carbon material as widely inclusive adsorbents. All things considered, Carbon material still illustrate possible in certain applications where just small quantities of adsorbents are required, which infers a lesser amount of material expense. These applications incorporate enhancing steps to expel unmanageable mixes or pre-convergence to follow organic pollutants for investigative uses (Qu et al. 2013).

6.8.2 Titanium-Based Materials

Previously, It was studied to determine the efficiency of titanium dioxide (TiO_2) toward the exclusion of arsenic in groundwater and photocatalytic oxidation process of As (III) species (Pena et al. 2005). The equilibrium condition of arsenic removal was achieved in 4 h using TiO_2 nanocrystalline materials, whereas, the equilibrium condition for commercial TiO_2 material was reached within 1 h. Moreover, the maximum adsorption limit was acquired utilizing TiO_2 nanocrystalline materials, which could be due to the maximum surface area than the commercial TiO_2 material. Using the nano adsorbent, more than 80% of arsenic species was removed at an equilibrium concentration of arsenic (45 g/L). As far as oxidation, TiO_2 nanocrystalline was likewise appeared as an effective photocatalyst and similar like arsenite was totally changed over to arsenate in 25 min under the light and dissolved oxygen completely.

Additionally, different titanium-based nanoadsorbents being utilized in the removal process of arsenic in which Hydrous Titania ($TiO_2 \times H_2O$) nanoparticles are very special. This hydrous nanoparticles offer the benefit of being compelling adsorbent for arsenite without the requirement for oxidation to arsenate or no need of pH change when the adsorption practice (Guan et al. 2012). In addition, hydrous Titania nanomaterials were studied for the removal of As (III) from the synthetic groundwater prepared at the laboratory and natural groundwater (Xu et al. 2010). 83 mg/g of arsenite was removed at neutral pH and 96 mg/g was removed at pH 9 which shows the use of $TiO_2 \times H_2O$ as a successful material with minimal effort, and single-step procedure for the arsenic-polluted water treatment. But, in view of their size, nanoparticle dispersion into the environment might happen. Thus, converting the nanoparticle to micron size particle using spray dry process or loading of these nanoparticles onto supporting material like porous material is required.

Previous report directed by Lee et al. (2015) demonstrated an improved arsenate expulsion in water utilizing Ti-incorporated basic yttrium carbonate (BYC) which showed the maximum adsorption capacity of 348.5 mg/g at pH 7 that is 25% higher than either titanium hydroxide or BYC. This higher adsorption capacity is due to the improved surface charge and specific surface area, i.e., point of zero charge (PZC): 8.4 and 82 m^2/g, respectively. In addition, Ti-incorporated BYC likewise showed high adsorption limits in a more extensive pH range (pH 3–11) and worked very well in presence of co-existing anions (e.g., bicarbonate, silicate, phosphate) without scarifying the adsorption limit. However, this examination didn't provide details regarding the capability of Ti-incorporated BYC to remove arsenite.

6.8.3 Iron-Based Materials

There are several nanomaterials used for the removal of arsenic- polluted water in which iron-based nanomaterials are extensively investigated, particularly iron oxide nanoparticles (i.e., Fe_2O_3 and Fe_3O_4) and zero-valent iron nanoparticles (nZVI). The valance state of iron in these materials impacts their capacity to remove heave metal contaminants (Tang and Lo 2013). A few mechanisms are engaged with these removal procedure (Fig. 6.5).

(a) Zero Valent Iron nanomaterial (nZVI)

A few examinations have exhibited that the utilization of nZVI is a compelling innovation for changing contaminations into their nontoxic structure (Xi et al. 2010). For example, dying components could be adsorbed adequately to functionalized nZVI that showed the greatest adsorption limit of 191.5 mg/g for one dye (Zhang et al. 2011). For this situation, adsorption was the consequence of donor and acceptor bonds happened in the reaction mixture between the – OH group on the objective component and the functional groups such as – NH_2 on the nZVI surface. With respect to heavy metals, co-precipitation and adsorption are commonly reported or accepted mechanism engaged with evacuation by nZVI (Habuda-Stanić and Nujić 2015). As schematically appeared in Fig. 6.5, these mechanisms happen due to the formation of iron oxide shell on nZVI in contact with water or air. Evacuation of arsenic is a very well-studied model (Kanel et al. 2005).

X-ray photoelectronic spectroscopy (XPS) was used to examine the As (III) immobilization system utilizing nZVI (Ramos et al. 2009). Because of the core-shell formation of nZVI material, it has been demonstrated that both oxidative and reductive mechanism happens in nZVI for the heavy metal removal application. It was described that the highly reducing core metal and thin amorphous surface layer of iron hydroxide that aides in the oxidation and coordination of As (III). Although, many reports indicated the advantage of nZVI usage in heavy metal removal treatment, other reports described the disadvantage of nZVI adsorbent with regards to the material synthesis (Litter et al. 2010).

(b) Iron oxide Nanomaterials

Iron oxide nanomaterials are progressively getting to be predominant in the field of arsenic expulsion due to the higher capacity to expel arsenic compare to the micron-sized materials This superior adsorption capacity with respect to metals is due their higher surface to volume rations (Mohmood et al. 2013). Additionally, iron oxide nanomaterials have magnetic properties that enable them to be advantageously isolated from water (Sharma et al. 2009).

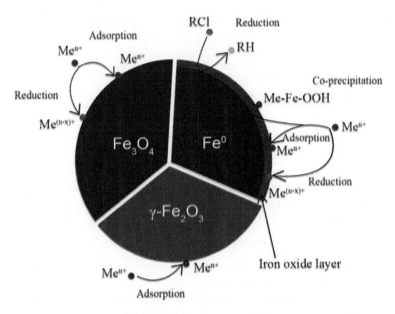

Fig. 6.5 Arsenic removal mechanism of Fe_2O_3, Fe_3O_4, and nZVI. (Reprinted with permission from (Tang and Lo 2013))

Tang et al. reported the utilization of ultrafine α-Fe_2O_3 nanoparticles to treat arsenic contaminated synthetic and natural ground water systems (Tang et al. 2011). Kinetic studies ascribed that As (V) and As (III) expulsion by α-Fe_2O_3 materials was very quick. With a loading of 0.04 g/L of α-Fe_2O_3 and initial As (III) concentration of 0.115 mg/L, about 74% of As (III) has been removed within the first 30 mins of interaction. In the case of As (V), the initial concentration was 0.095 mg/L, and 100% expulsion of As (V) has been accomplished when the α-Fe_2O_3 loading was just 50% of that utilized for As(III). The specific surface area of the synthesised material was around 162 m^2/g and the particle size was about 5 nm which showed the higher arsenic removal efficacy at neutral pH. Adsorption capacities with regards to As(V) and As (III) were resolved to be 47 mg/g and 95 mg/g, respectively. In addition, it was demonstrated that the competitive anions of NO_3^-, SO_4^{2-}, and Cl^- in the water has a negligible negative impact on arsenic removal process.

The action of magnetite nanoparticles (Fe_3O_4) to remove arsenic-polluted water was reported earlier (Chowdhury and Yanful 2011). The synthesised adsorbent had the average size of 20 nm and the specific surface area of 69.4 m^2/g. Results demonstrated that removal of arsenic by Fe_3O_4 nanomaterial is mainly depends on pH of the solution, initial concentration of arsenic, contact time, adsorbent concentration and PO_4^{3-} concentration. Maximum removal capacity for both arsenic species was accomplished by the initial concentration of 2 mg/L at pH 2. Arsenite adsorption did not vary by changing the pH from 2 to 9, whereas, arsenate adsorption decreased very quickly in higher pH (above 7). In addition, maximum arsenic adsorption limits were determined to be 8.8 mg/g and 8 mg/g for As(V) and As(III), respectively. The impact of phosphate ion on arsenic evacuation was contemplated and results demonstrated that expulsion rates reduced with expanding phosphate concentration. This outcome is as per the investigation directed by Roy et al. (2013a) the adsorption rate was decreased for As (V) and As(II) by 25% and 13%, respectively with the existence of phosphate at a concentration of 0.5 mg/L.

Mayo et al. additionally examined As(V) and As(III) expulsion utilizing magnetite nanocrystalline particles (Mayo et al. 2007). Their outcomes affirmed that the size of the nanoparticles significantly affects their adsorption conduct. Adsorption capacities with regards to both arsenic species expanded around 200% when size of the particle was reduced to 12 nm. Researchers (Hristovski et al. 2007) examined the As (V) adsorption through batch tests utilizing 16 commercially accessible nanoparticles. Many of these nanostructured materials adsorbed more than 90% of As (V) from the contaminated water, with ZrO_2, TiO_2, NiO, and Fe_2O_3 nanoparticles performance was greatest. These nanoparticles demonstrated the highest expulsion effectiveness, surpassing 98%, aside from ZrO_2 in contaminated water.

6.8.4 Other Metal-Based Materials

(a) Cerium oxide (CeO_2) nanoparticles

Feng et al. examined batch analyses for the removal of arsenic on CeO_2 nanomaterials (Feng et al. 2012). Obtained results demonstrated that arsenic evacuation by the CeO_2 nanomaterial is depends on the solution pH. For As (V), adsorption expanded as soon as the pH expanded upto 6, and after that reduced as pH kept on expanding past 6. Comparative patterns were detected for arsenite, even though arsenite removal rate was seen to persistently increases while increasing the pH from 1 to 8. Additionally, Langmuir adsorption isotherms exposed that the adsorption limits for the CeO_2 nanoparticle is 18.15 mg/g, and 17.08 mg/g at 50, and 10 °C, respectively, showing that arsenic adsorption is more favourable at higher temperatures.

(b) Zirconium oxide (ZrO_2) nanoparticles

Nontoxic, insoluble and chemically stable ZrO_2 nanoparticle is another option for the purification of drinking water (Cui et al. 2013). Cui et al. conducted few experiments with the group of nanoparticles (Cui et al. 2012). Amorphous ZrO_2 (am-ZrO_2) nanoparticles was prepared via hydrothermal synthesis method for arsenic expulsion. Through kinetic experiments, it was demonstrated that by utilizing very low dose (i.e., 0.10 g/L) of am-ZrO_2 nanoparticles, arsenic concentration could be decreased lower than EPA limits within 24 h for As (III) and 12 h for As (V). It was also demonstrated that the adsorption process is viable under neutral pH level and needn't bother with any pre-treatment or post-treatment. Highest adsorption limits for am-ZrO$_2$ nanoparticles were recorded as 32.5 mg/g and 83.2 mg/g for As (V) and As(III), respectively.

6.8.5 Metal Organic Framework

Metal organic frameworks (MOFs) are hybrid material with porous structure that are contained organic and inorganic building blocks associated with one another by coordination bonds (Hasan and Jhung 2015). Generally, the inorganic parts are a single or group of metal ions, in which the regularly utilized components are the transitional metals, for example, Zn^{2+}, Fe^{3+}, and Al^{3+}. Then organic parts, otherwise called linkers which are multidentate organic ligands, that could be cationic or anionic or electrically neutral, (Shen 2013). Carboxylates are the most generally utilized anionic linkers because of their capacity to make metal ionic groups, and therefore, more stable networks forms (Eddaoudi et al. 2001).

Due to their easy synthesis process, high surface areas, tuneable pore sizes and random shapes, coordinative unsaturated sites (CUS), and organic functional groups, MOFs have increased critical consideration in research and manufacturing in the recent years (Eddaoudi et al. 2001; Yaghi et al. 2003). Also, MOFs hybrid materials indicated potential in different fields including gas adsorption (He et al. 2012), hydrogen storage (Langmi et al. 2014), separation of synthetic compounds (He et al. 2012), biomedical and drug delivery (Huxford et al. 2010), catalytic application (Liu et al. 2014b), luminescence (Chandler et al. 2006), magnetism (Kurmoo 2009), and sensors (Chen et al. 2008).

Additionally, adsorption of dangerous substances like heavy metals from groundwater could be one of the potential utilizations of MOFs, despite the fact that their adsorption capacities have been less investigated when contrasted with different materials, for example, zeolites (Jia et al. 2013; Ungureanu et al. 2015). Less utilization in water purification field could be due to lack of stability in water for a more extended time (Low et al. 2009).

In contrast with nanoparticles, MOFs brings two important advantages in the field of adsorption that is, (1) open metal destinations in their structure that are promptly available, (2) Mechanical and thermal stability of MOFs causing them to withstand accumulation issues that are regular in nanoparticles (Zhu et al. 2012). These points of interest, together with their amazingly high pore volume, pore size and specific surface areas (10,450 m^2/g), cause MOFs accomplish better in expelling substantial metals from contaminated water than different permeable adsorbents (Furukawa et al. 2010; Jian et al. 2015).

Zhu et al. (2012) examined As (V) expulsion from groundwater utilizing Fe-BTC (iron and 1,3,5-benzenetricarboxylic) MOF. This MOF attributed high As (V) adsorption limit of 12.3 mg/g, that is around 11 times higher than commercial Fe$_2$O$_3$ and 2 times higher than Fe$_2$O$_3$ nanoparticles. In addition, Fe-BTC adsorbs arsenate in extensive pH range (pH 2–12). Ideal evacuation proficiency was seen under acidic conditions (pH 2-7). Expulsion effectiveness dropped considerably at pH levels over 12, as the stability of MOF is not strong in basic conditions. Also, it was shown that the adsorption of arsenic take place in the interior of MOFs and not on the exterior surface. This clarifies a higher adsorption limit of MOFs in contrast with Fe_2O_3 nanoparticles because it gives progressively inside space. Additionally, MIL-53 (Fe) MOFs was utilized for arsenate removal and showed the higher adsorption capacity (21.27 mg/g) than Fe-BTC (Vu et al. 2015). ZIF-8 (Zeolitic imidazolate framework-8) was also examined to remove arsenic species and showed the adsorption limits of 60.03 mg/g and 49.49 mg/g for As (V) and As (III), respectively, (Jian et al. 2015).

6.9 Disposal of Metal ion Contaminated Materials

The nanomaterials may require disposing when their absorption limit is saturated. For different metals and organics adsorbed nanoparticles, it might be recuperated through burning (Mohan and Pittman 2007). But, for arsenic-adsorbed materials, burning might not be perfect as oxide based arsenics are unpredictable and are effectively discharged to the environment during the ignition procedure, that makes another ecological hazard (Saiz et al. 2014). Hence, the most appealing choice to deal with arsenic-adsorbed nanoparticles is to encapsulate through stabilization-solidification process and transfer to safe landfill discarding (Bystrzejewska-Piotrowska et al. 2009). The initial process, stabilization-solidification, which is a mainstream strategy utilized to change over a possibly dangerous liquids or solids into a non-dangerous waste materials before it goes to the safe landfills (Leist et al. 2000).

6.10 Reusability

The economic situation is not favourable for the immediate disposal of hazardous waste materials is expensive process, hence, regeneration and reusability of the adsorbent is the favoured choice. Several investigations proposed that the greatest adsorption limit of metal oxide based adsorbents remains practically steady after multiple cycles of recovery and reuse (Tuutijärvi et al. 2012; Hu et al. 2005). Also, pH is seemed to be a significant factor in the desorption process of arsenic from the metal oxide adsorbents. The results showed the desorption qualities of As(V) and the regeneration of the maghemite (γ-Fe_2O_3) nanoparticle adsorbent (Tuutijärvi et al. 2012). 0.1 M NaOH alkaline solution demonstrated the most elevated desorption productivity of 90% compare to other alkaline solution such as Na_2CO_3, Na_2HPO_4, NaOAc, and $NaHCO_3$. Additionally, desorption was demonstrated to be influenced by the concentration of alkaline and pH of the solution, for example, when the concentration of NaOH was increased to 1M, complete desorption of As(V) was accomplished.

Nevertheless, other researchers found the reduction in the adsorption limit after recovery. Saiz et al. (2014) investigated the recovery and reusability of arsenate-adsorbed Fe_3O_4@SiO_2 composite materials. 0.01 M concentration of NaOH and HCl was used for the desorption process and found that NaOH performance much better in the desorption process. This alkaline treatment was further explored to assess the long-term execution of the recovery procedure, in which adsorbent materials were involved in functionalization steps (i.e., protonation of amino groups and coordination of Fe^{2+}) after each of the desorption process. After five cycles of adsorption-desorption process, 26% reduction in the desorption and 5.7% reduction in the re-adsorption process was observed. Deliyanni et al. (2003) also revealed the reduction in the adsorption capacity in an examination with respect to the adsorption of arsenate ions by akageneite-type nanoparticles. In addition, approximately 30% of akageneite's ability was lost in every recovery period, which implies that the adsorbent should be replaced after four recovery process.

Furthermore, metal oxide nanomaterials could be effectively recovered and recycled for the expulsion of arsenic provided that the adsorption limits are pretty much consistent even after a few cycles of recovery. This might be a benefit of utilizing nanoparticles as an adsorbent to reduce heavy metals in drinking water. Maybe some nanomaterials not holding their adsorption limit during recovery, this is probably not a drawback confining their potential usage as the preparation of these nano-adsorbent is very simple and the reagents used for their preparation is readily available and very cheap (Saiz et al. 2014). Additionally, these nanomaterials generally have high adsorption limits that could exceed the expenses expected to replace the absorbent after a few cycles.

6.11 Stability Issues

Nanomaterials was proved to be viable in the elimination of substantial metal ions from drinking water. Nevertheless, since they are typically exist as ultra-fine particles with low energy barriers which make them aggregate and accomplish a stabilized level (Petosa et al. 2010). Cluster formation of the nanoparticles reduces their specific surface area, thereby diminishing their adsorption limit and reactivity (Tang and Lo 2013). Additionally, the particle's mobility reduces, that further adds to decreasing their capability. To defeat the issues related with cluster formation, two solutions were accounted.

1. The impregnation of nanomaterials into permeable/porous materials or surface coatings (Chang and Chen 2005). Widely utilized host substrates are stimulated carbon (Kikuchi et al. 2006), bentonite (Eren et al. 2010), sand (Boujelben et al. 2010), alumina films (Hulteen et al. 1997), and resins (Pan et al. 2010). With respect to surface coating, there are many reports referencing that a thicker surface modifiers may decrease the response rate, although removal limit was upgraded because of an expanded number of active sites (Tang and Lo 2013). Therefore, trade off among reactivity and stability must be focused well.
2. Prepare micro- Nano hierarchically organized sorbents, that can adjust high adsorption limit and stability of nanoparticle (Zhu et al. 2012).

6.12 Conclusions

In the last decades, many examinations have demonstrated that consuming arsenic-polluted water ought to be the most significant worries for the strength of humanity. Along these lines, systems to evade the groundwater arsenic pollution and additionally to mitigate the effect of such tainting should be created trying to diminish the health dangers related to the admission of arsenic-polluted water. This chapter discussed in detail the chemistry and toxicity of arsenic and required conventional techniques to remove arsenic metal ions from drinking water. Focus was given to the conventional adsorption method which includes the mechanism, type of adsorption, and adsorption models. Analysis and modelling of small-scale column study is also discussed with various metal oxides being used in this study. Mainly, this chapter focused on two main concerns: (1) the possibility of metal oxide-based nanomaterials as an active adsorbent of arsenic expulsion for the utilization drinking water filtration systems, and (2) measuring the effects of variables such as arsenic initial concentration, adsorbents dosage, competing solute and solution pH. Additionally, a brief overview has also been given for the disposal of arsenic contaminated materials, regeneration and reuse of absorbents. In addition, other than the utilization of nanomaterials for the arsenic-expulsion water treatment, other novel permeable adsorbents have been explained which could go about as predominant adsorbent materials sooner rather than later because of their remarkable attributes

References

Abdus-Salam N, Adekola F (2005) The influence of pH and adsorbent concentration on adsorption of lead and zinc on a natural goethite. Afr J Sci Technol 6:55
Adepoju-Bello AA, Alabi O (2005) Heavy metals: a review. Nig J Pharm 37:41–45
Adepoju-Bello A, Ojomolade O, Ayoola G, Coker H (2009) Quantitative analysis of some toxic metals in domestic water obtained from Lagos metropolis. Nig J Pharm 42:57–60
Adewole AT (2009) Waste management towards sustainable development in Nigeria: a case study of Lagos state. Int NGO J 4:173–179
Adeyemi O, Oloyede O, Oladiji A (2007) Physicochemical and microbial characteristics of leachate-contaminated groundwater. Asian J Biochem 2:343–348

Ahmed MF (2001) An overview of arsenic removal technologies in Bangladesh and India. In: Proceedings of BUET-UNU international workshop on technologies for arsenic removal from drinking water, Dhaka, pp 5–7

Altundoğan HS, Altundoğan S, Tümen F, Bildik M (2000) Arsenic removal from aqueous solutions by adsorption on red mud. Waste Manag 20:761–767

Andjelkovic I, Tran DNH, Kabiri S, Azari S, Markovic M, Losic D (2015) Graphene aerogels decorated with alpha-FeOOH nanoparticles for efficient adsorption of arsenic from contaminated waters. ACS Appl Mater Interfaces 7:9758–9766. https://doi.org/10.1021/acsami.5b01624

Anjum A, Lokeswari P, Kaur M, Datta M (2011) Removal of As (III) from aqueous solutions using montmorillonite. J Anal Sci Methods Instrum 1:25

Bakare-Odunola M (2005) Determination of some metallic impurities present in soft drinks marketed in Nigeria. Nig J Pharm 4:51–54

Baldwin DR, Marshall WJ (1999) Heavy metal poisoning and its laboratory investigation. Ann Clin Biochem 36:267–300

Bang S, Patel M, Lippincott L, Meng X (2005) Removal of arsenic from groundwater by granular titanium dioxide adsorbent. Chemosphere 60:389–397

Bansal RC, Goyal M (2005) Activated carbon adsorption. CRC press

Basu A, Saha D, Saha R, Ghosh T, Saha B (2014) A review on sources, toxicity and remediation technologies for removing arsenic from drinking water. Res Chem Intermed 40:447–485

Benders R (2012) National institute for public health and environmental protection (RIVM). Integr Electr Resour Plan 261:123

Benjamin MM (2014) Water chemistry. Waveland Press, Long Grove

Benjwal P, Kumar M, Chamoli P, Kar KK (2015) Enhanced photocatalytic degradation of methylene blue and adsorption of arsenic(III) by reduced graphene oxide (rGO)-metal oxide (TiO2/Fe3O4) based nanocomposites. RSC Adv 5:73249–73260. https://doi.org/10.1039/c5ra13689j

Beolchini F, Pagnanelli F, De Michelis I, Vegliò F (2007) Treatment of concentrated arsenic (V) solutions by micellar enhanced ultrafiltration with high molecular weight cut-off membrane. J Hazard Mater 148:116–121

Bhardwaj V, Mirliss MJ (2005) Diatomaceous Earth filtration for drinking water. Water Encycl 1:174–177

Bottino A, Capannelli G, Comite A, Ferrari F, Firpo R, Venzano S (2009) Membrane technologies for water treatment and agroindustrial sectors. C R Chim 12:882–888

Boujelben N, Bouzid J, Elouear Z, Feki M (2010) Retention of nickel from aqueous solutions using iron oxide and manganese oxide coated sand: kinetic and thermodynamic studies. Environ Technol 31:1623–1634

Brandhuber P, Amy G (1998) Alternative methods for membrane filtration of arsenic from drinking water. Desalination 117:1–10

Bystrzejewska-Piotrowska G, Golimowski J, Urban PL (2009) Nanoparticles: their potential toxicity, waste and environmental management. Waste Manag 29:2587–2595

Calder L (1988) Chromium contamination of groundwater. In: Advances in environmental science and technology (USA). Wiley, New York

Carabante I (2012) Arsenic (V) adsorption on iron oxide: implications for soil remedeation and water purification. Thesis, Luleå tekniska universitet

Chandler BD, Cramb DT, Shimizu GK (2006) Microporous metal– organic frameworks formed in a stepwise manner from luminescent building blocks. J Am Chem Soc 128:10403–10412

Chandra V, Park J, Chun Y, Lee JW, Hwang I-C, Kim KS (2010) Water-dispersible magnetite-reduced graphene oxide composites for arsenic removal. ACS Nano 4:3979–3986

Chang Y-C, Chen D-H (2005) Preparation and adsorption properties of monodisperse chitosan-bound Fe3O4 magnetic nanoparticles for removal of Cu (II) ions. J Colloid Interface Sci 283:446–451

Chen B, Wang L, Zapata F, Qian G, Lobkovsky EB (2008) A luminescent microporous metal–organic framework for the recognition and sensing of anions. J Am Chem Soc 130:6718–6719

Chen ML, Sun Y, Huo CB, Liu C, Wang JH (2015) Akaganeite decorated graphene oxide composite for arsenic adsorption/removal and its proconcentration at ultra-trace level. Chemosphere 130:52–58. https://doi.org/10.1016/j.chemosphere.2015.02.046

Cheng RC, Liang S, Wang HC, Beuhler MD (1994) Enhanced coagulation for arsenic removal. J Am Water Works Assoc 86:79–90

Chiban M, Zerbet M, Carja G, Sinan F (2012) Application of low-cost adsorbents for arsenic removal: a review. J Environ Chem Ecotoxicol 4:91–102

Choong TS, Chuah T, Robiah Y, Koay FG, Azni I (2007) Arsenic toxicity, health hazards and removal techniques from water: an overview. Desalination 217:139–166

Chowdhury SR, Yanful EK (2011) Arsenic removal from aqueous solutions by adsorption on magnetite nanoparticles. Water Environ J 25:429–437

Chowdhury S, Chakraborty S, Saha PD (2013) Response surface optimization of a dynamic dye adsorption process: a case study of crystal violet adsorption onto NaOH-modified rice husk. Environ Sci Pollut Res 20:1698–1705

Cui H, Li Q, Gao S, Shang JK (2012) Strong adsorption of arsenic species by amorphous zirconium oxide nanoparticles. J Ind Eng Chem 18:1418–1427

Cui H, Su Y, Li Q, Gao S, Shang JK (2013) Exceptional arsenic (III, V) removal performance of highly porous, nanostructured ZrO2 spheres for fixed bed reactors and the full-scale system modeling. Water Res 47:6258–6268

De Anil K (2003) Environmental chemistry. New Age International, New Delhi

Deedar N, Aslam I (2009) Evaluation of the adsorption potential of titanium dioxide nanoparticles for arsenic removal. J Environ Sci 21:402–408

Deliyanni E, Bakoyannakis D, Zouboulis A, Matis K (2003) Sorption of As (V) ions by akaganeite-type nanocrystals. Chemosphere 50:155–163

Do DD (1998) Adsorption analysis: equilibria and kinetics. Imperial College Press, London

Dodd MC, Vu ND, Ammann A, Le VC, Kissner R, Pham HV, Cao TH, Berg M, Von Gunten U (2006) Kinetics and mechanistic aspects of As (III) oxidation by aqueous chlorine, chloramines, and ozone: relevance to drinking water treatment. Environ Sci Technol 40:3285–3292

Dubey SP, Nguyen TT, Kwon Y-N, Lee C (2015) Synthesis and characterization of metal-doped reduced graphene oxide composites, and their application in removal of Escherichia coli, arsenic and 4-nitrophenol. J Ind Eng Chem 29:282

Eddaoudi M, Moler DB, Li H, Chen B, Reineke TM, O'keeffe M, Yaghi OM (2001) Modular chemistry: secondary building units as a basis for the design of highly porous and robust metal–organic carboxylate frameworks. Acc Chem Res 34:319–330

Edition F (2011) Guidelines for drinking-water quality. WHO Chron 38:104–108

Eren E, Tabak A, Eren B (2010) Performance of magnesium oxide-coated bentonite in removal process of copper ions from aqueous solution. Desalination 257:163–169

Fauci A, Braunwald E, Isselbacher K, Wilson J, Martin J, Kasper D, Hauser S, Longo D (1998) Heavy Metal poisoning, principal of international medicine Harrison, vol 2. McGraw-Hill, New York, pp 2565–2566

Faust SD, Aly OM (2018) Chemistry of water treatment. CRC Press, Boca Raton

Feng Q, Zhang Z, Ma Y, He X, Zhao Y, Chai Z (2012) Adsorption and desorption characteristics of arsenic onto ceria nanoparticles. Nanoscale Res Lett 7:84

Figoli A, Cassano A, Criscuoli A, Mozumder MSI, Uddin MT, Islam MA, Drioli E (2010) Influence of operating parameters on the arsenic removal by nanofiltration. Water Res 44:97–104

Fred Lee G, Jones-Lee A (2005) Municipal solid waste landfills—water quality issues. Water Encycl 2:163–169

Furukawa H, Ko N, Go YB, Aratani N, Choi SB, Choi E, Yazaydin AÖ, Snurr RQ, O'Keeffe M, Kim J (2010) Ultrahigh porosity in metal-organic frameworks. Science 329:424–428

Garelick H, Dybowska A, Valsami-Jones E, Priest N (2005) Remediation technologies for arsenic contaminated drinking waters (9 pp). J Soils Sediments 5:182–190

Gecol H, Ergican E, Fuchs A (2004) Molecular level separation of arsenic (V) from water using cationic surfactant micelles and ultrafiltration membrane. J Membr Sci 241:105–119

Giles DE, Mohapatra M, Issa TB, Anand S, Singh P (2011) Iron and aluminium based adsorption strategies for removing arsenic from water. J Environ Manag 92:3011–3022

Guan X, Du J, Meng X, Sun Y, Sun B, Hu Q (2012) Application of titanium dioxide in arsenic removal from water: a review. J Hazard Mater 215:1–16

Guo X, Chen F (2005) Removal of arsenic by bead cellulose loaded with iron oxyhydroxide from groundwater. Environ Sci Technol 39:6808–6818

Guo H, Stüben D, Berner Z (2007) Adsorption of arsenic (III) and arsenic (V) from groundwater using natural siderite as the adsorbent. J Colloid Interface Sci 315:47–53

Guo L, Ye P, Wang J, Fu F, Wu Z (2015) Three-dimensional Fe 3 O 4-graphene macroscopic composites for arsenic and arsenate removal. J Hazard Mater 298:28–35

Gupta BS, Chatterjee S, Rott U, Kauffman H, Bandopadhyay A, DeGroot W, Nag N, Carbonell-Barrachina A, Mukherjee S (2009) A simple chemical free arsenic removal method for community water supply – a case study from West Bengal, India. Environ Pollut 157:3351–3353

Gupta A, Yunus M, Sankararamakrishnan N (2012) Zerovalent iron encapsulated chitosan nanospheres–A novel adsorbent for the removal of total inorganic Arsenic from aqueous systems. Chemosphere 86:150–155

Habuda-Stanić M, Nujić M (2015) Arsenic removal by nanoparticles: a review. Environ Sci Pollut Res 22:8094–8123

Han B, Runnells T, Zimbron J, Wickramasinghe R (2002) Arsenic removal from drinking water by flocculation and microfiltration. Desalination 145:293–298

Hasan Z, Jhung SH (2015) Removal of hazardous organics from water using metal-organic frameworks (MOFs): plausible mechanisms for selective adsorptions. J Hazard Mater 283:329–339

He Y, Zhou W, Krishna R, Chen B (2012) Microporous metal–organic frameworks for storage and separation of small hydrocarbons. Chem Commun 48:11813–11831

Hering JG, Chen PY, Wilkie JA, Elimelech M, Liang S (1996) Arsenic removal by ferric chloride. J Am Water Works Assoc 88:155–167

Hering JG, Chen P-Y, Wilkie JA, Elimelech M (1997) Arsenic removal from drinking water during coagulation. J Environ Eng 123:800–807

Hesami F, Bina B, Ebrahimi A, Amin MM (2013) Arsenic removal by coagulation using ferric chloride and chitosan from water. Int J Environ Health Eng 2:17

Ho Y-S, McKay G (2000) The kinetics of sorption of divalent metal ions onto sphagnum moss peat. Water Res 34:735–742

Hristovski K, Baumgardner A, Westerhoff P (2007) Selecting metal oxide nanomaterials for arsenic removal in fixed bed columns: from nanopowders to aggregated nanoparticle media. J Hazard Mater 147:265–274

Hu J, Chen G, Lo IM (2005) Removal and recovery of Cr (VI) from wastewater by maghemite nanoparticles. Water Res 39:4528–4536

Hu C, Liu H, Chen G, Jefferson WA, Qu J (2012) As (III) oxidation by active chlorine and subsequent removal of As (V) by Al13 polymer coagulation using a novel dual function reagent. Environ Sci Technol 46:6776–6782

Hua M, Zhang S, Pan B, Zhang W, Lv L, Zhang Q (2012) Heavy metal removal from water/wastewater by nanosized metal oxides: a review. J Hazard Mater 211:317–331

Hulteen J, Chen H, Chambliss C, Martin C (1997) Template synthesis of carbon nanotubule and nanofiber arrays. Nanostruct Mater 9:133–136

Huong PTL, Huy LT, Phan VN, Huy TQ, Nam MH, Lam VD, Le AT (2016) Application of graphene oxide-MnFe2O4 magnetic nanohybrids as magnetically separable adsorbent for highly efficient removal of arsenic from water. J Electron Mater 45:2372–2380. https://doi.org/10.1007/s11664-015-4314-3

Huxford RC, Della Rocca J, Lin W (2010) Metal–organic frameworks as potential drug carriers. Curr Opin Chem Biol 14:262–268

Igwilo IO, Afonne OJ, Maduabuchi UJ-M, Orisakwe OE (2006) Toxicological study of the Anam river in Otuocha, Anambra state, Nigeria. Arch Environ Occup Health 61:205–208

Iqbal J, Kim H-J, Yang J-S, Baek K, Yang J-W (2007) Removal of arsenic from groundwater by micellar-enhanced ultrafiltration (MEUF). Chemosphere 66:970–976

Jang M, Chen W, Cannon FS (2008) Preloading hydrous ferric oxide into granular activated carbon for arsenic removal. Environ Sci Technol 42:3369–3374

Jia S-Y, Zhang Y-F, Liu Y, Qin F-X, Ren H-T, Wu S-H (2013) Adsorptive removal of dibenzothiophene from model fuels over one-pot synthesized PTA@ MIL-101 (Cr) hybrid material. J Hazard Mater 262:589–597

Jian M, Liu B, Zhang G, Liu R, Zhang X (2015) Adsorptive removal of arsenic from aqueous solution by zeolitic imidazolate framework-8 (ZIF-8) nanoparticles. Colloids Surf A Physicochem Eng Asp 465:67–76

Johnston R, Heijnen H, Wurzel P (2001) Safe water technology, technologies for arsenic removal from drinking water, Matiar Manush, Dhaka, Bangladesh, pp 1–98

Kamsonlian S, Suresh S, Majumder C, Chand S (2012) Biosorption of arsenic from contaminated water onto solid Psidium guajava leaf surface: equilibrium, kinetics, thermodynamics, and desorption study. Biorem J 16:97–112

Kanel SR, Manning B, Charlet L, Choi H (2005) Removal of arsenic (III) from groundwater by nanoscale zero-valent iron. Environ Sci Technol 39:1291–1298

Katsoyiannis IA, Zouboulis AI (2004) Application of biological processes for the removal of arsenic from groundwaters. Water Res 38:17–26

Kikuchi Y, Qian Q, Machida M, Tatsumoto H (2006) Effect of ZnO loading to activated carbon on Pb (II) adsorption from aqueous solution. Carbon 44:195–202

Kim M-J, Nriagu J (2000) Oxidation of arsenite in groundwater using ozone and oxygen. Sci Total Environ 247:71–79

Kumari P, Sharma P, Srivastava S, Srivastava M (2005) Arsenic removal from the aqueous system using plant biomass: a bioremedial approach. J Ind Microbiol Biotechnol 32:521–526

Kurmoo M (2009) Magnetic metal–organic frameworks. Chem Soc Rev 38:1353–1379

Kwok CM (2009) Removal of arsenic from water using chitosan and nanochitosan. Hong Kong University of Science and Technology, Hong Kong

Lai C-H, Chen C-Y, Wei B-L, Yeh S-H (2002) Cadmium adsorption on goethite-coated sand in the presence of humic acid. Water Res 36:4943–4950

Lakshmanan D, Clifford D, Samanta G (2008) Arsenic removal by coagulation with aluminum, iron, titanium, and zirconium. J Am Water Works Assoc 100:76–88

Lambrechts C, Woodley B, Church C, Gachanja M (2003) Aerial survey of the destruction of the Aberdare Range forests. Division of Early Warning and Assessment, UNEP

Langmi HW, Ren J, North B, Mathe M, Bessarabov D (2014) Hydrogen storage in metal-organic frameworks: a review. Electrochim Acta 128:368–392

Lee B, De Haan M, Reddy S, Martindale A (2004) Evaluation of EBCT on arsenic adsorption performance and competitive preloading. In: Proceedings, AWWA water quality technology conference and exhibition

Lee S-H, Kim K-W, Lee B-T, Bang S, Kim H, Kang H, Jang A (2015) Enhanced arsenate removal performance in aqueous solution by yttrium-based adsorbents. Int J Environ Res Public Health 12:13523–13541

Leist M, Casey R, Caridi D (2000) The management of arsenic wastes: problems and prospects. J Hazard Mater 76:125–138

Lenoble V, Bouras O, Deluchat V, Serpaud B, Bollinger J-C (2002) Arsenic adsorption onto pillared clays and iron oxides. J Colloid Interface Sci 255:52–58

Li Y-H, Wang S, Luan Z, Ding J, Xu C, Wu D (2003) Adsorption of cadmium (II) from aqueous solution by surface oxidized carbon nanotubes. Carbon 41:1057–1062

Li R, Li Q, Gao S, Shang JK (2012) Exceptional arsenic adsorption performance of hydrous cerium oxide nanoparticles: Part A. Adsorption capacity and mechanism. Chem Eng J 185:127–135

Lin T-F, Wu J-K (2001) Adsorption of arsenite and arsenate within activated alumina grains: equilibrium and kinetics. Water Res 35:2049–2057

Litter MI, Morgada ME, Bundschuh J (2010) Possible treatments for arsenic removal in Latin American waters for human consumption. Environ Pollut 158:1105–1118

Liu YY, Leus K, Grzywa M, Weinberger D, Strubbe K, Vrielinck H, Van Deun R, Volkmer D, Van Speybroeck V, Van Der Voort P (2012) Synthesis, structural characterization, and catalytic performance of a vanadium-based metal–organic framework (COMOC-3). Eur J Inorg Chem 2012:2819–2827

Liu H, Zuo K, Vecitis CD (2014a) Titanium dioxide-coated carbon nanotube network filter for rapid and effective arsenic sorption. Environ Sci Technol 48:13871–13879

Liu J, Chen L, Cui H, Zhang J, Zhang L, Su C-Y (2014b) Applications of metal–organic frameworks in heterogeneous supramolecular catalysis. Chem Soc Rev 43:6011–6061

Logsdon GS, Kohne R, Abel S, LaBonde S (2002) Slow sand filtration for small water systems. J Environ Eng Sci 1:339–348

Lorenzen L, Van Deventer J, Landi W (1995) Factors affecting the mechanism of the adsorption of arsenic species on activated carbon. Miner Eng 8:557–569

Low JJ, Benin AI, Jakubczak P, Abrahamian JF, Faheem SA, Willis RR (2009) Virtual high throughput screening confirmed experimentally: porous coordination polymer hydration. J Am Chem Soc 131:15834–15842

Lu C, Liu C (2006) Removal of nickel (II) from aqueous solution by carbon nanotubes. J Chem Technol Biotechnol 81:1932–1940

Luo X, Wang C, Luo S, Dong R, Tu X, Zeng G (2012) Adsorption of As (III) and As (V) from water using magnetite Fe 3 O 4-reduced graphite oxide–MnO 2 nanocomposites. Chem Eng J 187:45–52

Luo XB, Wang CC, Wang LC, Deng F, Luo SL, Tu XM, Au CT (2013) Nanocomposites of graphene oxide-hydrated zirconium oxide for simultaneous removal of As(III) and As(V) from water. Chem Eng J 220:98–106. https://doi.org/10.1016/j.cej.2013.01.017

Mandal S, Sahu MK, Patel RK (2013) Adsorption studies of arsenic (III) removal from water by zirconium polyacrylamide hybrid material (ZrPACM-43). Water Resour Ind 4:51–67

Marcovecchio JE, Botté SE, Freije RH (2007) Heavy metals, major metals, trace elements. Handb Water Anal 2:275–311

Margat J, Van der Gun J (2013) Groundwater around the world: a geographic synopsis. CRC Press, Boca Raton/London/New York

Martinson CA, Reddy K (2009) Adsorption of arsenic (III) and arsenic (V) by cupric oxide nanoparticles. J Colloid Interface Sci 336:406–411

Mayo J, Yavuz C, Yean S, Cong L, Shipley H, Yu W, Falkner J, Kan A, Tomson M, Colvin V (2007) The effect of nanocrystalline magnetite size on arsenic removal. Sci Technol Adv Mater 8:71

McMurry J, Fay R (2004) In: Hamann KP (ed) Hydrogen, oxygen and water, McMurry Fay Chemistry, 4th edn. Pearson Education, Upper Saddle River, pp 575–599

Melia O, Coagulation C (1972) In: Weber WJ Jr (ed) Floccula-tion. Physicochemical processes for water quality control. Wiley Interscience, New York

Mendie U (2005) The nature of water. In: The theory and practice of clean water production for domestic and industrial use, vol 1. Lacto-Medals Publishers, Lagos, p 21

Mishra AK, Ramaprabhu S (2010) Magnetite decorated multiwalled carbon nanotube based supercapacitor for arsenic removal and desalination of seawater. J Phys Chem C 114:2583–2590. https://doi.org/10.1021/jp911631w

Misra, M. & Lenz, P. (2008). Removal of arsenic from drinking and process water: Google Patents.

Mohan D, Pittman CU Jr (2007) Arsenic removal from water/wastewater using adsorbents—a critical review. J Hazard Mater 142:1–53

Mohapatra D, Mishra D, Chaudhury GR, Das RP (2007) Arsenic adsorption mechanism on clay minerals and its dependence on temperature. Korean J Chem Eng 24:426–430

Mohmood I, Lopes CB, Lopes I, Ahmad I, Duarte AC, Pereira E (2013) Nanoscale materials and their use in water contaminants removal—a review. Environ Sci Pollut Res 20:1239–1260

Momodu M, Anyakora C (2010) Heavy metal contamination of ground water: the Surulere case study. Res J Environ Earth Sci 2:39–43

Mondal N, Roy P, Das B, Datta J (2011) Chronic arsenic toxicity and it's relation with nutritional status: a case study in Purabasthali-II, Burdwan, West Bengal, India. Int J Environ Sci 2:1103–1118

Mondal P, Bhowmick S, Chatterjee D, Figoli A, Van der Bruggen B (2013) Remediation of inorganic arsenic in groundwater for safe water supply: a critical assessment of technological solutions. Chemosphere 92:157–170

Moore JW, Ramamoorthy S (2012) Heavy metals in natural waters: applied monitoring and impact assessment. Springer, New York

Morgan JJ, Stumm W (1970) Aquatic chemistry. Wiley, New York

Namasivayam C, Ranganathan K (1995) Removal of Cd (II) from wastewater by adsorption on "waste" Fe (III) Cr (III) hydroxide. Water Res 29:1737–1744

Nigam S, Gopal K, Vankar PS (2013a) Biosorption of arsenic in drinking water by submerged plant: Hydrilla verticilata. Environ Sci Pollut Res 20:4000–4008

Nigam S, Vankar PS, Gopal K (2013b) Biosorption of arsenic from aqueous solution using dye waste. Environ Sci Pollut Res 20:1161–1172

Ohlinger K, Young TM, Schroeder E (1998) Predicting struvite formation in digestion. Water Res 32:3607–3614

Organization, W. H (2007) Quality assurance of pharmaceuticals: a compendium of guidelines and related materials. Good manufacturing practices and inspection. World Health Organization, Geneva

Pan B, Xing B (2008) Adsorption mechanisms of organic chemicals on carbon nanotubes. Environ Sci Technol 42:9005–9013

Pan B, Qiu H, Pan B, Nie G, Xiao L, Lv L, Zhang W, Zhang Q, Zheng S (2010) Highly efficient removal of heavy metals by polymer-supported nanosized hydrated Fe (III) oxides: behavior and XPS study. Water Res 44:815–824

Pandey PK, Choubey S, Verma Y, Pandey M, Chandrashekhar K (2009) Biosorptive removal of arsenic from drinking water. Bioresour Technol 100:634–637

Peak D, Sparks D (2002) Mechanisms of selenate adsorption on iron oxides and hydroxides. Environ Sci Technol 36:1460–1466

Pena ME, Korfiatis GP, Patel M, Lippincott L, Meng X (2005) Adsorption of As (V) and As (III) by nanocrystalline titanium dioxide. Water Res 39:2327–2337

Pennesi C, Vegliò F, Totti C, Romagnoli T, Beolchini F (2012) Nonliving biomass of marine macrophytes as arsenic (V) biosorbents. J Appl Phycol 24:1495–1502

Petosa AR, Jaisi DP, Quevedo IR, Elimelech M, Tufenkji N (2010) Aggregation and deposition of engineered nanomaterials in aquatic environments: role of physicochemical interactions. Environ Sci Technol 44:6532–6549

Pettine M, Campanella L, Millero FJ (1999) Arsenite oxidation by H2O2 in aqueous solutions. Geochim Cosmochim Acta 63:2727–2735

Pontius FW, Association, A. W. W (1990) Water quality and treatment: a handbook of community water supplies. AWWA, Norwich

Pous N, Casentini B, Rossetti S, Fazi S, Puig S, Aulenta F (2015) Anaerobic arsenite oxidation with an electrode serving as the sole electron acceptor: a novel approach to the bioremediation of arsenic-polluted groundwater. J Hazard Mater 283:617–622

Prasad KS, Ramanathan A, Paul J, Subramanian V, Prasad R (2013) Biosorption of arsenite (As+ 3) and arsenate (As+ 5) from aqueous solution by Arthrobacter sp. biomass. Environ Technol 34:2701–2708

Qu X, Alvarez PJ, Li Q (2013) Applications of nanotechnology in water and wastewater treatment. Water Res 47:3931–3946

Ramos MA, Yan W, Li X-Q, Koel BE, Zhang W-X (2009) Simultaneous oxidation and reduction of arsenic by zero-valent iron nanoparticles: understanding the significance of the core− shell structure. J Phys Chem C 113:14591–14594

Ranjan D, Talat M, Hasan S (2009a) Rice polish: an alternative to conventional adsorbents for treating arsenic bearing water by up-flow column method. Ind Eng Chem Res 48:10180–10185

Ranjan D, Talat M, Hasan S (2009b) Biosorption of arsenic from aqueous solution using agricultural residue 'rice polish'. J Hazard Mater 166:1050–1059

Rice EW, Baird RB, Eaton AD, Clesceri LS (2012) Standard methods for the examination of water and wastewater. American Public Health Association, Washington, DC

Roy P, Choudhury M, Ali M (2013a) As (III) and As (V) adsorption on magnetite nanoparticles: adsorption isotherms, effect of ph and phosphate, and adsorption kinetics. Int J Chem Environ Eng 4:55–63

Roy P, Mondal NK, Das B, Das K (2013b) Arsenic contamination in groundwater: a statistical modeling. J Urban Environ Eng 7:24–29

Roy E, Patra S, Madhuri R, Sharma PK (2016) Europium doped magnetic graphene oxide-MWCNT nanohybrid for estimation and removal of arsenate and arsenite from real water samples. Chem Eng J 299:244–254. https://doi.org/10.1016/j.cej.2016.04.051

Ruthven DM (2006) Adsorption and diffusion. Springer, Berlin, pp 1–43

Saiz J, Bringas E, Ortiz I (2014) New functionalized magnetic materials for As5+ removal: adsorbent regeneration and reuse. Ind Eng Chem Res 53:18928–18934

Sakthivel TS, Das S, Pratt CJ, Seal S (2017) One-pot synthesis of a ceria–graphene oxide composite for the efficient removal of arsenic species. Nanoscale 9(10):3367–3374

Salomons W, Förstner U, Mader P (2012) Heavy metals: problems and solutions. Springer, Berlin/Heidelberg

Samiey B, Cheng C-H, Wu J (2014) Organic-inorganic hybrid polymers as adsorbents for removal of heavy metal ions from solutions: a review. Materials 7:673–726

Sandoval R, Cooper AM, Aymar K, Jain A, Hristovski K (2011) Removal of arsenic and methylene blue from water by granular activated carbon media impregnated with zirconium dioxide nanoparticles. J Hazard Mater 193:296–303

Sato Y, Kang M, Kamei T, Magara Y (2002) Performance of nanofiltration for arsenic removal. Water Res 36:3371–3377

Sharma SK (2001) Adsorptive iron removal from groundwater. CRC Press, Leiden

Sharma R, Al-Busaidi T (2001) Groundwater pollution due to a tailings dam. Eng Geol 60:235–244

Sharma S, Petrusevski B, Schippers J (2002) Characterisation of coated sand from iron removal plants. Water Sci Technol Water Supply 2:247–257

Sharma YC, Srivastava V, Singh V, Kaul S, Weng C (2009) Nano-adsorbents for the removal of metallic pollutants from water and wastewater. Environ Technol 30:583–609

Shen L (2013) Synthesis, characterization and application of metal-organic frameworks. Thesis, University of Illinois at Urbana-Champaign

Sheng G, Li Y, Yang X, Ren X, Yang S, Hu J, Wang X (2012) Efficient removal of arsenate by versatile magnetic graphene oxide composites. RSC Adv 2:12400–12407

Shih M-C (2005) An overview of arsenic removal by pressure-drivenmembrane processes. Desalination 172:85–97

Shiklomanov I, Rodde J (2011) Summary of the monograph "world water resources at the beginning of the 21st century" prepared in the framework of ihp unesco

Shinde RN, Pandey A, Acharya R, Guin R, Das S, Rajurkar N, Pujari P (2013) Chitosan-transition metal ions complexes for selective arsenic (V) preconcentration. Water Res 47:3497–3506

Singh TS, Pant K (2004) Equilibrium, kinetics and thermodynamic studies for adsorption of As (III) on activated alumina. Sep Purif Technol 36:139–147

Singh R, Singh S, Parihar P, Singh VP, Prasad SM (2015) Arsenic contamination, consequences and remediation techniques: a review. Ecotoxicol Environ Saf 112:247–270

Smedley PL, Kinniburgh DG (2002) A review of the source, behaviour and distribution of arsenic in natural waters. Appl Geochem 17:517–568

Sorlini S, Gialdini F (2010) Conventional oxidation treatments for the removal of arsenic with chlorine dioxide, hypochlorite, potassium permanganate and monochloramine. Water Res 44:5653–5659

Sorlini S, Gialdini F, Stefan M (2014) UV/H 2 O 2 oxidation of arsenic and terbuthylazine in drinking water. Environ Monit Assess 186:1311–1316

Sun Y, Zhou G, Xiong X, Guan X, Li L, Bao H (2013a) Enhanced arsenite removal from water by Ti (SO4) 2 coagulation. Water Res 47:4340–4348

Sun Z, Yu Y, Pang S, Du D (2013b) Manganese-modified activated carbon fiber (Mn-ACF): Novel efficient adsorbent for Arsenic. Appl Surf Sci 284:100–106

Tang SC, Lo IM (2013) Magnetic nanoparticles: essential factors for sustainable environmental applications. Water Res 47:2613–2632

Tang W, Li Q, Gao S, Shang JK (2011) Arsenic (III, V) removal from aqueous solution by ultrafine α-Fe2O3 nanoparticles synthesized from solvent thermal method. J Hazard Mater 192:131–138

Tuutijärvi T (2013) Arsenate removal from water by adsorption with magnetic nanoparticles (γ-Fe2O3)

Tuutijärvi T, Vahala R, Sillanpää M, Chen G (2012) Maghemite nanoparticles for As (V) removal: desorption characteristics and adsorbent recovery. Environ Technol 33:1927–1936

Uddin MT, Mozumder MSI, Figoli A, Islam MA, Drioli E (2007) Arsenic removal by conventional and membrane technology: an overview. Indian J Chem Technol 14(5):441–450

Ungureanu G, Santos S, Boaventura R, Botelho C (2015) Arsenic and antimony in water and wastewater: overview of removal techniques with special reference to latest advances in adsorption. J Environ Manag 151:326–342

Urba J (1985) Impact of domestic and industrial wastes and agricultural activities on ground-water quality. Congr Int Assoc Hydrogeol 18:91–117. IAH

Uwamariya V (2013) Adsorptive removal of heavy metals from groundwater by iron oxide based adsorbents. IHE Delft Institute for Water Education

Van der Bruggen B, Vandecasteele C, Van Gestel T, Doyen W, Leysen R (2003) A review of pressure-driven membrane processes in wastewater treatment and drinking water production. Environ Prog 22:46–56

VanLoon GW, Duffy SJ (2017) Environmental chemistry: a global perspective. Oxford University Press

Veličković ZS, Bajić ZJ, Ristić MĐ, Djokić VR, Marinković AD, Uskoković PS, Vuruna MM (2013) Modification of multi-wall carbon nanotubes for the removal of cadmium, lead and arsenic from wastewater. Dig J Nanomater Biostruct (DJNB) 8:501

Velizarov S, Crespo JG, Reis MA (2004) Removal of inorganic anions from drinking water supplies by membrane bio/processes. Rev Environ Sci Biotechnol 3:361–380

Viet PH, Con TH, Ha CT, Van Ha H, Berg M, Giger W, Schertenleib R (2003) Arsenic exposure and health effects V. Elsevier, Amsterdam, pp 459–469

Vrba J, van der Gun J (2004) The world's groundwater resources, world water development report 2, Contribution to Chapter 4, Report IP 2004-1. System 2:1–10

Vu TA, Le GH, Dao CD, Dang LQ, Nguyen KT, Nguyen QK, Dang PT, Tran HT, Duong QT, Nguyen TV (2015) Arsenic removal from aqueous solutions by adsorption using novel MIL-53 (Fe) as a highly efficient adsorbent. RSC Adv 5:5261–5268

Wang S, Mulligan CN (2006) Occurrence of arsenic contamination in Canada: sources, behavior and distribution. Sci Total Environ 366:701–721

Wang LK, Tay J-H, Tay STL, Hung Y-T (2010) Environmental bioengineering. Springer, New York

Wang C, Luo HJ, Zhang ZL, Wu Y, Zhang J, Chen SW (2014) Removal of As(III) and As(V) from aqueous solutions using nanoscale zero valent iron-reduced graphite oxide modified composites. J Hazard Mater 268:124–131. https://doi.org/10.1016/j.jhazmat.2014.01.009

Warner NR, Christie CA, Jackson RB, Vengosh A (2013) Impacts of shale gas wastewater disposal on water quality in western Pennsylvania. Environ Sci Technol 47:11849–11857

Wen T, Wu XL, Tan XL, Wang XK, Xu AW (2013) One-pot synthesis of water-swellable Mg-Al layered double hydroxides and graphene oxide nanocomposites for efficient removal of As (V) from aqueous solutions. ACS Appl Mater Interfaces 5:3304–3311. https://doi.org/10.1021/am4003556

Xi Y, Mallavarapu M, Naidu R (2010) Reduction and adsorption of Pb2+ in aqueous solution by nano-zero-valent iron—a SEM, TEM and XPS study. Mater Res Bull 45:1361–1367

Xu Z, Li Q, Gao S, Shang JK (2010) As (III) removal by hydrous titanium dioxide prepared from one-step hydrolysis of aqueous TiCl4 solution. Water Res 44:5713–5721

Yaghi OM, O'Keeffe M, Ockwig NW, Chae HK, Eddaoudi M, Kim J (2003) Reticular synthesis and the design of new materials. Nature 423:705

Yoon Y, Park WK, Hwang TM, Yoon DH, Yang WS, Kang JW (2016) Comparative evaluation of magnetite-graphene oxide and magnetite-reduced graphene oxide composite for As(III) and As (V) removal. J Hazard Mater 304:196–204. https://doi.org/10.1016/j.jhazmat.2015.10.053

Yu L, Ma Y, Ong CN, Xie J, Liu Y (2015) Rapid adsorption removal of arsenate by hydrous cerium oxide–graphene composite. RSC Adv 5:64983–64990

Yuan T, Yong Hu J, Ong SL, Luo QF, Jun Ng W (2002) Arsenic removal from household drinking water by adsorption. J Environ Sci Health A 37:1721–1736

Zhang J, Liu Q, Ding Y, Bei Y (2011) 3-aminopropyltriethoxysilane functionalized nanoscale zero-valent iron for the removal of dyes from aqueous solution. Pol J Chem Technol 13:35–39

Zhu HJ, Jia YF, Wu X, Wang H (2009) Removal of arsenic from water by supported nano zero-valent iron on activated carbon. J Hazard Mater 172:1591–1596. https://doi.org/10.1016/j.jhazmat.2009.08.031

Zhu B-J, Yu X-Y, Jia Y, Peng F-M, Sun B, Zhang M-Y, Luo T, Liu J-H, Huang X-J (2012) Iron and 1, 3, 5-benzenetricarboxylic metal–organic coordination polymers prepared by solvothermal method and their application in efficient As (V) removal from aqueous solutions. J Phys Chem C 116:8601–8607

Zhu J, Lou ZM, Liu Y, Fu RQ, Baig SA, Xu XH (2015) Adsorption behavior and removal mechanism of arsenic on graphene modified by iron-manganese binary oxide (FeMnOx/RGO) from aqueous solutions. RSC Adv 5:67951–67961. https://doi.org/10.1039/c5ra11601e

Chapter 7
Earth Abundant Materials for Environmental Remediation and Commercialization

**J. Nimita Jebaranjitham, Adhimoorthy Prasannan,
K. Sankarasubramanian, K. S. Prakash, and Baskaran Ganesh Kumar**

Abstract Environmental remediation is a process of eliminating the toxic materials from living area of plants, animals and humans. The toxic materials contaminants air, water and land and made earth not habitable. The solutions for the environmental remediation should be renewable and sustainable in long term. The pressing issue is that water is heavily contaminated with industrials pollutants due to growing industrialization. Many technologies are being developed for the water remediation problem such as filtration, oxidation, reduction, catalysis and many more. Among that heterogenous photocatalysis found to be effective and it has potential of solving the water remediation problem which means the solution for water remediation should be commercialisable. Herein, we explain the importance of heterogenous catalysis in the environmental remediation and also explain the present status of state-of-art materials and current advanced materials for water remediation. For the viable commercialization, we predicted that material abundance being major challenge. The focus of the present work is to explore potential of photocatalytic

---

J. Nimita Jebaranjitham
P.G. Department of Chemistry, Women's Christian College (An Autonomous Institution Affiliated to University of Madras), Chennai, Tamil Nadu, India

A. Prasannan
Graduate Institute of Applied Science and Technology, National Taiwan University of Science and Technology, Taipei, Taiwan

K. Sankarasubramanian
School of Physics, Madurai Kamaraj University, Madurai, Tamil Nadu, India

K. S. Prakash
Department of Chemistry, Bharathidasan Government College for Women (Autonomous) (Affiliated to Pondicherry University, Pondicherry), Muthialpet, Puducherry U.T, India

B. G. Kumar (✉)
Department of Chemistry, P.S.R. Arts and College (Affiliated to Madurai Kamaraj University, Madurai), Sivakasi, Tamil Nadu, India

Department of Science and Humanities, P.S.R. Engineering College (Affiliated to Anna University, Chennai), Sivakasi, Tamil Nadu, India

© The Author(s), under exclusive license to Springer Nature Switzerland AG 2021
S. Rajendran et al. (eds.), *Metal, Metal-Oxides and Metal-Organic Frameworks for Environmental Remediation*, Environmental Chemistry for a Sustainable World 64, https://doi.org/10.1007/978-3-030-68976-6_7

materials for commercialization. We foresee and attempted to bring out the clear picture of the earth abundance crisis in the heterogenous catalytic based environmental remediation. The core concepts explained in this work could be extended to the other existing technologies.

Keywords Water remediation · Heterogenous catalysis · Materials abundance · Environment · Scale up · Commercialization · Toxicity · Material design · Novel materials · Sustainability · Photocatalyst

7.1 Introduction

7.1.1 Environmental Remediation

Water is often considered as the source of life and the water is essential and flow through entire earth get pure-impure cycle frequently. The current industrialization breaks the nature purification cycle due to the huge use of materials and its wastage (Dapeng and Jiuhui 2009; Dong et al. 2015; Khin et al. 2012; Mills et al. 1993; Santhosh et al. 2016; Tesh and Scott 2014; Umar and Aziz 2013). The impurities are contaminating the water and hence consecutively contaminate the land, air, plants and animals. The remediation aims for water purification till the consumption standards achieved both for drinking and domestic use. The increase in the world population and higher standards of people magnify the problem. Because water purification should be carried out at mass level which require higher quantity of materials and work force. Hence the toxic impurities are should be removed from the water to achieve best health for human and the living beings. Additionally, the technology for the water purification should directly convert toxic molecules to nontoxic products (Fujishima and Honda 1972; Ku and Hsieh 1992; Matthews 1987; Tanaka et al. 2000). The present work, review environmental remediation of water using heterogeneous catalyst with focus of earth abundance of materials.

Conventionally, water purified by heterogenous and homogenous catalysis (Chun et al. 2000; Corma et al. 2010; Descorme et al. 2012; Guillard et al. 2003; Hajiesmaili et al. 2010; Herrmann et al. 1997; Hu et al. 2011; Jia et al. 2017; Khanchandani et al. 2013; Malato et al. 2009; Mills and Le Hunte 1997; Ramos-Delgado et al. 2013; Sun et al. 2006; Thuy et al. 2014). Homogenous catalysis means catalyst is same phase with water; heterogenous catalysis means catalyst is different phase than water (mostly solid). Among that, heterogenous catalysis predicted to be most effective and scientifically proven method for destroying toxic chemicals. The photocatalyst is often the semiconductor materials (e.g TiO_2, ZnO, MgO) which oxidize organic and inorganic contaminates such as halogenated alkanes, alkenes, and aromatics (e.g Rhodamine B dye) using complex sequence of reaction (Mills et al. 1993; Mills and Le Hunte 1997; Xia et al. 2016). The mechanism of degradation is simple, semiconducting photocatalyst creates very reactive species in presence of sunlight. The reactive species ($OH°$ radicals) degrade the organic contaminants present in the water (Mills and Le Hunte 1997). The photocatalytic based degradation is effective and has tremendous potential for future with water remediation. Hence, researchers devoted considerable efforts towards new material development, new combination (sonolysis with catalysis), fabrication methods, performance enhancement, increase active sites using nano, deep understanding of mechanism of degradation, make it commercially viable and establishing mechanism (Bhatkhande et al. 2002; Dong et al. 2015; Khin et al. 2012; Qu et al. 2012; Thatai et al. 2014).

Heterogenous catalysis has potential to become the solution for environmental remediation. Since the water remediation is existing all over earth, the water remediation technology should be addressed commercially (Chirik and Morris 2015). World population is around 8 billion and to solve the problem at that sale requires tons of raw materials. If we assume water remediation problem can be solved by the heterogenous photocatalyst technology, then we should imagine in terms of material abundance. Otherwise, even after achieving the great efficiency, with the metal with less abundance does not make any impact in the technology or world. To put it simply, the water purification technology using heterogeneous catalyst was developed competitively but the scale up is often overlooked due to the lack of knowledge on mass production. The scale up-potential of every photocatalyst involving the water purification should be analyzed carefully. Then the material can be effectively and efficiently used for solve the environment remediation problem in global scale. Otherwise, developing the photocatalyst for the water remediation is mere a scientific interest not a realworld solution. Hence, the materials availability dictates the use and choice of the materials. Current state-of-art materials is TiO_2 where titanium has abundance of 0.57% and oxygen abundance is about 21%. It is apparent that oxygen is good choice of material but titanium is not a rational choice. Because, if we establish the commercialization based on TiO_2, we will end up in the situation of titanium depletion at the middle and eventually failure of technology. Therefore, the materials abundance is new scarcity and makes the technology costly for commercialization as well as implementation. Therefore, the analyzing the material abundance for the materials involved in environmental remediation is meaningful path for commercialization of the technologies.

The problem of earth abundance in environmental remediation is too great to be overcome by conventional hit-and-trial approach. That is by taking the random materials and improving it to the best performance and mostly researchers choose the readily available materials. By this approach, we end up with least-abundant materials as state-of-art industrial materials e.g. Ru, Rh, Pd, Pt and more and all those technology metals are rare and precious metals. It is clear that importance of abundance is poorly understood among the scientific community and the earth abundant materials environmental remediation not been widely studied to date. These materials amount in the earth crust is constant, if the demand increased, the materials cannot serve intended purpose with the time. In general, research work starts with milligrams scale and eventually converted to multi-ton chemical manufacturing. Hence material abundance limits the expansion of technology and scaling up for the mass market. This situation has the potential to create real hard challenges for water remediation technologies. An open consideration is minimizing the use of materials which individual elements earth abundance is very low. Therefore, we should start with earth abundant materials and tune it to the best performance. The abundance of materials seriously analyzed and should be integrated during the material design and engineering. The earth abundant materials are cheaper, less prone to supply fluctuation and industrially more feasible. There have been recent discussions on only use the new and carbon-based materials for the photocatalysis but in different context (Arvidsson and Sandén 2017; Chirik and Morris 2015; Johnson et al. 2007; Sunada and Nagashima 2017). There are no present studies concerning in terms of materials abundance in environmental radiation technologies for high volume commercial application.

Photocatalyst based environmental remediation is perceived as next generation water remediation technology. To realize full potential, we discussed drivers and barriers for commercialization of environmental remediation technology. We emphasized that sustainable commercialization requires the use of non-precious metals or earth abundant materials. The materials abundance is grand challenge it should be solved make the photocatalyst technology commercially viable. As long as use of scarce metals for environmental remediation problem of sustainable solution will arise. we believe the present work informs the researchers and industry the importance of earth abundance based materials design. By taking longer time perspective, this work discourages the use of scarce materials for the future materials design for any application. By choosing earth abundant materials based design, the scope, selectivity, performance, stability, reactivity characteristics and electron flow of reactions of photocatalyst can be improved and hence the overall competency of the catalyst among the existing one. The present work focuses the representative synthetic methods that could potentially address the material abundance issues. We suggested four possible solutions for current precious materials based research namely recycling, use of earth abundant materials as photocatalyst, increasing efficiency of the catalyst and develop two or more state-of-art materials. We argue that still possible to shift the direction of water remediation technology to the sustainability and affordability. The work also provides actionable insights to solve the materials abundance issue and develop the effective technology for water remediation. Overall, we suggest that if we controllably design and prepare the earth abundant photocatalyst, we could realize photocatalyst based water remediation technology. We concluded with challenges associated with implementation of earth abundant materials in the heterogeneous catalysis based environmental remediation.

7.1.2 Water Remediation and Heterogenous Catalysis

The ever-increasing world population demand industrial revolution and mass manufacturing. The industries not only positive impact but also generate the negative impact due to the industrial contaminants. The water standards permit only about 40 ppb of toxic chemicals. The contaminants seriously pollute the eco-systems including land, air and water which in turn harm to human, animals and plants. Since human body contains most of water, the water is considered as source of life. Hence much consideration given to the purification of water. Providing high pure water is pressing problem is many developing, undeveloped and developed countries such as India and Africa. The contaminants are many types and it can be removed by physical, chemical and biological methods. Among the various purification/remediation technologies of water, heterogenous technology is proven to be cost-effective, efficient and often rated as most practical. Since the method does not require any energy to purify the chemicals it attracted wide interest in industry and academia. Simply, the catalyst is continuously irradiated with light, which converted the toxic impurities to the less toxic ones (Fig. 7.1). Many state-of-art efforts have been made in the water purification technologies using heterogenous catalysis (Bessekhouad et al. 2004; Bessekhouad et al. 2005; Cao et al. 2011; Chen et al. 2012; Furukawa et al. 2011; Herrmann et al. 1997; Li et al. 2009; Li et al. 2011; Matthews 1987; Matthews 1991). The heterogenous catalyst can be categorizes in to type-I and type-II based on its band types. The type-I catalyst band gap found to be a straddle band gap between two co-existing semiconductors materials (e.g GaAs/GaAlAs, CdSe/ZnSe) (Furukawa et al. 2011; Huang et al. 2013; Jeon et al. 2011; Nasr et al. 1997; Ostermann et al. 2009; Schultz et al. 2012; Siedl et al. 2009; Su et al. 2011; Thibert et al. 2011). The type-II catalyst band gap found to be a staggered band gap between

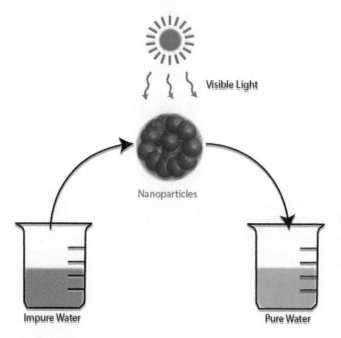

Fig. 7.1 Visible light driven environmental remediation. The advanced materials like nanoparticles are converted the impure water into purified water

two co-existing semiconductors materials (Ag_2O/Bi_2WO_6, CdS/TiO_2) (Bessekhouad et al. 2004; Cao et al. 2011; Chen et al. 2012; Khan et al. 2012; Li et al. 2012; Liu et al. 2012; Nasr et al. 1997; Shi et al. 2012; Wang et al. 2012; Yu et al. 2012). To be effective, both type of catalysts is associated with two difference semiconducting materials to create the heterojunction which provides unprecedented capability of light absorption and charge separation. Both type-I and type-II proven to be effective and it has been mere choice of type while choosing the environmental materials. In this section, we have given general overview of technology, degradation mechanism and factor influencing catalytic activity.

7.1.2.1 The Contaminants of Domestic Water

The industrial waste containing many type of contaminants including biological, chemical and physical materials. Within the scope of the present work, chemical contaminants in water remediation is analyzed. Chemical contaminants can be organic and inorganic type chemicals with micro or macro size. As by the definition, water remediation method should completely mineralize the organic impurities without any side-product toxins. The purification process is essentially convert the toxic chemicals in to the less toxic ones. Organic contaminants are synthetic compounds namely 1. Dyes: dyes are colorful materials often released by textile industry e.g. rhodamine B, congo red, methylene blue and methylene blue. The dyes are homogenous impurities and major cause of the water pollution, 2. Aliphatic compounds: They are often come from side product in mass production of chemicals e.g. carbon tetrachloride, chloroform, bromoform and glycolic acid. Aromatic compounds are inevitable in industry often found in the plastic and emulsion industries as solvent and reactant. The major aromatic compounds are 2,4-dichlorophenol, toluene, 4-nitrophenol, chlorobenzene and benzoic acid. The aromatic compounds are miscible or partially miscible with water. But aromatic compounds have very good affinity towards the catalytic surfaces due to anchoring group and hence the heterogenous catalyst is effective. Other very common form water pollutants are pesticides parathion, dicarzol, diuron, alachlor, trichlorfon and turbophos. Since most of the second world countries are depend on agriculture, the mass scale pollution is raised from the pesticides. There is other type of contaminants which can contribute less towards water pollution such as drugs (e.g paracetamol, proguanil, chloromycetin and explosives (tear gas chloropicrin). As per the catalytic theory, the contaminants adsorbed on the active surface of the catalyst. The contaminants form elastic bond with the surface of the catalyst through the anchoring groups (e.g –OH group of the phenol anchor with the catalyst through co-ordinate bond). The contaminants may photodegrade at the surface of the photocatalyst where critical bonds are broken and the catalyst become less toxic.

7.1.2.2 The Mechanism of Photocatalytic Water Remediation

In the photocatalysis reaction, both the catalysis and photolysis occur simultaneously. In the catalysis reaction, the catalyst is absorbent and contaminants are adsorbate. During the degradation process, contaminants adsorb at the active sites of the catalytic surface and degrades gradually. The entire catalytic process happens at the surface of the photocatalyst. During the irradiation, catalyst absorb the irradiated light based on its absorption region such UV or visible. The UV region is 200–400 nm where pi-pi transition occurs and visible region is 400–750 nm where n-pi transition occurs. The band gap of the catalyst is determined by separation of valance-conduction band and different materials have different types of bandgap. When the energy of the absorbed light is equal or higher than bandgap of the photocatalyst, the electron excitation occurs in the catalyst. The excitation process generates the hole and electron pair (Fig. 7.2). Any adsorbent contaminants come in contact with surface undergoes chemical modification. The photogenerated the electron-hole pair and promoted the degradation process through the various photochemical pathways. The detailed description of all photochemical pathways is beyond the scope of the present work. Because catalysis widely depend on structure and transition states and difficult to characterize the primary events. The electron-hole pair promote the redox reaction and generate the active free radicals. The free radicals are very active and aggressively facilitate breaking of structure any type of contaminants. Eventually, the adsorption process weakens the inter-atomic bonds in the contaminant molecules results in the rupture and degradation respectively. The photocatalyst accelerate the degradation process through the effective pathway of

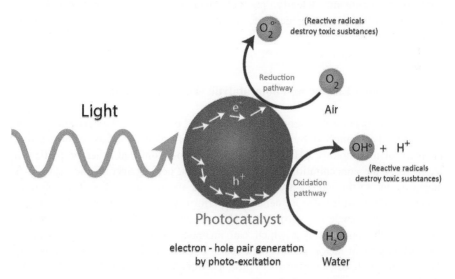

Fig. 7.2 Absorption process of the photocatalyst. While irradiation, photocatalyst generate the electron-hole pair and it further promotes free radical formation. The free radicals are very active and destroy the contaminants

lowering the energy which is possible by the lowering the active energy of the reaction. Hence, when compared to the uncatalyzed reaction, rate of the reaction is much higher.

$$\text{Photocatalyst} \xrightarrow{h\nu} e^{\ominus} \xrightarrow{O_2} O_2^{\bullet} \xrightarrow{\text{Dyes}} H_2O_2 \xrightarrow{\text{Dyes}} H_2O \quad (7.1)$$

$$\text{Photocatalyst} \xrightarrow{h\nu} h^{\oplus} \xrightarrow{H_2O} OH^{\bullet} \xrightarrow{\text{Dyes}} CO_2 + H_2O \quad (7.2)$$

The degradation of contaminants occurs through two of the following process. The photoexcitation generates the electron and the photoelectron react with the oxygen molecules (Eq. 7.1). The reaction produces the oxygen free radicals then it is converted to oxidizing agent hydrogen peroxide. The peroxide degrades the any types of the dyes or contaminants present in the solution. Indirectly, the photogenerated electron transfer it energy to the contaminated molecule and decompose it. The overall reduction of contaminants proceeds through the reduction pathway by using the electron. Another path way of degradation of contaminants occurs through the following process. The photoexcitation generates the holes and the photo-hole react with the water molecules (Eq. 7.2). The reaction produces very reactive hydroxyl free radicals. The hydroxyl radicals degrade the any types of the dyes or contaminants present in the contaminant solution. Indirectly, the photogenerated hole transfer it energy to the contaminated molecule and decompose it. The overall reduction of contaminants proceeds through the oxidation pathway by using the holes. Practically, different types of photocatalyst is required to different types of contaminants. Clearly, photochemical processes are generated by the photocatalyst which in turn generate the free radicals which further degrade the contaminants.

7.1.2.3 Factors Enhancing Photocatalytic Water Remediation

Since the photocatalysis is sensitive to operational parameters, the performance can be effectively controlled by the concentration of catalyst, nature of impurities, intensity of the photons, pH of the solution, temperature and other properties. If we increase the concentration of the photocatalyst, performance of the catalysis is increases due to the more active sites and more absorption centers. The nature of contaminant impurities also influences the performance of the catalyst. The contaminants with the anchoring group, which can tentatively bond with photoactive surface of the photocatalyst which can increase the performance considerably. Another important factor to be considered is intensity of irradiated photon used in the catalytic process. The intensity of proton is directly proportional to number of the photogenerated electron and holes. Hence intensity of irradiated photon is key factor for the effective catalytic performance. Further, controllable parameter is pH of the reaction medium. The reaction medium with low pH has positive charge on the surface of the catalyst due to the excess of the protons. Similarly, low pH values have high concentration of OH⁻ ions and induce the negative charge on the surface of the catalyst. If the contaminants are charged, or Lewis base or acid, the charge on the surface can attached or detach contaminants effectively due to electrostatic interaction. Hence the performance of the catalyst can be tuned to the maximum level by adjusting the pH of the solution. The charge on the surface also prevent the aggregation of the surface of the catalysts due to same charge on the surface and hence electrostatic repulsion. The temperature is a conventional key parameter enhance the performance of the catalyst. By increasing the temperature, the collision between the photocatalyst and contaminant adsorbents increases which reflected in catalyst-contaminant binding efficiency. Hence performance of the catalyst is linearly related with the temperature of the reaction. The other parameters also influence the performance of catalyst such as shape of catalyst, nano size of photocatalyst, morphology of the catalyst and band alignment of photocatalyst. Relating performance with those parameters are beyond the scope of the present work. In summary, by controlling above mentioned parameters, effectiveness and efficiency of the catalyst can be improved to desired level.

7.1.3 State of Art Materials for Water Remediation

The photocatalytic based water remediation is an exciting area and considerable efforts put forward and numerous state of art materials were developed (Dong et al. 2015; Malato et al. 2009). The materials which can co-coordinatively bond with more number of contaminants were found to be effective catalysis such as TiO_2 and ZnO. Among that TiO_2, (Fujishima and Honda 1972) is very first compound utilized for the photocatalytic degradation and it currently fine-tuned to highest performance (Dapeng and Jiuhui 2009; Sin et al. 2012). Along with photochemical performance, TiO_2 gained it position by very good stability towards photons and less corrosive to most of the chemical contaminants. Hence, TiO_2 is ideal model system for photocatalytic water degradation and by following TiO_2 many associated catalysts was developed. Most of the cations are the transition metals and such as Ti, Fe, Ni, Zn, Cd, Cu, Ag and more (Ajmera et al. 2002; Bessekhouad et al. 2004; Bessekhouad et al. 2005; Bi and Xu 2013; Choi and Hoffmann 1996; Kondo and Jardim 1991; Liu et al. 2013; Srinivas et al. 2015; Thuy et al. 2014; Tongying et al. 2012; Wenhua et al. 2000). Since the transition metals are having partially filled d-orbitals, high charge forming ability and have good coordinating tendency make them good candidates for the photocatalysis process. The high charge of the co-ordination metals (Ti^{4+}, Mn^{5+}) helps the ligand molecules (adsorbents) by electrostatic attraction. The bigger size of the transition metals also provides the possibility of high coordination number and hence one metal on photocatalytic surface can adsorb up to six molecules. High charge and high co-ordination number of transition metals synergistically promote adsorption of the contaminants. The anionic species in photocatalyst is mostly chalcogenide elements namely sulfide, selenide and telluride. Chalcogenides have dominant covalent bond forming ability and made photocatalyst covalent in nature. It is to be mentioned that anions bond with cation is covalent form whereas contaminants bond with cation though co-ordination induced adsorption process. Novel photocatalyst includes multicomponent oxides, metals, nanocomposites, core-shell nanoparticles and photocatalyst with active supports. Among the heterogenous catalyst, type-I catalyst has heterojunctions with straddle band gap. Some notable type-I catalysts are CdSe/CdS, Bi_2S_3/CdS, Fe_2O_3/$SrTiO_3$ and many more (Fang et al. 2011; Schultz et al. 2012; Thibert et al. 2011). Among the heterogenous catalyst, type-II catalyst have heterojunctions with staggered band gap. Some notable type-II catalysts are CdS/TiO_2, Bi_2S_3/TiO_2, In_2O_3/TiO_2 and many more(Bessekhouad et al. 2004; Chen et al. 2012; Liu et al. 2012).

7.2 Earth Abundance: Matter of Importance

Heterogenous catalysis for water remediation has a huge potential to become commercially viable technology. The water remediation occurs all over the earth, the real situation requires tonnes of materials. Hence choice of earth abundant material for the photocatalytic water remediation have the paramount importance. Because even if we achieve the highly effective catalyst with precious/non-abundant materials, the materials will deplete over commercialisation before it solves the real-world remediation problem. Therefore, the analysing the material abundance for the materials involved in environmental remediation is meaningful path for commercialization of the technologies. Hence, we represented abundance of cations and anions based on its earth abundance (Fig. 7.3). Among the anions, oxygen (46%), sulphur (0.042%) and phosphorus (0.099%) are best bet on photocatalytic materials. It is to be mentioned that selenium is being favourable choice of photocatalytic materials. But the abundance is very low 5×10^{-6}% and it would not be viable material for scale up and commercialisation. Similarly, most abundant cations are

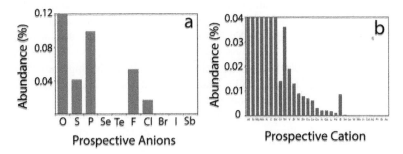

Fig. 7.3 Abundance of photocatalytic materials: (**a**) Abundance of prospective anions. Oxygen and sulfur is favourable choice of anion (**b**) Abundance of prospective cation and Aluminum and magnesium favorable choice of cation

aluminium (8.1%), silicon (27%) and unfortunately, the photocatalyst research developments did not integrate the materials for the water remediation. The popular choice the materials are titanium (0.66%), cadmium (1.5×10^{-5}%), zinc (0.0078%), copper (0.0068%), silver (7.9×10^{-6}%), bismuth (2.5×10^{-6}%), tin (0.00022%) and Nickel (0.009%). Ironically, state-of-art material titanium is existing in earth about 0.66%. The materials also have the issue of extraction technology where primary and daughter materials extraction (Graedel 2011). The daughter materials are materials extracted along with side of primary materials which cannot be extracted separately. Notable primary materials copper, nickel, zinc, lead and aluminium. For example, copper is extracted as primary materials and the daughter materials of the copper are cobalt, selenium, tellurium. Explaining the influence of extraction technologies in abundance is beyond the scope of present work. Hence, the research direction of photocatalytic water remediation towards the high abundant materials, ideally aluminium and oxygen. The materials abundance is new scarcity and makes the photocatalytic technology costly for commercialization as well as implementation. Therefore, we suggest that if we controllably design and prepare the earth abundant photocatalyst, we could realize photocatalyst based water remediation technology.

unconsciously, very few excellent efforts addressed the environmental issue in the current literature. Our intention is not to present exhaustive review of the reported work on abundant materials, but rather to note new research direction and reported the representative examples. We presented mostly oxides of high abundant and moderately abundant materials.

7.2.1 Aluminium Compounds as Earth Abundant Materials

If we focus about the earth abundant material research, aluminium is an ideal choice of cation. But the aluminium is suffered by its high band gap and not been the favorable choice of materials. But the novel compounds like metal organic frame works (MOFs) may integrate the aluminum in photocatalytic research. Fujita et al. initially revealed potential use of the MOFs as heterogeneous catalyst (Fujita et al. 1994). MOFs have active metal ion at the surface and inside cavity provides high regioselectivity, stereoselectivity, and shape selectivity for the photocatalytic applications. MOFs have wide attention due to the fact that, the cavity shape, size and surface metal atoms of MOFs can be tuned to the required level of photocatalysis. The MOFs is combination of organic molecules with inorganic metals and hence it provides excellent chemical tunability to photocatalyst which help to optimise the catalyst for best performance. After that, Martín et al. reported that ruthenium complexation with aluminium metal–organic framework and its application in alcohol oxidation catalysis with high selectivity (Carson et al. 2012). Along with the aluminium, bipyridine and ruthenium in the metal organic frame work. The aluminium was introduced in the metal organic frame works through the post synthetic modification. The introduction of the aluminium provide the stability, crystallinity and commercial viability to metal organic frame work (Kang et al. 2011). Moreover, incorporation of aluminum provides, excellent recyclability for metal organic frame work up to six times. The novel catalyst oxidized variety of secondary alcohols to ketones in mild reaction conditions with turn over number of 1980 and yield up to 99%. The attractive results are due to its effective charge separation of metal organic frame works. Since metal attached with linker organic molecules, one the electron-hole pair is generated and its flowed and separated through linkers. The advanced application of the MOF is demonstrated by oxidation of cholestanol to cholestanone successfully. Over all, integration of aluminium is challenging and the MOFs provided an opportunity to develop commercially viable photocatalyst.

7.2.2 Metal Oxides and Sulfides as Earth Abundant Materials

Oxygen is a most abundant material on earth 46% and have excellent photochemical stability. Hence it been a favorite choice and numerous number of efforts made by the researchers. To main the focus of the present work, herein we discussed the earth abundant metal oxides with photocatalytic application. After titanium, copper is moderately abundant material when compare to cadmium or gallium. Copper is narrow band gap semiconductor and p-type and have the potential for the commercially viable catalyst. Unlike other photocatalytic material, copper is bio-friendly and can be implemented to any public technology without any regulation. Pathania et al. (Katwal et al. 2015) reported Electrochemically synthesized copper oxide nanoparticles with photocatalytic activity. The prepare nanoparticle is in nano-regime and they control the size and shape electrochemically. The particle size is around 4 nm and nano surface provided excellent surface to volume ratio and made the materials effective for the photocatalysis. The CuO had degraded the cationic dyes namely methylene blue, methyl red and congo red and analyzed through absorption spectra. Within 2 hours, the color of the solution disappears and dyes were decomposed completely from water.

After oxygen, the favorable choice of the photocatalytic material is sulfur. Since the titanium sulfide is conductive nature it cannot be effectively used for the photocatalytic purpose. Hence the copper can be used instead of titanium and has potential commercial value. Srinivas et al. reported that photocatalytic degradation of rhodamine B using CuS nanostructures (Srinivas et al. 2015). The method is known as hexmethyldisilazane assisted synthesis and gave ligand free surfaces which is ideal for the photocatalytic activity. The method nanoflowers as nanostructures with the dimensions of 9–35 nm. They have demonstrated that CuS flowers rhodamine dye in just 12 minutes. The bare clean surface of the nanoparticles influences the photocatalytic activity and 64% catalyst is degraded in just 2 minutes. The catalyst performance can be increased up to 4.2 time by increasing light intensity. Similarly, performance of the catalyst can be increased up to 3.3 times by increasing the concentration of the catalyst.

7.2.3 Carbon Based Materials as Earth Abundant Materials

Carbon abundance is 0.18% is better than sulfur and selenium. Moreover, carbon is chemically versatile and have excellent synthetic handles because of its rich chemistry knowledge. But carbon-based nanomaterials long been discouraged due to it homogenous nature and its challenging to recycle. But the new materials like graphene provided a revival of carbon based heterogenous catalysis. Recently many efforts emerging based on the carbon or carbon integrated catalysts. Typical example is reported by Hui et al. and they reported metal free C_3N_4 photocatalyst for the phenol degradation (Zhang et al. 2016a).The C_3N_4 photocatalyst has band gap of 2.7% and has high photochemical stability, makes the ideal candidate for the photocatalysis. The C_3N_4 has the layered structure and due to Vander Waals interaction between the adjacent layers and layer structure provided the effective light capture, charge separation and optimized for the degradation. The band alignment in the regime of the photocatalyst is about CdS and Cu_2O and could be viable replacement for conventional non-abundant photocatalytic materials. The nano regime of 3.6 nm provided another handle to tuned the photocatalyst to high performance. The C_3N_4 catalyst degrade the phenol efficiently when compared to dark conditions. About, 0.5% of the C_3N_4 catalyst degraded the phenol in 3 hours and follow the pseudo first order kinetics.

Recently, more number of publication using graphene and graphene based materials for photocatalytic water purification (Upadhyay et al. 2014). The graphene attained wide attention due its high absorption coefficient, tunable optical behavior, high surface area, mechanical strength, stability and cost-effectiveness. It is notable that adsorption capability of the graphene is high and made it ideal for the photocatalyst studies. Lee. et al. reported that hydrothermal synthesis of carbon based nanomaterial multiwall carbon nanotubes (MWCNT) and its photochemical degradation of the Rhodamine B was analyzed (Pawar et al. 2015). They reported the composite Fe_2O_3/MWCNT has high performance than the individual Fe_2O_3 nanoparticles. The demonstrated that Fe_2O_3/MWCNT can degrade rhodamine dye in 2 h and it was followed by UV analysis. The claimed that Fe_2O_3/MWCNT catalyst removed the methyl groups from the rhodamine B and remove consecutively. The improved photocatalytic activity is due to high surface area and high density of catalytic centers in the surfaces of the catalyst which enhance the absorption and charge separation. They reported that the catalyst may be part of green and clean technology for the purification of the water. Overall, the carbon and nano regime of the Fe_2O_3/MWCNT photocatalyst provides the considerable photocatalytic performance.

7.3 The Solving the Problem of Abundancy

Any world problem must have various parallel solution with immediate, middle and long-term solution. The solutions for water remediation problem aspired to be permanent and sustainable Herein, we described the four-possible solution for earth abundance materials based research and development (Table 7.1). The solution suggested here is representative but not exhaustive. The expectation is all the solution of sustainable water remediation should be executed in parallel to address the problem in world scale.

7.3.1 High-Efficient Material Design During Pre-Synthesis

Material efficiency means making more out of the catalyst by using less. This is possible by increasing the photocatalytic efficiency of the catalyst (Ma et al. 2013; Mushtaq et al. 2016; Tanwar et al. 2017; Torres-Martínez et al. 2001; Wang et al. 2011). Hence the degradation process requires nanogram of catalyst instead of grams of catalyst. The solution is temporary, by increasing efficiency, we just postponing the materials depletion for few years and the end of material abundance eventually will come. Moreover, the process of increasing efficiency requires tremendous effort from academia and industry. Because fundamentals of photocatalysis such as charge

Table 7.1 Solving the problem of non-earth abundant materials in photocatalytic materials research

S. No	Possible solution for use of non-abundance of materials	How it works	Issue
1	High efficient material design	Increasing the efficiency of the state-of-art materials. Nano gram of materials can be used instead of the grams.	Increase the efficiency of current materials is a short-term solution. Eventually non-abundance will come, but delayed. Hence the solution is temporary
2	Recycling	Recycling the existing state-of-art heterogenous materials after remediation process.	Recycling process is cost intensive process. The collection of materials, purification included in the recycle cost.
3	Using two or more efficient materials	Directing research from less abundant materials to two or more moderately abundant materials e.g. Sulphur, copper	Since two materials involved, various skills required for manufacturing and scale up.
4	Earth abundant materials made material design	Directing research to earth materials e.g. aluminium and oxide	Band gap engineering of new materials could be difficult. But the solution is permanent and sustainable.

separation and charge injection not understood for the many materials. The energy management and energy-economy should be maintained in every step of the photocatalytic degradation process. The efficiency also refers to the production efficiency, such as minimization use of the raw materials and all operations. Current trial and error approach may not be best approach to attain the high efficiency. Hence the research direction requires serious collaboration with the chemists, physicist, material scientist, electronics and photo-chemists which research direction utilize every possible opportunity. Alternatively, the nanomaterials are providing wonderful opportunity for the enhancing the efficiency (Farnesi Camellone and Marx 2013; Jabbari et al. 2016; Savage and Diallo 2005). Since the nanomaterials has high surface to volume ratio, it provided extra catalytic centers on the photocatalytic surfaces. Moreover, nanoparticles band gap is depending on the size of the nanoparticles which provides the unexplored opportunity for the photocatalyst bandgap engineering. Due to the small surface area charge separation and charge transfer also easy in nano photocatalyst. Hence nanotechnology can empower the photocatalytic technology and have the promise of high efficiency. In summary, the solution of high efficient material design is short term and requires cooperation from material design as well as research and development.

7.3.2 Recycling the Utilized Photocatalytic Materials

Recycling means utilizing the used catalyst instead of mining the materials and this may address the earth abundant materials problem. The photocatalyst is heterogenous in nature and it can be recycled numerous times with negligible loss photocatalytic activity which is applicable to both powder and the thin films. It is to be mentioned that photocatalyst must be extracted from the catalytic medium after the catalysis. Then it must clean up, reactivated and reused for the next cycle for the catalysis. The solution is permanent but the technology for recycling process should be developed separately. It is extremely unlikely to implement the recycling method due to economics involved in the process. From the industrial perspective, the recycling is money intense process and involving collection, separation, recycling of photocatalyst, time and labor cost. Recently, magnetically recoverable catalysts gain wide attention due to the simplicity in the process and zero loss during the recovery. The process improves the separation efficiency and recovery. It is believed that recycling technology completely dominated by the magnetic materials (Ambashta and Sillanpää 2010; Hu et al. 2016; Lu et al. 2007; Zhang et al. 2010; Zhang et al. 2016b). But the water remediation process restricts the photocatalytic water remediation process to few magnetic materials (Mamba and Mishra 2016). Recycling process is more of engineering and technology than the science. Hence the problem of recycling associated with the industries and solution should industry conducive. Another important factor to consider the while recycling process is stability of active materials. The materials must be stable towards photons, charge transfer, oxidation, reduction and leaching process. Due to this factor materials recycling tendency is different for different materials and all materials cannot be recycled many times. In summary, recycling the photocatalytic materials indirectly maximize the life time of the catalyst. But the balance of economics should be achieved to realize the sustainable next generation technology.

7.3.3 Using Two or more Efficient Materials by Optimised Combination

Earth abundance issue can be solved by using two or more moderately earth abundant materials in photocatalysis. For example, develop the TiO_2 and CuS based photocatalyst and use it as two different technology for water remediation. Hence, total abundance of materials used in the photocatalyst meets the world-scale demand. The solution is also temporary and eventually postpone the materials depletion. The solution can be simply explained as selective materials for the selective water remediation. One another similar solution is using the composite materials, the composite materials have the advantages of the both components and enhance the efficiency (Pawar et al. 2015). Hence, composting the scarce materials with the earth abundant materials is another meaningful direction. Another similar direction is using core-shell nanoparticles for the water remediation (Kaur et al. 2014; Khanchandani et al. 2013; Wang et al. 2013; Zhu et al. 2010). Because the core and shell is different materials and different function can be integrated. Another important advantage of core-shell materials is charge separation through the band gap engineering. That is if we coat the low band gap material than core, the excited electron is free flowing and can be utilized effectively for photocatalysis (Balet et al. 2004; Dorfs et al. 2008; Ivanov et al. 2007). The solution of using two or different materials for water remediation is scientific and not favorable in terms of technological point view. Because two different technology requires different set of the skills to operate and different type of training required for labor. Another issue is that research community generally chooses the materials at will, hence convincing them to adopt two different materials for research and development is impractical. Overall, using two or more efficient materials research is temporary and provide immediate solution.

7.3.4 Earth Abundant Materials Based Material Design

The meaningful solution for the less abundant materials used in the photocatalytic water remediation process is directing the research towards more abundant materials such as aluminium and oxygen. To the best of our knowledge, the solution is permanent, feasible and sustainable in the water remediation. By our analysis ideal choice of cations are aluminium, silicon, copper, magnesium and barium. Similarly, ideal choice for the anions are oxygen, sulfur and phosphorus. Unfortunately, there are no present studies concerning in terms of materials abundance in environmental radiation technologies for high volume commercial application. The technology is nascent, still possible to shift the direction of water remediation technology to the sustainability and affordability. Hence new synthetic methods should be encouraged to create new photocatalytic materials with high abundance elements. The research direction can be advanced by developing theories and computation models in new photocatalytic materials. The main challenge associated with the solution is understanding the fundamental of the new catalyst and exploit them. Because new catalyst must be with optimal band gap for absorption and should have more catalytic surface centers which will reflect in photon collection and charge separation. Hence, focused and committed research efforts required in this direction of material abundance based research approach. This is possible by new type of material regime such as nanoscale materials and MOFs. Another issue is that selectivity of photocatalyst towards variety of contaminants present in water. The catalyst expected to be universal in terms of anchoring contaminants but further it need to be identified and developed. Analytically, the approach of using earth abundant materials provides expansion of photocatalytic materials based water remediation technology and industrial level scale up. In summary, the integrating earth abundant materials will provide the permanent solution and we could realize photocatalyst based water remediation technology.

7.4 Conclusion and Outlook

Our work suggests that world problems like environmental remediation not solved by merely considering research and development but abundance also should be taken in to consideration. The materials abundance based research should be taken as research approach for industry and academia. This approach not only solve the problem of material abundance but also integrate the solution of future problems. Heterogenous catalysis is ideal solution for the environmental remediation. The new heterogenous catalysis technologies and products should have the vision of sustainability and hence materials abundance.

The ever-increasing demands of sustainability forces the mass production also in the sustainable way such as abundance of raw materials. The present work informs the importance of abundance while designing heterogenous photocatalytic materials for water treatment process. The problem of abundance can be solved in terms of the four categories. First is to increase the efficiency of current running horse materials. While increase the efficiency, the less amount of the catalyst required to the cleanup water. But at some point of time, end of the material abundance will come. Second is materials used in the heterogenous catalysis can be recycled. But the recycling at the global scale itself a new technology and not possible at current scenarios. Third is use two or more less earth abundant materials for the heterogenous catalyst based water remediation. That is, for few occasions aluminium can be used and some places silicon can be used. But the issue with this approach is different materials requires different skills to operate and hence technology. Hence scaling up is not the meaningful solution. Fourth method is directing the research and development to the earth abundant materials like iron or magnesium and its oxides. There are few unconscious efforts made one this direction, however, efforts still for from addressing the issue at global scale. Hence focused and committed research efforts required in this direction of material abundance based research approach. Finally, the problem of earth abundant materials could be solved by using above one solution or combinations for suitable water remediation problem.

Our humble warning is to use earth abundant materials for sustainable water remediation. Otherwise the water remediation process ends up like current petroleum-depletion and new renewable solutions required in future. It is to be mentioned that the main challenge to use earth abundant materials would be making the materials high efficient like current state of art materials. It is not always possible to make the materials high efficient, but the problem can be addressed by material engineering such as nanoscale preparation, doping etc. By choosing earth abundant materials based design, the scope, selectivity, performance, stability, reactivity characteristics and electron flow of reactions of photocatalyst can be improved and hence the overall competency of the catalyst among the existing one. We foresee and attempted to bring out the clear picture of the earth abundance crisis in the heterogenous catalytic based environmental remediation. The emphasis mainly given to material abundance in scale up and expect that it will be useful to selection of materials for environmental remediation. The water quality is most important issue, and it is directly reflecting the health of mankind and meticulous analysis of long term impact of new earth abundant photocatalysts should be analyzed before implementation. The core concepts explained in this work could be extended to the other existing technologies.

Acknowledgements We thank the editor Dr. Saravanan Rajendran for valuable suggestions, feedbacks, and discussions. B.G.K Thank the P.S.R group for institutions for infrastructure and funding.

References

Ajmera AA, Sawant SB, Pangarkar VG, Beenackers AA (2002) Solar-assisted photocatalytic degradation of benzoic acid using titanium dioxide as a photocatalyst chemical engineering & technology. 25:173–180. https://doi.org/10.1002/1521-4125(200202)25:2<173::AID-CEAT173>3.0.CO;2-C

Ambashta RD, Sillanpää M (2010) Water purification using magnetic assistance: a review. J Hazard Mater 180:38–49. https://doi.org/10.1016/j.jhazmat.2010.04.105

Arvidsson R, Sandén BA (2017) Carbon nanomaterials as potential substitutes for scarce metals. J Clean Prod 156:253–261. https://doi.org/10.1016/j.jclepro.2017.04.048

Balet L, Ivanov S, Piryatinski A, Achermann M, Klimov VI (2004) Inverted core/shell nanocrystals continuously tunable between type-I and type-II localization regimes. Nano Lett 4:1485–1488. https://doi.org/10.1021/nl049146c

Bessekhouad Y, Robert D, Weber J (2004) Bi2S3/TiO2 and CdS/TiO2 heterojunctions as an available configuration for photocatalytic degradation of organic pollutant. J Photochem Photobiol A Chem 163:569–580. https://doi.org/10.1016/j.jphotochem.2004.02.006

Bessekhouad Y, Robert D, Weber J-V (2005) Photocatalytic activity of Cu2O/TiO2, Bi2O3/TiO2 and ZnMn2O4/TiO2 heterojunctions. Catal Today 101:315–321. https://doi.org/10.1016/j.cattod.2005.03.038

Bhatkhande DS, Pangarkar VG, Beenackers AACM (2002) Photocatalytic degradation for environmental applications–a review journal of Chemical Technology & Biotechnology: international research in process. Environ Clean Technol 77:102–116. https://doi.org/10.1002/jctb.532

Bi D, Xu Y (2013) Synergism between Fe2O3 and WO3 particles: Photocatalytic activity enhancement and reaction mechanism. J Mol Catal A Chem 367:103–107. https://doi.org/10.1016/j.molcata.2012.09.031

Cao T, Li Y, Wang C, Shao C, Liu Y (2011) A facile in situ hydrothermal method to SrTiO3/TiO2 nanofiber heterostructures with high photocatalytic activity. Langmuir 27:2946–2952. https://doi.org/10.1021/la104195v

Carson F, Agrawal S, Gustafsson M, Bartoszewicz A, Moraga F, Zou X, Martín-Matute B (2012) Ruthenium complexation in an aluminium metal–organic framework and its application in alcohol oxidation catalysis. Chem Eur J 18:15337–15344. https://doi.org/10.1002/chem.201200885

Chen Y-C, Pu Y-C, Hsu Y-J (2012) Interfacial charge carrier dynamics of the three-component In2O3–TiO2–Pt heterojunction system. J Phys Chem C 116:2967–2975. https://doi.org/10.1021/jp210033y

Chirik P, Morris R (2015) Getting down to earth: the renaissance of catalysis with abundant metals. ACS Publications. https://doi.org/10.1021/acs.accounts.5b00385

Choi W, Hoffmann MR (1996) Novel photocatalytic mechanisms for CHCl3, CHBr3, and CCl3CO2-degradation and the fate of photogenerated trihalomethyl radicals on TiO2 Environmental science & technology 31:89–95 doi:https://doi.org/10.1021/es960157k

Chun H, Yizhong W, Hongxiao T (2000) Destruction of phenol aqueous solution by photocatalysis or direct photolysis. Chemosphere 41:1205–1209. https://doi.org/10.1016/S0045-6535(99)00539-1

Corma A, García H, Llabrés i, Xamena F (2010) Engineering metal organic frameworks for heterogeneous catalysis. Chem Rev 110:4606–4655. https://doi.org/10.1021/cr9003924

Dapeng L, Jiuhui Q (2009) The progress of catalytic technologies in water purification: a review. J Environ Sci 21:713–719. https://doi.org/10.1016/S1001-0742(08)62329-3

Descorme C, Gallezot P, Geantet C, George C (2012) Heterogeneous catalysis: a key tool toward sustainability. ChemCatChem 4:1897–1906. https://doi.org/10.1002/cctc.201200483

Dong S et al (2015) Recent developments in heterogeneous photocatalytic water treatment using visible light-responsive photocatalysts: a review. RSC Adv 5:14610–14630. https://doi.org/10.1039/C4RA13734E

Dorfs D, Franzl T, Osovsky R, Brumer M, Lifshitz E, Klar TA, Eychmüller A (2008) Type-I and type-II nanoscale Heterostructures based on CdTe nanocrystals: a comparative study. Small 4:1148–1152. https://doi.org/10.1002/smll.200800287

Fang Z et al (2011) Epitaxial growth of CdS nanoparticle on Bi2S3 nanowire and photocatalytic application of the heterostructure. J Phys Chem C 115:13968–13976. https://doi.org/10.1021/jp112259p

Farnesi Camellone M, Marx D (2013) On the impact of solvation on a au/TiO2 nanocatalyst in contact with water. J Phys Chem Lett 4:514–518. https://doi.org/10.1021/jz301891v

Fujishima A, Honda K (1972) Electrochemical photolysis of water at a semiconductor electrode. Nature 238:37. https://doi.org/10.1038/238037a0

Fujita M, Kwon YJ, Washizu S, Ogura K (1994) Preparation, clathration ability, and catalysis of a two-dimensional square network material composed of cadmium (II) and 4, 4′-bipyridine. J Am Chem Soc 116:1151–1152. https://doi.org/10.1021/ja00082a055

Furukawa S, Shishido T, Teramura K, Tanaka T (2011) Photocatalytic oxidation of alcohols over TiO2 covered with Nb2O5. ACS Catal 2:175–179. https://doi.org/10.1021/cs2005554

Graedel T (2011) On the future availability of the energy metals. Annu Rev Mater Res 41:323–335. https://doi.org/10.1146/annurev-matsci-062910-095759

Guillard C et al (2003) Solar efficiency of a new deposited titania photocatalyst: chlorophenol, pesticide and dye removal applications. Appl Catal B Environ 46:319–332. https://doi.org/10.1016/S0926-3373(03)00264-9

Hajiesmaili S, Josset S, Bégin D, Pham-Huu C, Keller N, Keller V (2010) 3D solid carbon foam-based photocatalytic materials for vapor phase flow-through structured photoreactors. Appl Catal A Gen 382:122–130. https://doi.org/10.1016/j.apcata.2010.04.044

Herrmann J-M, Tahiri H, Ait-Ichou Y, Lassaletta G, Gonzalez-Elipe A, Fernandez A (1997) Characterization and photocatalytic activity in aqueous medium of TiO2 and ag-TiO2 coatings on quartz. Appl Catal B Environ 13:219–228. https://doi.org/10.1016/S0926-3373(96)00107-5

Hu Y et al (2011) BiVO4/TiO2 nanocrystalline heterostructure: a wide spectrum responsive photocatalyst towards the highly efficient decomposition of gaseous benzene. Appl Catal B Environ 104:30–36. https://doi.org/10.1016/j.apcatb.2011.02.031

Hu L et al (2016) Fabrication of magnetic water-soluble hyperbranched polyol functionalized graphene oxide for high-efficiency water remediation. Sci Rep 6:28924. https://doi.org/10.1038/srep28924

Huang L, Wang X, Yang J, Liu G, Han J, Li C (2013) Dual cocatalysts loaded type I CdS/ZnS core/shell nanocrystals as effective and stable photocatalysts for H2 evolution. J Phys Chem C 117:11584–11591. https://doi.org/10.1021/jp400010z

Ivanov SA et al (2007) Type-II core/shell CdS/ZnSe nanocrystals: synthesis, electronic structures, and spectroscopic properties. J Am Chem Soc 129:11708–11719. https://doi.org/10.1021/ja068351m

Jabbari V, Hamadanian M, Shamshiri M, Villagrán D (2016) Band gap and Schottky barrier engineered photocatalyst with promising solar light activity for water remediation. RSC Adv 6:15678–15685. https://doi.org/10.1039/C5RA24096D

Jeon TH, Choi W, Park H (2011) Photoelectrochemical and photocatalytic behaviors of hematite-decorated titania nanotube arrays: energy level mismatch versus surface specific reactivity. J Phys Chem C 115:7134–7142. https://doi.org/10.1021/jp201215t

Jia X, Cao J, Lin H, Zhang M, Guo X, Chen S (2017) Transforming type-I to type-II heterostructure photocatalyst via energy band engineering: a case study of I-BiOCl/I-BiOBr. Appl Catal B Environ 204:505–514. https://doi.org/10.1016/j.apcatb.2016.11.061

Johnson J, Harper E, Lifset R, Graedel TE (2007) Dining at the periodic table: metals concentrations as they relate to recycling. Environ Sci Technol 41:1759–1765. https://doi.org/10.1021/es060736h

Kang IJ, Khan NA, Haque E, Jhung SH (2011) Chemical and thermal stability of isotypic metal–organic frameworks: effect of metal ions. Chem Eur J 17:6437–6442. https://doi.org/10.1002/chem.201100316

Katwal R, Kaur H, Sharma G, Naushad M, Pathania D (2015) Electrochemical synthesized copper oxide nanoparticles for enhanced photocatalytic and antimicrobial activity. J Ind Eng Chem 31:173–184. https://doi.org/10.1016/j.jiec.2015.06.021

Kaur R, Hasan A, Iqbal N, Alam S, Saini MK, Raza SK (2014) Synthesis and surface engineering of magnetic nanoparticles for environmental cleanup and pesticide residue analysis: a review. J Sep Sci 37:1805–1825. https://doi.org/10.1002/jssc.201400256

Khan Z, Khannam M, Vinothkumar N, De M, Qureshi M (2012) Hierarchical 3D NiO–CdS heteroarchitecture for efficient visible light photocatalytic hydrogen generation. J Mater Chem 22:12090–12095. https://doi.org/10.1039/C2JM31148H

Khanchandani S, Kundu S, Patra A, Ganguli AK (2013) Band gap tuning of ZnO/In2S3 core/shell nanorod arrays for enhanced visible-light-driven photocatalysis. J Phys Chem C 117:5558–5567. https://doi.org/10.1021/jp310495j

Khin MM, Nair AS, Babu VJ, Murugan R, Ramakrishna S (2012) A review on nanomaterials for environmental remediation. Energy Environ Sci 5:8075–8109. https://doi.org/10.1039/c2ee21818f

Kondo MM, Jardim WF (1991) Photodegradation of chloroform and urea using ag-loaded titanium dioxide as catalyst. Water Res 25:823–827. https://doi.org/10.1016/0043-1354(91)90162-J

Ku Y, Hsieh C-B (1992) Photocatalytic decomposition of 2, 4-dichlorophenol in aqueous TiO2 suspensions. Water Res 26:1451–1456. https://doi.org/10.1016/0043-1354(92)90064-B

Li G-S, Zhang D-Q, Yu JC (2009) A new visible-light photocatalyst: CdS quantum dots embedded mesoporous TiO2. Environ Sci Technol 43:7079–7085. https://doi.org/10.1021/es9011993

Li X, Hou Y, Zhao Q, Chen G (2011) Synthesis and photoinduced charge-transfer properties of a ZnFe2O4-sensitized TiO2 nanotube array electrode. Langmuir 27:3113–3120. https://doi.org/10.1021/la2000975

Li X, Huang R, Hu Y, Chen Y, Liu W, Yuan R, Li Z (2012) A templated method to Bi2WO6 hollow microspheres and their conversion to double-shell Bi2O3/Bi2WO6 hollow microspheres with improved photocatalytic performance. Inorg Chem 51:6245–6250. https://doi.org/10.1021/ic300454q

Liu S, Zhang N, Tang Z-R, Xu Y-J (2012) Synthesis of one-dimensional CdS@ TiO2 core–shell nanocomposites photocatalyst for selective redox: the dual role of TiO2 shell. ACS Appl Mater Interfaces 4:6378–6385. https://doi.org/10.1021/am302074p

Liu D, Zheng Z, Wang C, Yin Y, Liu S, Yang B, Jiang Z (2013) CdTe quantum dots encapsulated ZnO nanorods for highly efficient photoelectrochemical degradation of phenols. J Phys Chem C 117:26529–26537. https://doi.org/10.1021/jp410692y

Lu AH, Salabas EL, Schüth F (2007) Magnetic nanoparticles: synthesis, protection, functionalization, and application. Angew Chem Int Ed 46:1222–1244. https://doi.org/10.1002/anie.200602866

Ma Z et al (2013) Luffa-sponge-like glass–TiO2 composite fibers as efficient photocatalysts for environmental remediation. ACS Appl Mater Interfaces 5:7527–7536. https://doi.org/10.1021/am401827k

Malato S, Fernández-Ibáñez P, Maldonado MI, Blanco J, Gernjak W (2009) Decontamination and disinfection of water by solar photocatalysis: recent overview and trends. Catal Today 147:1–59. https://doi.org/10.1016/j.cattod.2009.06.018

Mamba G, Mishra A (2016) Advances in magnetically separable photocatalysts: smart, recyclable materials for water pollution mitigation. Catalysts 6:79. https://doi.org/10.3390/catal6060079

Matthews RW (1987) Solar-electric water purification using photocatalytic oxidation with TiO2 as a stationary phase. Sol Energy 38:405–413. https://doi.org/10.1016/0038-092X(87)90021-1

Matthews RW (1991) Photooxidative degradation of coloured organics in water using supported catalysts. TiO2 on sand. Water Res 25:1169–1176. https://doi.org/10.1016/0043-1354(91)90054-T

Mills A, Le Hunte S (1997) An overview of semiconductor photocatalysis. J Photochem Photobiol A Chem 108:1–35. https://doi.org/10.1016/S1010-6030(97)00118-4

Mills A, Davies RH, Worsley D (1993) Water purification by semiconductor photocatalysis. Chem Soc Rev 22:417–425. https://doi.org/10.1039/CS9932200417

Mushtaq F, Asani A, Hoop M, Chen XZ, Ahmed D, Nelson BJ, Pané S (2016) Highly efficient coaxial TiO2-PtPd tubular nanomachines for photocatalytic water purification with multiple locomotion strategies. Adv Funct Mater 26:6995–7002. https://doi.org/10.1002/adfm.201602315

Nasr C, Hotchandani S, Kim WY, Schmehl RH, Kamat PV (1997) Photoelectrochemistry of composite semiconductor thin films. Photosensitization of SnO2/CdS coupled nanocrystallites with a ruthenium polypyridyl complex. J Phys Chem B 101:7480–7487. https://doi.org/10.1021/jp970833k

Ostermann R, Sallard S, Smarsly BM (2009) Mesoporous sandwiches: towards mesoporous multilayer films of crystalline metal oxides. Phys Chem Chem Phys 11:3648–3652. https://doi.org/10.1039/b820651c

Pawar RC, Choi D-H, Lee CS (2015) Reduced graphene oxide composites with MWCNTs and single crystalline hematite nanorhombohedra for applications in water purification. Int J Hydrog Energy 40:767–778. https://doi.org/10.1016/j.ijhydene.2014.08.084

Qu X, Brame J, Li Q, Alvarez PJ (2012) Nanotechnology for a safe and sustainable water supply: enabling integrated water treatment and reuse. Acc Chem Res 46:834–843. https://doi.org/10.1021/ar300029v

Ramos-Delgado N, Hinojosa-Reyes L, Guzman-Mar I, Gracia-Pinilla M, Hernández-Ramírez A (2013) Synthesis by sol–gel of WO3/TiO2 for solar photocatalytic degradation of malathion pesticide. Catal Today 209:35–40. https://doi.org/10.1016/j.cattod.2012.11.011

Santhosh C, Velmurugan V, Jacob G, Jeong SK, Grace AN, Bhatnagar A (2016) Role of nanomaterials in water treatment applications: a review. Chem Eng J 306:1116–1137. https://doi.org/10.1016/j.cej.2016.08.053

Savage N, Diallo MS (2005) Nanomaterials and water purification: opportunities and challenges. J Nanopart Res 7:331–342. https://doi.org/10.1007/s11051-005-7523-5

Schultz AM, Salvador PA, Rohrer GS (2012) Enhanced photochemical activity of α-Fe$_2$O$_3$ films supported on SrTiO$_3$ substrates under visible light illumination. Chem Commun 48. https://doi.org/10.1039/c2cc16715h

Shi Y, Li H, Wang L, Shen W, Chen H (2012) Novel α-Fe2O3/CdS cornlike nanorods with enhanced photocatalytic performance. ACS Appl Mater Interfaces 4:4800–4806. https://doi.org/10.1021/am3011516

Siedl N, Elser MJ, Bernardi J, Diwald O (2009) Functional interfaces in pure and blended oxide nanoparticle networks: recombination versus separation of photogenerated charges. J Phys Chem C 113:15792–15795. https://doi.org/10.1021/jp906368f

Sin J-C, Lam S-M, Mohamed AR, Lee K-T (2012) Degrading endocrine disrupting chemicals from wastewater by TiO2 Photocatalysis: a review. Int J Photoenergy 2012. https://doi.org/10.1155/2012/185159

Srinivas B, Kumar BG, Muralidharan K (2015) Stabilizer free copper sulphide nanostructures for rapid photocatalytic decomposition of rhodamine B. J Mol Catal A Chem 410:8–18. https://doi.org/10.1016/j.molcata.2015.08.028

Su J et al (2011) Macroporous V2O5− BiVO4 composites: effect of heterojunction on the behavior of photogenerated charges. J Phys Chem C 115:8064–8071. https://doi.org/10.1021/jp200274k

Sun J, Wang X, Sun J, Sun R, Sun S, Qiao L (2006) Photocatalytic degradation and kinetics of Orange G using nano-sized Sn (IV)/TiO2/AC photocatalyst. J Mol Catal A Chem 260:241–246. https://doi.org/10.1016/j.molcata.2006.07.033

Sunada Y, Nagashima H (2017) Design and development of iron-based non-precious metal catalyst systems. Yuki Gosei Kagaku Kyokaishi/J Synth Organic Chem 75:1253–1263. https://doi.org/10.5059/yukigoseikyokaishi.75.1253

Tanaka K, Padermpole K, Hisanaga T (2000) Photocatalytic degradation of commercial azo dyes. Water Res 34:327–333. https://doi.org/10.1016/S0043-1354(99)00093-7

Tanwar R, Kaur B, Mandal UK (2017) Highly efficient and visible light driven Ni0. 5Zn0. 5Fe2O4@ PANI modified BiOCl heterocomposite catalyst for water remediation. Appl Catal B Environ 211:305–322. https://doi.org/10.1016/j.apcatb.2017.04.051

Tesh SJ, Scott TB (2014) Nano-composites for water remediation: a review. Adv Mater 26:6056–6068. https://doi.org/10.1002/adma.201401376

Thatai S, Khurana P, Boken J, Prasad S, Kumar D (2014) Nanoparticles and core–shell nanocomposite based new generation water remediation materials and analytical techniques: a review. Microchem J 116:62–76. https://doi.org/10.1016/j.microc.2014.04.001

Thibert A, Frame FA, Busby E, Holmes MA, Osterloh FE, Larsen DS (2011) Sequestering high-energy electrons to facilitate photocatalytic hydrogen generation in CdSe/CdS nanocrystals. J Phys Chem Lett 2:2688–2694. https://doi.org/10.1021/jz2013193

Thuy UTD, Liem NQ, Parlett CM, Lalev GM, Wilson K (2014) Synthesis of CuS and CuS/ZnS core/shell nanocrystals for photocatalytic degradation of dyes under visible light. Catal Commun 44:62–67. https://doi.org/10.1016/j.catcom.2013.07.030

Tongying P, Plashnitsa VV, Petchsang N, Vietmeyer F, Ferraudi GJ, Krylova G, Kuno M (2012) Photocatalytic hydrogen generation efficiencies in one-dimensional CdSe heterostructures. J Phys Chem Lett 3:3234–3240. https://doi.org/10.1021/jz301628b

Torres-Martínez CL, Kho R, Mian OI, Mehra RK (2001) Efficient photocatalytic degradation of environmental pollutants with mass-produced ZnS nanocrystals. J Colloid Interface Sci 240:525–532. https://doi.org/10.1006/jcis.2001.7684

Umar M, Aziz HA (2013) Photocatalytic degradation of organic pollutants in water. In: Organic pollutants-monitoring, risk and treatment. InTech. https://doi.org/10.5772/53699

Upadhyay RK, Soin N, Roy SS (2014) Role of graphene/metal oxide composites as photocatalysts, adsorbents and disinfectants in water treatment: a review. RSC Adv 4:3823–3851. https://doi.org/10.1039/C3RA45013A

Wang TC, Lu N, Li J, Wu Y (2011) Plasma-TiO2 catalytic method for high-efficiency remediation of p-nitrophenol contaminated soil in pulsed discharge. Environ Sci Technol 45:9301–9307. https://doi.org/10.1021/es2014314

Wang J et al (2012) Highly efficient oxidation of gaseous benzene on novel Ag3VO4/TiO2 nanocomposite photocatalysts under visible and simulated solar light irradiation. J Phys Chem C 116:13935–13943. https://doi.org/10.1021/jp301355q

Wang D, Hisatomi T, Takata T, Pan C, Katayama M, Kubota J, Domen K (2013) Core/Shell photocatalyst with spatially separated co-catalysts for efficient reduction and oxidation of water. Angew Chem Int Ed 52:11252–11256. https://doi.org/10.1002/anie.201303693

Wenhua L, Hong L, Sao'an C, Jianqing Z, Chunan C (2000) Kinetics of photocatalytic degradation of aniline in water over TiO2 supported on porous nickel. J Photochem Photobiol A Chem 131:125–132. https://doi.org/10.1016/S1010-6030(99)00232-4

Xia Y, Wang J, Chen R, Zhou D, Xiang L (2016) A review on the fabrication of hierarchical ZnO nanostructures for photocatalysis application. Crystals 6:148. https://doi.org/10.3390/cryst6110148

Yu H, Liu R, Wang X, Wang P, Yu J (2012) Enhanced visible-light photocatalytic activity of Bi2WO6 nanoparticles by Ag2O cocatalyst. Appl Catal B Environ 111:326–333. https://doi.org/10.1016/j.apcatb.2011.10.015

Zhang D et al (2010) Carbon-stabilized iron nanoparticles for environmental remediation. Nanoscale 2:917–919. https://doi.org/10.1039/C0NR00065E

Zhang H, Zhao L, Geng F, Guo L-H, Wan B, Yang Y (2016a) Carbon dots decorated graphitic carbon nitride as an efficient metal-free photocatalyst for phenol degradation. Appl Catal B Environ 180:656–662. https://doi.org/10.1016/j.jcis.2019.04.027

Zhang Q, Teng J, Zou G, Peng Q, Du Q, Jiao T, Xiang J (2016b) Efficient phosphate sequestration for water purification by unique sandwich-like MXene/magnetic iron oxide nanocomposites. Nanoscale 8:7085–7093. https://doi.org/10.1039/C5NR09303A

Zhu Q, Tao F, Pan Q (2010) Fast and selective removal of oils from water surface via highly hydrophobic core− shell Fe2O3@ C nanoparticles under magnetic field. ACS Appl Mater Interfaces 2:3141–3146. https://doi.org/10.1021/am1006194

Chapter 8
Arsenic Contamination: Sources, Chemistry and Remediation Strategies

Pankaj K. Parhi, Snehasish Mishra, Ranjan K. Mohapatra, Puneet K. Singh, Suresh K. Verma, Prasun Kumar, and Tapan K. Adhya

Abstract Growing industrialisation, urbanisation and technological advancements have been endlessly increasing the environmentally contaminating heavy metals load. Arsenic contamination as an environmental pollutant has transcended as a major global concern to address. Arsenic contamination of air, soil, water, sediment and crops due to the various anthropogenic (agricultural) and geogenic (geochemical) sources is a major global threat, including India, owing to its hazardous and toxic nature. Primarily, the three and five valency arsenic cause severe human health concerns at an elevated concentration (>0.05 mg/l) affecting millions of people worldwide year-after-year. Generally non-biodegradable, arsenic can be transformed into less toxic forms by adopting chemical, biological and/or composite techniques involving oxidation-reduction, methylation, complexation, precipitation, immobilisation through sorption, etc. Microbial and phyto-remediation of arsenic through adsorption, absorption, extracellular entrapment, precipitation and oxidation-reduction reactions are gaining global attention due to their greater advantages. While phytoremediation includes phytoextraction and phytovolatilisation, microbial biomass remediates through active/passive/combined arsenic binding. The chapter embodies the underlying arsenic toxicity and bioremediation

---

P. K. Parhi
Department of Chemistry, Fakir Mohan University, Balasore, Odisha, India

S. Mishra (✉) · P. K. Singh
Bioenergy Lab and BDTC, School of Biotechnology, KIIT (Deemed University), Bhubaneswar, Odisha, India
e-mail: smishra@kiitbiotech.ac.in

R. K. Mohapatra · T. K. Adhya
School of Biotechnology, KIIT (Deemed University), Bhubaneswar, Odisha, India

S. K. Verma
Division of Molecular Toxicology, Institute of Environmental Medicine, Karolinska Institute, Stockholm, Sweden

P. Kumar
Department of Chemical Engineering, Chungbuk National University, Cheongju, Chungbuk, Republic of Korea

© The Author(s), under exclusive license to Springer Nature Switzerland AG 2021
S. Rajendran et al. (eds.), *Metal, Metal-Oxides and Metal-Organic Frameworks for Environmental Remediation*, Environmental Chemistry for a Sustainable World 64,
https://doi.org/10.1007/978-3-030-68976-6_8

mechanisms for a cleaner and healthier environment. It details the chemical and biological remediation of arsenic highlighting the advantages of biological approaches.

Keywords Pollution · Arsenic contamination · Methylation · Bioremediation · Phytoextraction

8.1 Introduction

The delicate environmental balance has been upset due to excess load of numerous inorganics including heavy metals generated through mining, rapid industrialisation, and urbanisation. Amongst the heavy metals, environmental pollution caused due to arsenic contamination is considered as vital for its severe toxicity and potential associated health risks. Arsenic could generate multiple adverse health effects because of its many (inorganic and organic) chemical forms, the most common inorganic trivalent arsenic forms being arsenic trioxide, sodium arsenite, and arsenic trichloride (Adeniji 2004). The acute symptoms of arsenic poisoning, as a consequence of consuming it above the maximum tolerable limit of 0.05 mg/l, are skin discoloration, skin thickening and ultimately skin cancer (Khan et al. 2000; Dey et al. 2017; Banerjee et al. 2011). Arsenic is a metalloid (considered a heavy metal) under group 'V' element of the periodic table (Satyapal et al. 2016). It is found in four different oxidation states, i.e., $+5$, $+3$, 0, and -3 in nature. Of these, the pentavalent (As^V; Arsenate) and trivalent (As^{III}; Arsenite) exist mostly in inorganic forms (Satyapal et al. 2018). Although both pentavalent and trivalent forms are poisonous but As^{III} due to its high mobility is 1000-fold more toxic than its counterpart (Satyapal et al. 2018; Dey et al. 2016). Organic arsenicals derived from pesticides, herbicides and preservatives are also encountered in the environment (Satyapal et al. 2018).

Heavy metals like lead, arsenic, nickel, mercury, and cadmium are not at all beneficial to plants, affecting the plant wellbeing negatively through reduced photosynthesis, nutrient uptake and certain enzymatic malfunctioning (Lim et al. 2014). Low concentration of heavy metals leads to cytotoxicity and higher concentration to cancer (Tak et al. 2013). It occurs due to contamination in food chain at a point and bioaccumulation inside the living organisms (Tak et al. 2013). The cellular damage happens due to the reactive oxygen species ROS, mainly oxygen radicals, damaging the DNA (Chibuike and Obiora 2014). The sources of arsenic contamination, both natural and anthropogenic, have resulted in wide arsenic contamination of the soil, water, air and crops (Satyapal et al. 2016). In certain geographical regions, the animals and humans are constantly exposed to high arsenic concentrations through contaminated drinking water and food crops (Satyapal et al. 2018; Dey et al. 2016). As per World Health Organisation (WHO), the maximum permissible limit for arsenic contaminant in drinking water is 0.01 mg/l (WHO 2011; Aksornchu et al. 2008).

Arsenic's notoriety is most predominant in groundwater systems as toxic metalloid. The greatest mass poisoning in human history was reportedly due to the drinking of arsenic contaminated groundwater in millions of people (Dey et al. 2017). Various anthropogenic activities have further accelerated arsenic contamination, especially in the South East Asia region. It was estimated that more than six million people in West Bengal, India (Dey et al. 2016; Anyanwu and Ugwu 2010) and 46 million people in Bangladesh are at a risk of arsenic poisoning due to drinking water (Dey et al. 2016; Bachate and L Cavalca 2009). Arsenic contamination in groundwater is at an alarming situation in Indian states of West Bengal, Chhattisgarh, Bihar, Telangana and Uttar Pradesh.

Removing the contaminating heavy metals through (micro)biological means is widely accepted as the most efficient, cost effective, and eco- and health-friendly (Ekperusi and Aigbodion 2015; Ayangbenro and Babalola 2017). The capacity of microbes to remove heavy metals and metalloids is dependent on the suitability of the abiotic factors, such as, temperature, pH, physical and chemical properties of soil, and moisture (Verma and Jaiswal 2016). The chapter deliberates on the chemical and biological remediation of arsenic from contaminated soils, and the phytological/microbial mechanisms involved in heavy metals decontamination.

8.2 Sources of Arsenic Contamination

Arsenic in the environment (air, soil, water, and sediment) originates from both natural (geogenic) and anthropogenic sources (Satyapal et al. 2018; Fig. 8.1). The key origin points of arsenic flow from the natural geogenic sources include volcanic eruptions, weathering, fossil fuels, minerals, parent/sedimentary rock bearing arsenic (Mohapatra et al. 2017a). Various anthropogenic activities like agriculture, mining, smelting, refining, electroplating, coal combustion, painting and chemical manufacturing have added to the arsenic release to the environment (Mohapatra et al. 2017a; Dey et al. 2016; Akhtar et al. 2013). Manufacturing of agricultural chemicals (e.g. pesticides, herbicides, fertilisers, wood preservatives, etc.), dying materials and medical products, are other major sources of arsenic contamination (Mohapatra et al. 2017a; Vishnoi and Singh 2014).

8.3 Arsenic Toxicity

Presence of Arsenic in the environment beyond the permissible limit of 0.01 mg/l can generate multiple acute and chronic health disorders (Dey et al. 2016). As^{III} compound such as arsenic trioxide, sodium arsenite and arsenic trichloride could cause neurotoxicity of both the peripheral and the central nervous system (Adeniji 2004). On the other hand, As^V inorganic forms such as arsenic pentoxide, arsenic acid and Arsenates could also affect the enzyme activity of human metabolism (Klaassen and Watkins III 2003). Trivalent or pentavalent organoarsenic, specially

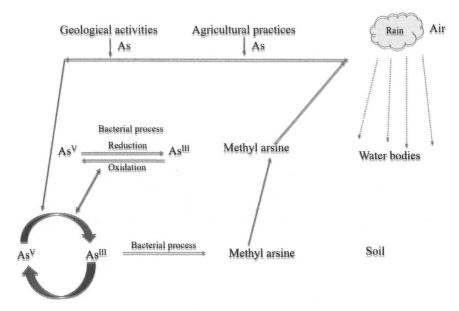

Fig. 8.1 Schematic representation of Biogeochemical cycle of arsenic in the atmosphere

the methylated, forms cause biomethylation and are potential health hazards in humans, animals and other higher living organisms as they enter the food chain (Mateos et al. 2006; Adeniji 2004). Excess exposure to arsenic even at a low concentration could create acute health issues including skin itching, skin discoloration, hyperkeratosis, skin thickening leading to skin cancer, fever, anorexia, melanosis, weight loss, appetite loss, gastrointestinal disorders (e.g. nausea, anorexia, stomach irritation, abdominal pain, enlarged liver and spleen), anaemia, weakness, lethargy, granulocyopenia cardiac arrhythmia, and cardiovascular failure (Mohapatra et al. 2017a; Taran et al. 2013; Cavalca et al. 2013; Dey et al. 2016). Long-term arsenic exposure could cause chronic respiratory disorders, lungs irritation, immune-suppression, arsenicosis, sensory loss, changes in skin epithelium, ultimately leading to cancer due to DNA damage (Mohapatra et al. 2017a; Adeniji 2004). High arsenic intake could cause infertility, fatal health issues, miscarriages in women, type-II diabetes, brain damage, cardiovascular problems including hypertension, coronary artery diseases, peripheral vascular disease and atherosclerosis (Mohapatra et al. 2017a). As^{III} could also cease protein functions by binding to the sulfhydryl groups of cysteine residues (Cavalca et al. 2013).

8.4 Soil Reclamation Strategies

The various remediation measure for heavy metal contamination are functional now days (Fig. 8.2). The measures for removal of arsenic from the contaminated environment should follow several minimal technical standards that consists

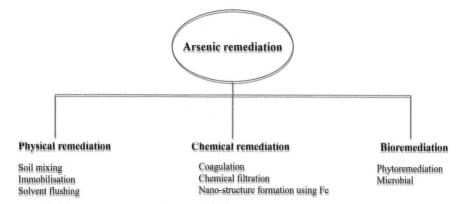

Fig. 8.2 Schematic representation of scalable technologies (physical, chemical and biological) currently available for arsenic remediation from contaminated surfaces

effectiveness, no adverse impact on the environment, and no health hazard for the neighbouring organism or the system. Presently, three main reclamation practices are functional namely, physical, chemical, and biological/bioremediation (Lim et al. 2014; Duarte et al. 2009). There are varieties of conventional methods have been used for removal of arsenic from the contaminated aqueous system such as membrane filtration, precipitation, reverse osmosis, coagulation, oxidation-reduction, adsorption etc. (Dey et al. 2017; Dey et al. 2016; Bahar et al. 2012). In addition to these physicochemical remediation processes, microbial bioremediation is also proven as a potential eco-friendly approach for clean-up of the arsenic contaminated area because of its low operation cost, less energy requirements, non and high removal efficiency (Voica et al. 2016; Dey et al. 2017; Das and Dash 2014).

Complex make of the microbial cell and their metabolic specialities make them potential agents to address arsenic toxicity by transforming it to a less-toxic form, and/or remove them during the cellular metabolic process, biosorption and accumulation phenomena (Mohapatra et al. 2017a; Dey et al. 2016). Table 8.1 describes the advantages and disadvantages of some of the reported physicochemical and biological arsenic remediation approaches.

8.4.1 Arsenic Removal from Contaminated Waters

Conventional and advanced successful treatment approaches to remove arsenic from groundwater under both laboratory and field conditions have been reported, which include: (i) coagulation/flocculation, (ii) adsorption, (iii) ion-exchange and (iv) membrane processes (Mondal et al. 2013). A potentially scalable bioremediation technology involving green synthesis of nano-adsorbents using bacteria, yeasts, fungi and plant extracts is also reported (Mondal et al. 2006; Bahar et al. 2012).

Table 8.1 Various physicochemical and biological techniques in arsenic remediation

Technique	Approach (es) employed	Speciality of approach (es)	Reference(s)
Physical	Coagulation, precipitation, sedimentation, etc.	Well accepted; high operational cost, useful in small-scale operations. Up to 30–90% As^{III} and > 95% of As^V removal efficiency	Mahimairaja et al. (2005) and Fazi et al. (2016)
	Adsorption by activated carbon and/or alumina		
	Ion exchange using anionic resins		
	Membrane filtration		
Chemical	Coagulation, complexation and precipitation using ferric chloride, sulphates of aluminium, copper and ammonia	Economical but could be expensive to remediate a larger area. Up to 30% As^{III} and 90–95% As^V removal efficiency	Duarte et al. (2009), Komárek et al. (2013) and Lim et al. (2014)
	Adsorption using granular iron hydroxide, iron impregnated polymer resins, iron oxide impregnated activated alumina, etc.	Widely applicable and economical. Up to 30–60% As^{III} and > 95% As^V removal efficiency	Shrivastava et al. (2015) and Fazi et al. (2016)
Phytoremediation	Phytoremediation using plants	Widely accepted ecofriendly approach useful primarily in large field applications	Porter and Peterson (1975), Chakraborti et al. (2001), Mishra et al. (2000), Silva et al. (2006) and Yang et al. (2012a, 2012b)
Microbial biosorption	Immobilisation of As in the solid phase using microbial (bacterial, fungal and algal) biomass	Cost-effective and ecofriendly; cellular and microbiological/molecular analyses needed	Mahimairaja et al. 2005, Ahmed et al. (2005) and Lim et al. (2014)
Microbial (RedOx) transformation	Microbial transformation of toxic arsenic to lesser toxic forms through oxidation-reduction, by heterotrophs and chemoautotrophs; arsenate can be reduced to arsenite by microbial dissimilatory reduction mechanism	For controlled environmental condition. Arsenic reduction is carried out in anaerobic condition using facultative or obligate anaerobes	Xiong et al. (2006), Chipirom et al. (2012) and Leiva et al. (2014)
Microbial methylation	Biomethylation of arsenic by microbes with cellular enzymes like As(III)-S-adenosyl methionine methyltransferase	An effective and highly efficient biological process to remediate arsenic contaminated aquatic bodies	Mahimairaja et al. (2005) and Lim et al. (2014)

8.4.2 Arsenic Removal from Contaminated Soils

There are many arsenic removal approaches that could be divided primarily into three categories, physical, chemical, and biological (Lim et al. 2014).

8.4.3 Physical Approach

One of the popular approaches is, mixing both the uncontaminated and contaminated soils together till the arsenic concentration reaches an acceptable level (Lim et al. 2014; Mahimairaja et al. 2005). Soil washing is a physicochemical approach whereby the soil contaminated with arsenic is washed in presence of chemicals such as sulphuric/nitric/phosphoric acids, and/or hydrogen bromide (Lim et al. 2014).

8.4.4 Chemical Approach

The chemical approach employed for the purpose as extractant is costly and often is restricted to soil washing at smaller-scale operations (Mahimairaja et al. 2005). Cement could also immobilise soluble Arsenites and has been successfully used to stabilise arsenic-rich sludge (Sullivan et al. 2010). Furthermore, additives, such as, surfactants, cosolvents, etc. could also enhance the soil flushing efficiencies using aqueous solutions. Surfactant alone was about 80–85% efficient in laboratory conditions, while more complex processes such as polymer injection enhanced the efficiency (Atteia et al. 2013). Available chemical remediation approaches involve methods such as adsorption by using specific media, immobilisation, modified coagulation along with filtration, precipitations, immobilisations and complexation (Duarte et al. 2009; Mahimairaja et al. 2005). Coagulation along with filtration for arsenic removal is quite economical but often displayed lower (<90%) efficiencies (Lim et al. 2014).

8.4.5 Biological Approach

Biological measures are broadly distinguished as phytoremediation and microbially-mediated remediation. Plants and microbes, especially the ones thriving in arsenic-rich environment, have evolved themselves to sustain and metabolise arsenic and their metalloids. Some bacteria convert the inorganic and organic arsenic to trimethyl-arsine (less toxic gaseous arsenic), particularly under anaerobic conditions. This could be accomplished through various ways including biomethylation,

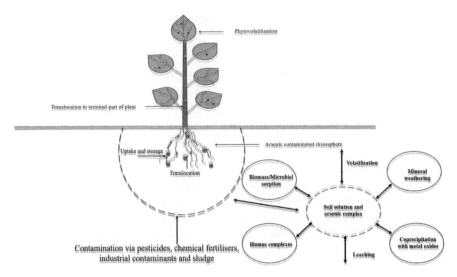

Fig. 8.3 Schematic representation of process involved in arsenic decontamination using plant/plant–microbe interactions

biotransformation and biooxidation (Srivastava et al. 2011; Liu et al. 2009; Casarett et al. 2008). With several limitations of non-biological approaches, biological approach is gaining popularity, particularly due to its cost effectiveness. Biological remediation approach is primarily divided into two subcategories, intrinsic and engineered. Intrinsic bioremediation is mainly meant to address low-level contamination by specialised natural/wild microbes, whereas engineered bioremediation is useful in addressing critically contaminated soils by engineered microbes (Lim et al. 2014) (Fig. 8.3).

8.4.5.1 Phytoremediation

Phytoremediation is an efficient way to bioremediate contaminated soils and water bodies (Mishra et al. 2000). Several hyperaccumulating plant varieties (1 kg biomass accumulating up to one-gram arsenic) are reported. The cheapest technology for heavy metal removal, this approach is time saving and also decreases the volume of the contaminated biomass (Chattopadhyay et al. 2017). Phosphorus helped in mobilising and enhancing the uptake capacity of arsenic in sunflower which could sustain 250 mg of arsenic/kg plant biomass in soil, whereas Chinese brake (*Pteris vittata* L.) could tolerate up to 22,600 mg of arsenic/kg plant biomass on a dry weight basis (Jang et al. 2016). The nonprotein thiols, phytochelatins, phytochelatins and glutathione produced by plants as a defense mechanism help in decontaminating arsenic-rich soil (Dixit et al. 2016). The mechanisms involved arsenic decontamination involve phytoextraction, rhizofiltration and phytovolatilization (Fig. 8.4).

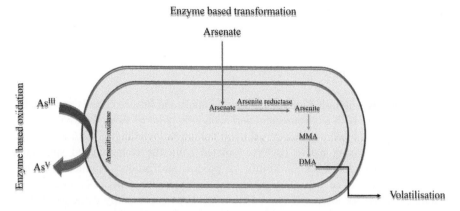

Fig. 8.4 Schematic representation of bacterial remediation of Arsenic through oxidation-reduction pathway (*MMA* Monomethylarsinic acid, *DMA* Dimethylarsinic acid)

Microbial associations with plants often play an important role as the facilitator in arsenic cycling.

The internal resistant mechanisms to reduce metal toxicity in plants include sequestration of metals and phytochelatins. Phytochelatins are cysteine rich peptides formed by glutathione at high arsenic concentration (Mesa et al. 2017). Further, plants are divided into the following groups on the basis of their metal removal efficiencies.

Excluders

These group of plants restrict the uptake and translocation of arsenic on the terminal parts by tolerating the existing high concentration of arsenic through intracellular chelators. The excess arsenic segregated and is stored in the non-sensitive plant parts, a phenomenon known as compartmentalisation (Sun et al. 2009).

Accumulators

Accumulator plant performs remediation via the uptake and translocation arsenic into the terminal parts without any discernible plant symptoms. These can uptake upto 1000 mg As/kg dry shoot.

8.4.5.2 Microbially-Mediated Remediation

Microbes (bacteria, fungi and algae) are very effective and efficient bioremediating agents. Their small life span and adaptative abilities help thrive well even in harsh environments. Below is an account of the usefulness of each group as bioremediation agents.

Bacterial Remediation

Bacteria possess multiple bioremediation potentials and hence certainly are beneficial agents, from both environmental and economic point of view, for toxic pollutants cleanup. Such bioremediation (of toxic pollutants including metals/metalloids) is achieved by using native bacteria isolated from the contaminated sites and stimulating their detoxification ability by process and product engineering (Das and Dash 2014). The use of suitable non-native and/or genetically engineered microbes suited for arsenic bioremediation has been successfully demonstrated at least at research-scale if not at field-scale (Das and Dash 2014).

Mechanism of Arsenic Bioremediation

The mechanisms in arsenic bioremediation are majorly biotransformation and biosorption.

Biotransformation Mechanism

In this, the microbes could decrease the toxicity of the contaminants by using them as energy sources while transforming them through the energy-yielding oxidation-reduction reactions utilising oxygen, carbon dioxide, nitrates, sulphate acetate, lactate and glucose as electron acceptors/donors during metabolism (Dey et al. 2017; Akhtar et al. 2013). Dey et al. (2017) reported that the bacteria having the capacity to resist toxic metals can chemically transform heavy metals/metalloids through their common cellular metabolism through oxidation, reduction, methylation, demethylation, precipitation etc. Bacteria could exploit arsenic in their metabolic process either as an electron acceptor as in case of anaerobic respiration or as an electron donor as in case of chemoautotrophic fixation of CO_2 into cell carbon (Akhtar et al. 2013). Dissimilatory arsenate-reducing bacteria use arsenate as an electron acceptor and reduce it to arsenite. Chemoautotrophic arsenite oxidising bacteria use CO_2 as the carbon source and arsenite as an electron acceptor, oxidising it to arsenate for energy, whereas heterotrophic arsenite oxidisers use oxygen as an electron acceptor to oxidise arsenite to arsenate (Fig. 8.3; Akhtar et al. 2013).

Biosorption of Arsenic

Microbial arsenic biosorption involves the sorption of arsenate ions (sorbate) present in aqueous form on to the surface of a solid microbial biomass (biosorbent). Due to their higher affinity towards charged ions which further dependent on the chemical constituents of the cell wall, biosorbents facilitate the binding of the contaminant ions. Further, the degree of biosorption differs according to the distribution of arsenic ions between the solid biosorbent and the liquid (aqueous) phase. This process continues till establishment of equilibrium between the amount of contaminant-bound biosorbent and the free ions in the solution. The extracellular polymeric substances (EPSs) such as the peptidoglycan, phospholipids, lipopolysaccharides, proteins, teichoic and teichuronic acids of the bacterial cell, primarily having a role in quorum sensing, play a key role in binding and adsorption of the toxic arsenic ions. Several (carboxylic, amino, thiol, hydroxyl and hydrocarboxylic) functional groups present in the biomass also actively participate in the binding process (Mohapatra et al. 2017a, b).

Biochemistry of Bacterial Arsenic Removal

Studies have demonstrated that plant could sustain in high arsenic contaminated soils when the phosphorus concentration was also high (Rosen et al. 2011). The arsenic in soil and water occurring naturally generally enters the plant via phosphate transporters and facilitate bacterial survival under arsenic stress condition. A common bacterial defence mechanism is the three detoxifying operons, ArsR, ArsC, and ArsB (Musingarimi et al. 2010; Yang et al. 2012a, b). The transportation of arsenate to cell and its reduction to arsenite is accomplished by ArsC gene and the outward transportation of arsenite from cell by ArsB gene (Musingarimi et al. 2010).

Arsenic-tolerant bacteria *Acinetobacter* from the rhizospheric soil of *Pteris vittata*, a fern, oxidises As^{III}, whereas a few others (*Flavobacterium, Pseudomonas,* and *Staphylococcus*) could oxidise as well as reduce arsenic (Wang et al. 2012). *Agrobacterium radiobacter* in the roots of *Populus deltoids* makes the plant tolerant to 300 mg/kg Arsenic in soil, with a 54% removal efficiency (Wang et al. 2011). Reports suggest volatilisation (*Sphingomonas desiccabilis* and Cyanobacteria), adsorption (*Ralstonia eutropha*) and oxidation (*Rhodococcus equi, Thiomonas arsenivorans* and *Ensifer adhaerens*) of arsenic from the contaminated soil (Table 8.2; Liu et al. 2011; Mondal et al. 2008; Yin et al. 2011; Bag et al. 2010; Dastidar and Wang 2012; Ito et al. 2012). A few of the siderophore-producing arsenic-tolerant bacteria are *Pseudomonas fluorescens, Micrococcus luteus* and *Bacillus licheniformis*. They are also active in solubilising phosphorus and fixing nitrogen (Ivan et al. 2017). A genetically modified *Rhizobium leguminosarum* incorporated with As^{III} S-adenosylmethionine methyltrasnferase gene (CrarsM) from *Chlamydomonas reinhardtii* was useful in arsenic detoxification through the methylation of As^{III} (Zhang et al. 2017). Some microbial mechanisms enhance the plant growth by producing indole-3-acetic and other organic acids. These metabolites metabolise the heavy metal through the bacterial 1-ammino-cyclopropane-1-carboxylic acid deaminase (Ma et al. 2011).

Algal Remediation

Algae from the groups Cyanophyta and Chlorophyta help in absorption and accumulation of arsenic from contaminated water (Mitra et al. 2017). The prokaryotic

Table 8.2 Reported bacterial species and their modes of action on arsenic-contaminated soil

Sl. No.	Microorganism	Mechanism	Reference
1.	*Sphingomonas desiccabilis*	Volatilisation	Liu et al. (2011)
2.	*Ralstonia eutropha*	Adsorption	Mondal et al. (2008)
3.	*Cyanobacteria*	Volatilisation	Yin et al. (2011)
4.	*Rhodococcus equi*	Oxidation	Bag et al. (2010)
5.	*Thiomonas arsenivorans*	Oxidation	Dastidar and Wang et al. (2012)
6.	*Ensifer adhaerens*	Oxidation	Ito et al. (2012)

(such as cyanobacteria) and eukaryotic (such as *Chlorella*) algae usually bioremediate arsenic (arsenate and arsenite) via the phosphate transportation and plasma membrane-based hexose permeases and aqua-glyceroporins pathways (Zhang et al. 2014). Arsenate is transported by competitive inhibition of phosphate due to their chemical similarity (between AsO_4^{3-} and PO_4^{3-}). The different functional groups present on the cell wall of algae help in adsorbing the metal. Wang et al. (2013) and Zhang et al. (2013) reported more than 60% of arsenic removal by algae from contaminated water through adsorption. With regard the biochemistry of arsenic biotransformation to reduce its toxicity, the two biochemical conversion pathways occurring inside the algal cells, viz. oxidation and methylation, are discussed below.

Oxidation of Arsenic

Few algae such as *Synechocystis* and *Cynidiales* could oxidise As^{III} to As^{IV} inside the cells. Zhang et al. (2011) reported a process of detoxification through the uptake, accumulation and transformation of arsenic in *Synechocystis* sp. inside the cell. A few other reports confirm that the oxidation of As^{III} happens outside with the help of extracellular phosphatases (Mitra et al. 2017). The role of the enzyme involved in the process particularly in the oxidation process is hitherto obscure (Mitra et al. 2017; Zhang et al. 2014).

Methylation of Arsenic

This mechanism involves the conversion of toxic As^{III} arsenic to a less toxic monomethyl and dimethyl arsenates with the help of arsenite methyltransferases (Ye et al. 2012). Qin et al. (2009) confirmed that *Cyanidioschyzon* sp. (an extremophilic alga) could alone oxidise As^{III} to As^{V}, reduce As^{V} to As^{III} and methylate As^{V} to monomethyl arsenate and dimethyl arsenate.

Fungal Remediation

In terms of bioactive compound production, fungi are the most prominent and potent biomass in soil. The fungal cell wall is made up of polysaccharide molecules and proteins with hydroxyl, phosphate, sulphate and amino functional groups that could bind to the metal ions and metalloids relatively easily (Maheswari and Murugesan 2011). Most fungi, viz., *Trichoderma, Candida, Aspergillus, Fusarium* and *Penicillium*, help in methylating inorganic arsenic to its organic counterpart (Upadhyay et al. 2018). The advantages of fungi over bacteria as bioremediation agents are their longer life-span, higher biomass content and a complex hyphal network (Singh et al. 2016). Additionally, metal savouring fungi can compete with native bacteria in relatively inhospitable conditions (Sun et al. 2012).

Trichoderma is another filamentous Ascomycete fungus of great significance in plant growth promotion (Waghunde et al. 2016). It improves soil fertility and has the ability to induce stress-tolerance, a peculiar characteristic unlike the competing neighbouring rhizospheric microbes. It could promote hormone production, nutrient release from the soil, and rhizosphere development (de Souza et al. 2017). It contains a variety of functional groups on the outer layer of cell wall that could bind to metal ions and metalloids (Tripathi et al. 2017). *Westerdykella aurantiaca*, a soil fungus, bears arsenic methyl-transferase (WaarsM) gene which could be expressed in *Saccharomyces cerevisiae* (Verma and Jaiswal 2016). Such bioengineered yeasts capable of expressing the WaarsM gene demonstrated a higher arsenic methylation property. Laboratory studies confirmed an enhanced arsenic tolerance in paddy when such yeast cells were cocultured/inoculated in paddy (Verma and Jaiswal 2016).

8.5 Approach Involving Plant-Microbe Associations

Phytoremediation is a selective way used by plants to clean heavy metals from the environment through modified rhizospheric PGPR and PGPM. Several studies have been performed to select hyper-accumulating plants to assess the consequence of metal stress on the useful rhizospheric microbes (PGPMs) that can further facilitate the development of a more promising bioremediation strategy (Tak et al. 2013). The efficacy of phytoremediation is limited by the major factors, such as, tolerance level for the contaminant by the plant, selection of the plant variety to be employed for bioremediation, and its capacity to uptake and translocate the heavy metals (Jutsz and Gnida 2015). Phytoremediation, as indicated earlier, is an economically feasible bioremediation strategy as it produces the utilisable biomass while removing the toxic metals (Angelova et al. 2016).

Most plant species harbour vesicular-arbuscular mycorrhizae (VAM) that primarily help in phosphate solubilisation and uptake thereby enhancing their stress tolerance ability (Sharma et al. 2017). Upadhyay et al. (2018) reported that VAM supplementation helped overcome arsenic-induced phosphate deficiency in wheat. VAM also helps in maintaining a good ratio of arsenic and phosphate by translocating arsenic to inside the plant cells, particularly in soils with low arsenic contamination. In a similar study, Li et al. (2016) observed a decrease in the inorganic and organic ratio of arsenic in seeds when rice was inoculated with *Rhizophagus irregularis*.

It is important to note that arsenic volatilisation and methylation depends on the structure, organic content, the degree of the contamination and the chemical status of the soil (Mestrot et al. 2011). Upadhyay et al. (2018) recorded an annual 0.002–0.13% of net arsenic biovolatilisation in rice fields, with an about 4 μg/kg/year rate of volatilisation.

8.6 Challenges in Field-Scale Replication of the Strategy

The challenges met particularly by the translational (lab-to-land) researchers are manifold. These challenges include ecological, environmental, biotic and abiotic. For instance, in situ bioremediation could be a huge challenge when the arsenic concentration and the soil characteristics are adversely positioned. As every technology has an associated risk so is the bioremeation. For example, useful more efficient genetically modified microbes and plants could be employed to remediate arsenic contamination but its on-field application remains a topic of concern with biosafety consequences. The pollens of the genetically engineered plants and the plasmid of the genetically modified microbes could be major challenges to address the biosafety concern.

8.7 Future Research Directions

The role of genetically modified microbes in expediting the removal and remediation of contaminating arsenic and their survival when transferred for in situ bioremediation need to be addressed. Factors like temperature, lesser available nutrients and other related factors that are not easy to restore, may impact bioremediation potential negatively (Freitas et al. 2013). Furthermore, access to the genetically engineered plants and microbes to evaluate their role in heavy metals decontamination needs to be more focussed. Hyperaccumulative plants producing high biomass must be identified and could be further improved genetically to enhance their remediation efficiency. Similarly, the bioremediation ability of the these microbes to compete with the indigenous microbiota for efficient bioremediation by demonstrating an upper hand in the competitive-exclusion ecological principle calls for technological insights.

8.8 Conclusion

Several mechanisms and biochemical interactions, and their role in bioremediation have been detailed, with an attempt to expose the practicality of chemical- and bio-remediation strategies and their effect on the scaled-up remediation processes. As it is time consuming, generates harmful byproducts, and a costly proposition, chemical approach has slowly taken a backseat in the recent technological advancements. In that place, biological strategies for arsenic removal are slowly gaining popularity as they are eco-friendly. Plants and microbes have their own adaptive mechanisms to survive and sustain in contaminated soils and waters. Microbes bioremediate by oxidation, reduction, biosorption, and degradation of metals with the help of extracellular transformation and phytoremediation is based on phytoextraction and phytovolatilisation which have been very useful. Nevertheless, a bioremediation to be better accomplished would require friendly environmental conditions. Phytoremediation depends on the concentration of the contaminant, and the physical and chemical properties of soil.

Acknowledgements The technical, logistics and administrative support received by the authors from their respective affiliated organisations are duly acknowledged.

Declaration Authors declare no conflict of interest.

References

Adeniji A (2004) Bioremediation of arsenic, chromium, lead, and mercury. National network of environmental management studies fellow for US Environmental Protection Agency Office of Solid Waste and Emergency Response Technology Innovation Office, Washington, DC, pp 14–19

Ahmed I, Hayat S, Pichtel J (2005) Heavy metal contamination of soil: problems and remedies. Science Publishers Inc, Enfield

Akhtar MS, Chali B, Azam T (2013) Bioremediation of arsenic and lead by plants and microbes from contaminated soil. Res Plant Sci 1(3):68–73

Aksornchu P, Prasertsan P, Sobhon V (2008) Isolation of arsenic-tolerant bacteria from arsenic-contaminated soil. Sonklanakarin J Sci Technol 30(1):95–102

Angelova VR, Perifanova-Nemska M, Uzunova G, Ivanov K, Lee H (2016) Potential of sunflower (Helianthus annuus L.) for phytoremediation of soils contaminated with heavy metals. World. J Sci Eng Technol 10:1–11. https://doi.org/10.5281/zenodo.1126371

Anyanwu CU, Ugwu CE (2010) Incidence of arsenic resistant bacteria isolated from a sewage treatment plant. Int J Basic Appl Sci 10:64–78

Atteia O, Estrada EDC, Bertin H (2013) Soil flushing: a review of the origin of efficiency variability. Rev Environ Sci Biotechnol 12:379–389. https://doi.org/10.1007/s11157-013-9316-0

Ayangbenro AS, Babalola OO (2017) A new strategy for heavy metal polluted environments: a review of microbial biosorbents. Int J Environ Res Public Health 14:94. https://doi.org/10.3390/ijerph14010094

Bachate SP, L Cavalca V (2009) Andreoni, arsenic-resistant bacteria isolated from agricultural soils of Bangladesh and characterization of arsenate-reducing strains. J Appl Microbiol 107:145–156. https://doi.org/10.1111/j.1365-2672.2009.04188

Bag P, Bhattacharya P, Chowdhury R (2010) Bio-detoxification of arsenic laden ground water through a packed bed column of a continuous flow reactor using immobilized cells. Soil Sediment Contam 19:455–466. https://doi.org/10.1080/15320383.2010.486050

Bahar MM, Megharaj M, Naidu R (2012) Arsenic bioremediation potential of a new arsenite-oxidizing bacterium Stenotrophomonas sp. MM-7 isolated from soil. Biodegradation 23:803–812. https://doi.org/10.1007/s10532-012-9567-4

Banerjee S, Datta S, Chattyopadhyay D, Sarkar P (2011) Arsenic accumulating and transforming bacteria isolated from contaminated soil for potential use in bioremediation. J Environ Sci Health Part A 4:1736–1747. https://doi.org/10.1080/10934529.2011.623995

Casarett LJ, Doull J, Klaassen CD (2008) Casarett and Doull's toxicology: the basic science of poisons, 7th edn. McGraw-Hill, New York, p 1331

Cavalca L, Corsini A, Zaccheo P, Andreoni V, Muyzer G (2013) Microbial transformations of arsenic: perspectives for biological removal of arsenic from water. Future Microbiol 8 (6):753–768. https://doi.org/10.2217/fmb.13.38

Chakraborti D, Basu GK, Biswas BK, Chowdhury UK, Rahman MM, Paul K, Roy Chowdhury T, Chanda CR, Lodh L, Ray SL (2001) Characterization of arsenic bearing sediments in Gangetic Delta of West Bengal-India. In: Chappell WR, Abernathy CO, Calderon RL (eds) Arsenic exposure and health effects 4. Elsevier Science, New York, pp 27–52

Chattopadhyay A, Singh AP, Rakshit A (2017) Bioreclamation of arsenic in the soil-plant system: a review. Sci Int 5(1):30–41. https://doi.org/10.3923/sciintl.2017.30.41

Chibuike G, Obiora S (2014) Heavy metal polluted soils: effect on plants and bioremediation methods. Appl Environ Soil Sci 2014:752708–752712. https://doi.org/10.1155/2014/752708

Chipirom K, Tanasupawat S, Akaracharanya A, Leepepatpiboon N, Prange A, Kim KW, Lee KC, Lee JS (2012) Comamonas terrae sp. nov., an arsenite-oxidizing bacterium isolated from agricultural soil in Thailand. J Gen Appl Microbiol 58(3):245–251

Das S, Dash HR (2014) Microbial bioremediation: a potential tool for restoration of contaminated areas. In: Microbial biodegradation and bioremediation. Elsevier, pp 1–21

Dastidar A, Wang YT (2012) Modelling arsenite oxidation by chemoautotrophic Thiomonas arsenivorans strain b6 in a packed-bed bioreactor. Sci Total Environ 432:113–121. https://doi.org/10.1016/j.scitotenv.2012.05.051

de Souza Vandenberghe LP, Garcia LM, Rodrigues C, Camara MC, de Melo Pereira GV, de Oliveira J, Soccol CR (2017) Potential applications of plant probiotic microorganisms in agriculture and forestry. AIMS Microbiol 3(3):629

Dey U, Chatterjee S, Mondal NK (2016) Isolation and characterization of arsenic-resistant bacteria and possible application in bioremediation. Biotechnol Rep 10:1–7. https://doi.org/10.1016/j.btre.2016.02.002

Dey U, Chatterjee S, Mondal NK (2017) Investigation of bioremediation of arsenic by Bacteria isolated from an arsenic contaminated area. Environ Process 4(1):183–199

Dixit G, Singh AP, Kumar A, Mishra S, Dwivedi S, Kumar S, Trivedi PK, Pandey V, Tripathi RD (2016) Reduced arsenic accumulation in rice (Oryza sativa L.) shoot involves sulfur mediated improved thiol metabolism, antioxidant system and altered arsenic transporters. Plant Physiol Biochem 99:86–96. https://doi.org/10.1016/j.plaphy.2015.11.005

Duarte AALS, Cardoso SJA, Alcada AJ (2009) Emerging and innovative techniques for arsenic removal applied to a small water supply system. Sustainability 1:1288–1304. https://doi.org/10.3390/su1041288

Ekperusi O, Aigbodion F (2015) Bioremediation of petroleum hydrocarbons from crude oil-contaminated soil with the earthworm: Hyperiodrilus africanus. 3 Biotech 5:957–965. https://doi.org/10.1007/s13205-015-0298-1

Fazi S, Amalfitano S, Casentini B, Davolos D, Pietrangeli B, Crognale S, Lotti F, Rossetti S (2016) Arsenic removal from naturally contaminated waters: a review of methods combining chemical and biological treatments. Rendiconti Lincei 27(1):51–58

Freitas EV, Nascimento CW, Souza A, Silva FB (2013) Citric acid-assisted phytoextraction of lead: a field experiment. Chemosphere 92:213–217. https://doi.org/10.1016/j.chemosphere.2013.01.103

Ito A, Miura JI, Ishikawa N, Umita T (2012) Biological oxidation of arsenite in synthetic groundwater using immobilised bacteria. Water Res 46:4825–4831. https://doi.org/10.1016/j.watres.2012.06.013

Ivan FP, Salomon MV, Berli F, Bottini R, Piccoli P (2017) Characterization of the As(III) tolerance conferred by plant growth promoting rhizobacteria to in vitrogrown grapevine. Appl Soil Ecol 109:60–68. https://doi.org/10.1016/j.apsoil.2016.10.003

Jang YC, Somanna Y, Kim H (2016) Source, distribution, toxicity and remediation of arsenic in the environment–a review. Int J Appl Environ Sci 11(2):559–581

Jutsz AM, Gnida A (2015) Mechanisms of stress avoidance and tolerance by plants used in phytoremediation of heavy metals. Arch Environ Prot 2015(41):104–114. https://doi.org/10.1515/aep-2015-0045

Khan AH, Rasul SB, Munir A, Habibuddowla M, Alauddin M, Newaz SS, Hussan A (2000) Appraisal of a simple arsenic removal method for groundwater of Bangladesh. J Environ Sci Health A 35:1021–1041. https://doi.org/10.1080/10934520009377018

Klaassen CD, Watkins JB III (2003) Absorption, distribution, and excretion of toxicants. Karl K Rozman Essen Toxicol

Komárek M, Vaněk A, Ettler V (2013) Chemical stabilization of metals and arsenic in contaminated soils using oxides–a review. Environ Pollut 172:9–22. https://doi.org/10.1016/j.envpol.2012.07.045

Leiva ED, dP Rámila C, Vargas IT, Escauriaza CR, Bonilla CA, Pizarro GE, Regan JM, Pasten PA (2014) Natural attenuation process via microbial oxidation of arsenic in a high Andean watershed. Sci Total Environ 466:490–502. https://doi.org/10.1016/j.scitotenv.2013.07.009

Li H, Chen XW, Wong MH (2016) Arbuscular mycorrhizal fungi reduced the ratios of inorganic/organic arsenic in rice grains. Chemosphere 145:224–230

Lim KT, Shukor MY, Wasoh H (2014) Physical, chemical, and biological methods for the removal of arsenic compounds. Biomed Res Int 2014. https://doi.org/10.1155/2014/503784

Liu Y, Zheng BH, Fu Q, Meng W, Wang YY (2009) Risk assessment and management of arsenic in source water in China. J Hazard Mater 170:729–734. https://doi.org/10.1016/j.jhazmat.2009.05.006

Liu S, Zhang F, Chen J, Sun G (2011) Arsenic removal from contaminated soil via biovolatilization by genetically engineered bacteria under laboratory conditions. J Environ Sci 23:1544–1550. https://doi.org/10.1016/S1001-0742(10)60570-0

Ma Y, Prasad MNV, Rajkumar M, Freitas H (2011) Plant growth promoting rhizobacteria and endophytes accelerate phytoremediation of metalliferous soils. Biotechnol Adv 29:248–258. https://doi.org/10.1016/j.biotechadv.2010.12.001

Maheswari S, Murugesan AG (2011) Removal of arsenic(III) ions from aqueous solution using Aspergillus flavus isolated from arsenic contaminated site. Ind J Chem Technol 18:45–52

Mahimairaja S, Bolan NS, Adriano DC, Robinson B (2005) Arsenic contamination and its risk management in complex environmental settings. Adv Agron 86:1–82

Mateos LM, Ordóñez E, Letek M, Gil JA (2006) Corynebacterium glutamicum as a model bacterium for the bioremediation of arsenic. Int Microbiol 9:207–215

Mesa V, Navazas A, González-Gil R, González A, Weyens N, Lauga B, Peláez AI (2017) Use of endophytic and rhizosphere bacteria to improve phytoremediation of arsenic-contaminated industrial soils by autochthonous Betula celtiberica. Appl Environ Microbiol 83(8):03411–03416. https://doi.org/10.1128/AEM.03411-16

Mestrot A, Feldmann J, Krupp EM, Hossain MS, Roman-Ross G, Meharg AA (2011) Field fluxes and speciation of arsines emanating from soils. Environ Sci Technol 45:1798–1804. https://doi.org/10.1021/es103463d

Mishra S, Barik SK, Ayyappan S, Mohapatra BC (2000) Fish bioassays for evaluation of raw and bioremediated dairy effluent. Bioresour Technol 72(3):213–218

Mitra A, Chatterjee S, Gupta DK (2017) Uptake, transport, and remediation of arsenic by algae and higher plants. In: Arsenic contamination in the environment. Springer, Cham, pp 145–169

Mohapatra RK, Parhi PK, Patra JK, Panda CR, Thatoi HN (2017a) Biodetoxification of toxic heavy metals by marine metal resistant Bacteria-a novel approach for bioremediation of the polluted saline environment. Microb Biotechnol 1:343–376

Mohapatra RK, Parhi PK, Thatoi H, Panda CR (2017b) Bioreduction of hexavalent chromium by Exiguobacterium indicum strain MW1 isolated from marine water of Paradip port, Odisha, India. Chem Ecol 33(2):114–130

Mondal P, Majumder CB, Mohanty B (2006) Laboratory based approaches for arsenic remediation from contaminated water: recent developments. J Hazard Mater 137:464–479. https://doi.org/10.1016/j.jhazmat.2006.02.023

Mondal P, Majumder CB, Mohanty B (2008) Treatment of arsenic contaminated water in a batch reactor by using Ralstonia eutropha MTCC2487 and granular activated carbon. J Hazard Mater 153:588–599. https://doi.org/10.1016/j.jhazmat.2007.09.028

Mondal P, Bhowmic S, Chatterjee D, Figoli A, Bruggen B (2013) Remediation of inorganic arsenic in groundwater for safe water supply: a critical assessment of technological solutions. Chemosphere 92:157–170. https://doi.org/10.1016/j.chemosphere.2013.01.097

Musingarimi W, Tuffin M, Cowan D (2010) Characterisation of the arsenic resistance genes in Bacillus sp. UWC isolated from maturing fly ash acid mine drainage neutralised solids. S Afr J Sci 106(1–2):59–63. https://doi.org/10.4102/sajs.v106i1/2.17

Porter EK, Peterson PJ (1975) Arsenic accumulation by plants on mine waste (United Kingdom). Sci Total Environ 4(4):365–371. https://doi.org/10.1016/0048-9697(75)90028-5

Qin J, Lehr CR, Yuan CG, Le XC, Mc Dermott TR, Rosen BP (2009) Biotransformation of arsenic by a Yellowstone thermoacidophilic eukaryotic alga. Proc Natl Acad Sci U S A 106:5213–5217. https://doi.org/10.1073/pnas.0900238106

Rosen BP, Ajees AA, McDermott TR (2011) Life and death with arsenic: arsenic life: an analysis of the recent report "A bacterium that can grow by using arsenic instead of phosphorus". BioEssays 33(5):350–357. https://doi.org/10.1002/bies.201100012

Satyapal GK, Rani S, Kumar M, Kumar N (2016) Potential role of arsenic resistant bacteria in bioremediation: current status and future prospects. J Microb Biochem Technol 8(3):256–258. https://doi.org/10.4172/1948-5948.1000294

Satyapal GK, Mishra SK, Srivastava A, Ranjan RK, Prakash K, Haque R, Kumar N (2018) Possible bioremediation of arsenic toxicity by isolating indigenous bacteria from the middle Gangetic plain of Bihar, India. Biotechnol Rep 17:117–125. https://doi.org/10.1016/j.btre.2018.02.002

Sharma S, Anand G, Singh N, Kapoor R (2017) Arbuscular mycorrhiza augments arsenic tolerance in wheat (Triticum aestivum L.) by strengthening antioxidant defense system and thiol metabolism. Front Plant Sci 8:906

Shrivastava A, Ghosh D, Dash A, Bose S (2015) Arsenic contamination in soil and sediment in India: sources, effects, and remediation. Current Pollution Reports 1(1):35–46

Silva GMI, Gonzaga SJA, Qiying ML (2006) Arsenic phytoextraction and hyperaccumulation by fern species. Sci Agric. https://doi.org/10.1590/S0103-90162006000100015

Singh N, Marwa N, Mishra SK, Mishra J, Verma PC, Rathaur S, Singh N (2016) Brevundimonas diminuta mediated alleviation of arsenic toxicity and plant growth promotion in Oryza sativa L. Ecotoxicol Environ Saf 125:25–34. https://doi.org/10.1016/j.ecoenv.2015.11.020

Srivastava PK, Vaish A, Dwivedi S, Chakrabarty D, Singh N, Tripathi RD (2011) Biological removal of arsenic pollution by soil fungi. Sci Total Environ 409:2430–2442. https://doi.org/10.1016/j.scitotenv.2011.03.002

Sullivan C, Tyrer M, Cheeseman CR, Graham NJD (2010) Disposal of water treatment wastes containing arsenic—a review. Sci Total Environ 408(8):1770–1778. https://doi.org/10.1016/j.scitotenv.2010.01.010

Sun Y, Zhou QX, Liu WT, Wang L (2009) Joint effects of arsenic, cadmium on plant growth and metal bioaccumulation: a potential Cd hyperaccumulator and As-excluder Bidens pilosa. L. J Hazard Mater 161(2–3):808–814. https://doi.org/10.1016/j.jhazmat.2008.10.097

Sun J, Zou X, Ning Z, Sun M, Peng J, Xiao T (2012) Culturable microbial groups and thallium-tolerant fungi in soils with high thallium contamination. Sci Total Environ 441:258–264. https://doi.org/10.1016/j.scitotenv.2012.09.053

Tak HI, Ahmad F, Babalola OO (2013) Advances in the application of plant growth-promoting rhizobacteria in phytoremediation of heavy metals. In: Whitacre DM (ed) Reviews of environmental contamination and toxicology, vol 223. Springer, New York, pp 33–52

Taran M, Safari M, Monaza A, Reza JZ, Bakhtiyari S (2013) Optimal conditions for the biological removal of arsenic by a novel halophilic archaea in different conditions and its process optimization. Pol J Chem Technol 15(2):7–9

Tripathi P, Singh PC, Mishra A, Srivastava S, Chauhan R, Awasthi S, Mishra S, Dwivedi S, Tripathi P, Kalra A, Tripathi RD (2017) Arsenic tolerant Trichoderma sp. reduces arsenic induced stress in chickpea (Cicer arietinum). Environ Pol 223:137–145

Upadhyay MK, Yadav P, Shukla A, Srivastava S (2018) Utilizing the potential of microorganisms for managing arsenic contamination: a feasible and sustainable approach. Front Environ Sci 6:24. https://doi.org/10.3389/fenvs.2018.00024

Verma JP, Jaiswal DK (2016) Book review: advances in biodegradation and bioremediation of industrial waste. Front Microbiol 6:1555. https://doi.org/10.3389/fmicb.2015.01555

Vishnoi N, Singh DP (2014) Biotransformation of arsenic by bacterial strains mediated by oxidoreductase enzyme system. Cell Mol Biol (Noisy-le-Grand, France) 60(5):7–14

Voica DM, Bartha L, Banciu HL, Oren A (2016) Heavy metal resistance in halophilic bacteria and archaea. FEMS Microbiol Lett 363(14):146. https://doi.org/10.1093/femsle/fnw146

Waghunde RR, Shelake RM, Sabalpara AN (2016) Trichoderma: A significant fungus for agriculture and environment. African J Agric Res 11(22):1952–1965

Wang Q, Xiong D, Zhao P, Yu X, Tu B, Wang G (2011) Effect of applying an arsenic-resistant and plant growth-promoting rhizobacterium to enhance soil arsenic phytoremediation by Populus deltoides LH05-17. J Appl Microbiol 111:1065–1074. https://doi.org/10.1111/j.1365-2672.2011.05142.x

Wang X, Rathinasabapathi B, Oliveira LMD, Guilherme LR, Ma LQ (2012) Bacteria-mediated arsenic oxidation and reduction in the growth media of arsenic hyperaccumulator Pteris vittata. Environ Sci Technol 46:11259–11266. https://doi.org/10.1021/es300454b

Wang ZH, Luo ZX, Yan CZ (2013) Accumulation, transformation, and release of inorganic arsenic by the freshwater cyanobacterium Microcystis aeruginosa. Environ Sci Pollut Res 20:7286–7295. https://doi.org/10.1007/s11356-013-1741-7

WHO (2011) Guidelines for drinking-water quality. WHO Chron 38:104–108

Xiong J, Wang W, Fan H, Cai L, Wang G (2006) Arsenic resistant bacteria in mining wastes from Shangrao coal mine of China. Environ Sci Technol 1:535–540

Yang HC, Fu HL, Lin YF, Rosen BP (2012a) Pathways of arsenic uptake and efflux. In: Current topics in membranes, vol 69. Academic Press, pp 325–358

Yang Q, Tu S, Wang G, Liao X, Yan X (2012b) Effectiveness of applying arsenate reducing bacteria to enhance arsenic removal from polluted soils by Pteris vittata L. Int J Phytoremediation 14(1):89–99

Ye J, Rensing C, Rosen BP, Zhu YG (2012) Arsenic biomethylation by photosynthetic organisms. Trends Plant Sci 17:155–162. https://doi.org/10.1016/j.tplants.2011.12.003

Yin XX, Chen J, Qin J, Sun GX, Rosen BP, Zhu YG (2011) Biotransformation and volatilization of arsenic by three photosynthetic cyanobacteria. Plant Physiol 156:1631–1638. https://doi.org/10.1104/pp.111.178947

Zhang B, Wang LH, Xu YX (2011) Study on absorption and transformation of arsenic in blue alga (Synechocystis sp. PCC6803). Asian J Ecotoxicol 6:629–633

Zhang JY, Ding TD, Zhang CL (2013) Biosorption and toxicity responses to arsenite (As(III)) in Scenedesmus quadricauda. Chemosphere 92:1077–1084. https://doi.org/10.1016/j.chemosphere.2013.01.002

Zhang SY, Rensing C, Zhu YG (2014) Cyanobacteria-mediated arsenic redox dynamics is regulated by phosphate in aquatic environments. Environ Sci Technol 489:994–1000. https://doi.org/10.1021/es403836g

Zhang J, Xu Y, Cao T, Chen J, Rosen BP, Zhao FJ (2017) Arsenic methylation by a genetically engineered Rhizobium-legume symbiont. Plant Soil 416(1):259–269

Chapter 9
Mycoremediation: An Elimination of Metal and Non-metal Inclusions from Polluted Soil

Jegadeesh Raman, Jang Kab-Yeul, Hariprasath Lakshmanan, Kong Won-Sik, and Babu Gajendran

Abstract Soil is a natural resource of the ecosystem, comprising of solids (minerals), living (organic materials), air and water. Soil is the base for land plants and all other habitats. In recent years, the soil ecosystem has been contaminated by a high concentration of metal and non-metal deposition. For ecological concern, the elimination and detoxification of pollutants are warranted. Several technologies have been developed to tackle soil remediation. Mycoremediation is a new technology in the bioremediation process, which can be practiced with fungus for the management of polluted soil and water. Fungi based approach is considered as eco-friendly, cost-effective and one of the promising methods. Fungi are eukaryotic organisms with heterotrophic food habits and absorb food from the environment/host through digestive enzymes. They can eliminate/detoxify the environmental pollutants, like polycyclic aromatic hydrocarbons, plastic, heavy metals, and other inclusions. Most of the saprophytic fungi are involved in the bioremediation process, and they obtain nutrients by degradation/decomposing. Accordingly, they can be classified as brown-rot and white-rot and litter decomposing fungi. Mycoremediation relies on the efficient enzymes, produced by ligninolytic and non-ligninolytic fungi, for the degradation of various substrates and pollutants. Besides, the white-rot fungi and arbuscular mycorrhizal fungi can be used to treat organic/inorganic contaminants,

---

J. Raman (✉)
Department of Biochemical and Polymer Engineering, Chosun University, Gwangju, Republic of Korea

Mushroom Research Division, National Institute of Horticultural and Herbal Science, Rural Development Administration, Eumsung, Republic of Korea

J. Kab-Yeul · K. Won-Sik
Mushroom Research Division, National Institute of Horticultural and Herbal Science, Rural Development Administration, Eumsung, Republic of Korea

H. Lakshmanan
Department of Biochemistry, Karpagam Academy of Higher Education, Coimbatore, India

B. Gajendran
State Key Laboratory of Functions and Applications of Medicinal Plants and Chinese Academy of Sciences, Guizhou Medical University, Guiyang, China

© The Author(s), under exclusive license to Springer Nature Switzerland AG 2021
S. Rajendran et al. (eds.), *Metal, Metal-Oxides and Metal-Organic Frameworks for Environmental Remediation*, Environmental Chemistry for a Sustainable World 64, https://doi.org/10.1007/978-3-030-68976-6_9

remove metals from polluted soil and detoxify the environmental chemicals. Heavy metals uptake by fungi produces unusual metabolites, which might increase their toxicity and stress level. Even though naturally present fungi can mineralize the organic/inorganic pollutions in the soil and make this planet a safe habitat. This chapter deals with the mycoremediation process in detail, types and the mechanisms behind the remediation process.

Keywords Mycoremediation · Fungi · Soli · Mushroom

9.1 Introduction

The bioremediation process, whereby the fungi degrade organic/inorganic wastes, is termed mycoremediation. The American mycologist Paul Stamets coined the term 'mycoremediation'. Fungi are hyperaccumulators and fungal based remediation is both cost and yield effective and can be used as an alternative approach in traditional remediation of the polluted environment (Gadd 2001). In this chapter, we discuss the types of fungal involved in the remediation process and their effectiveness against the contamination. Diverse plant and microbes like bacterial and fungal species have been tested against the metal and nonmetal inclusions (Pouli and Agathos 2011). The recent global census highlights that fungi are holding the third position with 2% of the biomass. Fungi can decompose the complex organic/inorganic materials and accumulate the heavy metal toxins from the polluted environment. Xenobiotic and aromatic compounds are degrading by the filamentous fungi, particularly lignicolous white-rot fungi. The lignocellulose materials are usually degraded by the fungi/bacteria and may enrich soil fertility through the composting process. Most of the decontamination processes were performed through wood-degrading basidiomycetes fungi. The fungi constitute the order Agaricales and Polyporales were the ones mostly involved in the mycoremediation. In addition to that, other lower fungi like Ascomycota, Zygomycota and Oomycetes are fungi also involved in the degradation process (Fig. 9.1). Fungi exhibit a high ability to immobilize toxic heavy metals by the formation of insoluble metal oxalate. Heavy metals regulate extracellular ligninolytic and cellulolytic enzymes at the level of transcription. White rot fungi can degrade and accumulate the aromatic hydrocarbons (PAHs), dioxins, many synthetic dyes and pesticides from the polluted ecosystem (Johannes and Majcherczyk 2000; Kanaly and Hur 2006). During the degradation process, heavy metals interfere with both the enzyme activity and fungal colonization. Even though extracellular enzymes break down the complex polymers into monomers and finally get metabolized.

The resemblance of several environmental pollutions might be compared with plant complex structures (Harvey and Thurston 2001). In fungi, the substrate specified enzymes were encoded during their growth on different substrates but remain vague (Chigu et al. 2010; Syed et al. 2010). D'Annibale et al. 2006 reported that metagenomics and specific gene identification facilitate the analysis and fungi intensely that decrease the soil toxicity. The most suitable solid-phase treatment by ex-situ/substrate-specific mycoremediation was paramount. Identification of the

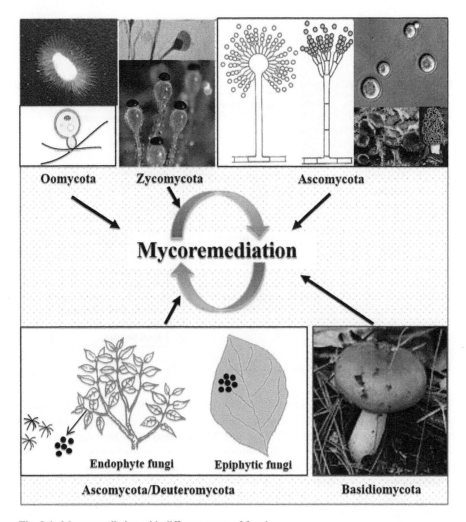

Fig. 9.1 Mycoremediation with different group of fungi

specific fungal gene regulated and expressed differently during their growth on substrates may be used to identify fungal diversity.

9.2 Heavy Metal Mycoremediation

Heavy metals are carcinogenic and persistent pollutants in the soil and water ecosystem. Soil is considered a long term reservoir of heavy metals, those metals entering through human activity. In recent years, heavy metals pollution has brought scientists and researchers the attention to human health and ecosystem concern.

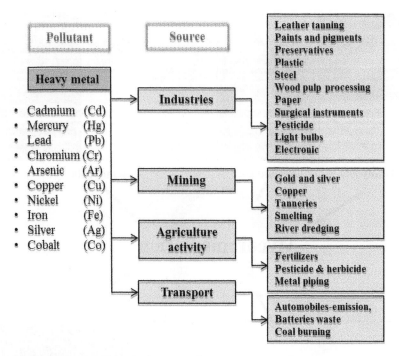

Fig. 9.2 Source of soil pollution by heavy metals

Heavy metals are released to the soil and water ecosystem from a different activity like leather tanning, wood and pulp processing steel industrial, mining, agricultural activities (Congeevaram et al. 2007) Fig. 9.2. The previous study revealed that waste disposal, waste incineration, traffic density, fertilizers, pesticides and long-term infiltration of sewage water leads to soil pollution (Bilos et al. 2001). Cadmium (Cd), mercury (Hg), lead (Pb), chromium (Cr) and arsenic (As) are causing cellular damage in animals and plants and are a significant threat to the ecosystem and soil health (Zafar et al. 2007). Silver and gold mining and extraction are profoundly affecting soil diversity.

Therefore, industrial effluent treatment processes are by conventional methods are cost-effective and time-consuming processes (Kadirvelu et al. 2002). Several technologies developed for the remediation process from heavy metal polluted soil. *In situ* treatment process on surface soil and is identified as a low cost, rapid and straightforward technique (USEPA 1997). The strategy was achieved through the microbiota associated agronomic practices (Cunningham and Ow 1996). Whereas in the *ex-situ* method, the soil was removed from the polluted site, this method is also equally low cost, fast and straightforward applicability on heavily polluted soil (Salt et al. 1998). Among the conventional methods are considered as challenging and strictest processes. The fungi detoxify heavy metals through natural processes such as biosorption, chemical precipitation and volatilization (Lone et al. 2008). High accumulation in soil has severe effects on the microbe population, decreases tin he bio-composting process, soil respiration, and reduction in the enzyme activity and affects the fungal spore formation (Tyler 1974; Hepper and Smith 1976).

In the soil ecosystem, metal ions are immobilized and mobilized by bacteria/fungi and final uptake by the plants (Birch and Bachofen 1990). Soil is an essential source for soil-dwelling microorganisms, as it includes various essential nutrients for growth. Approximately one billion colony forming units (CFUs) of diverse microorganisms were present in the soil, including fungi comprising of 10^5 to 10^8 numbers/gram (Lenart-Boron and Boron 2014).

Fungal cell wall chemical composition and structure may trigger metal ions. Their active binding sites were uptake during binding the heavy metals from the water and soil ecosystem. Mycoremediation may be done through mechanisms such as elemental transformation, precipitation and active uptake by enzymes and organic acid secretion (Fig. 9.3).

The presence of plant-associated fungal communities in heavy metal contaminated soil triggers the plant nutrient uptake and regulated the phosphorous cycle. Particularly the filamentous fungi (white-rot fungi) may speed up the bacteria bioaugmentation (Sasek 2003). Arbuscular mycorrhizal fungi can metabolize and detoxify cadmium, copper and zinc metals in the soil (Christie et al. 2004). The most common groups of fungi found in soil are Zygomycetes, mycorrhizal associated Ascomycota and Basidiomycota. The different groups of fungus involved in the mycoremediation process have been listed in Table 9.1. The important heavy metal

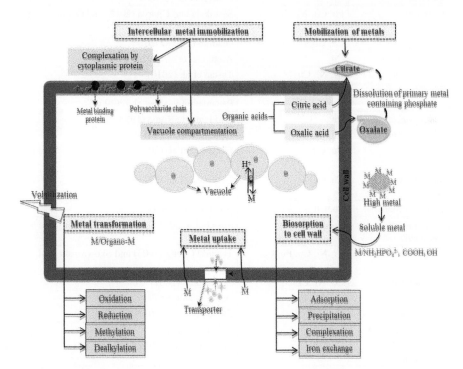

Fig. 9.3 Schematic representation of fungal biomass and functional groups involved in metal interaction. (Adapted from Siddiquee et al. 2015)

Table 9.1 Heavy metals mycoremediation by different group of fungi. (Adapted and modified from Archana and Jaitly 2015)

Fungi	Metal remediate	References
Zygomycota		
Circinella sp.	Ni	Alpat et al. (2010)
Cunnighamella echinulata	Pb, Cu	El-Morsy (2004)
Mucor hiemalis	Cd, Cr	Cabuk et al. (2005)
M. rouxii	Pb, Cd, Ni	Majumdar et al. (2010)
Rhizopus arrhizus	Pb, Cr, Cd, Ni	Fourest and Roux (1992)
R. cohnii	Cd	Luo et al. (2010)
R. nigricans	Pb, Cr, Zn	Bai and Abraham (2001)
R. oryzae	Cu, Cd	Mishra and Malik (2014)
Ascomycota		
Aspergillus flavus	Cu, Ni, Pb	Thippeswamy et al. (2012)
Aspergillus foetidus	Cr	Prasenjit and Sumathi (2005)
A. fumigatus	Cu, Cd, Co, Ni, Pb	Rao et al. (2005) and Oladipo et al. (2016)
A. luchuensis	Cu, Cd	Hassanein et al. (2012)
A. nidulans	Ar, Cu, Cd, Pb	Maheswari and Murugesan (2009) and Oladipo et al. (2016)
A. niger	Cd, Pb, Cu, Cr, Ar	Pal et al. (2010)
A. oryzae	Cr	Nasseri et al. (2002)
A. terreus	Pb, Cu, Ni, Cr	Seshikala and Charya (2012) and Oladipo et al. (2016)
A. tubingensis	Cu, Cd, Pb, Ar	Oladipo et al. (2016)
A. ustus	Cu	Alothman et al. (2020)
A. versicolor	Cr, Ni, Cu	Tastan et al. (2010)
Candida species	Cu, Cr, Pb, Ar, Ag, Co, Cd, Hg	Acosta-Rodriguez et al. (2018) and Pattanapipitpaisal et al. (2001)
Cladosporium sp.	Cu	Gadd and de Rome (1988)
Curvularia lunata	Cu, Cd, Cr	El-Gendy et al. (2011)
Drechslera hawaiiensis	Cu, Cd	El-Gendy et al. (2011)
Monacrosporium elegans	Cu, Cd	El-Gendy et al. (2011)
Neurospora crassa	Pb, Cu	Ismail et al. (2005)
Penicillium canescens	Cd, Pb, As, Hg	Say et al. (2003)
P. chrysogenum	Cr, Ni, Cu, Pb, Cd	Tan and Cheng (2003) and Skowronski et al. (2001)
P. decumbens	Cd, Ni, Cr	Levinskaite (2001)
P. digitatum	Cd, Cu, Pb	Galun et al. (1987)
P. lilacium	Cu, Cd	El-Gendy et al. (2011)
P. simplicissimum	Cd, Pb	Fan et al. (2008)
Pestalotiopsis clavispora	Cu, Cd	El-Gendy et al. (2011)

(continued)

Table 9.1 (continued)

Fungi	Metal remediate	References
Saccharomyces cerevisae	Cd, Ni, Pb, Cr, Cu	Thippeswamy et al. (2012)
Trichoderma harzianum	Cu, Ni, Cr	Shoaib et al. (2012) and Sarkar et al. (2010)
T. virde	Pb, Ni, Cd, Cr	Prasad et al. (2013), Levinskaite (2001) and Hala and Eman (2009)
Verticillium fungicola	Cu, Cd	El-Gendy et al. (2011)
Basidiomycota		
Agaricus biosporous	Ni, Cu, Pb, Cd, Hg, Fe	Ita et al. (2008) and Nagy et al. (2014)
A. bitorquis	Cu, Fe, Cd, Pb, Ni	Lamrood and Ralegankar (2013)
Armillariella mellea	Ni, Cu, Pb, Cd	Ita et al. (2008)
Calocybe indica	Cr, Ni	Kuzhali et al. (2012)
Ganoderma lucidum	Cu	Muraleedharan et al. (1995)
Lentinus edodes	Cd, Pb, Cr	Chen et al. (2005)
Phanerochaete chysosporium	Cu, Ni, Pb, Cd, Fe	Mihova and Godjevargova (2001) and Mamun et al. (2011)
Pleurotus florida	Cr, Zn, Ni, Pb	Kuzhali et al. (2012) and Prasad et al. (2013)
P. floridianus	Cu, Fe, Cd, Pb, Ni	Lamrood and Ralegankar 2013
P. ostreatus	Ni, Cu, Pb, Cr, Cd, Fe, Hg	Ita et al. (2008) and Arbanah et al. (2012)
P. sajor-caju	Hg, Pb, Cd, Fe, Pb, Ni	Arica et al. (2003) and Lamrood and Ralegankar (2013)
P. sapidus	Ni, Cu, Pb, Cd	Ita et al. (2008)
Polyporus frondosis	NI, Cu, Pb, Cd	Ita et al. (2008)
P. sulphureus	Ni, Cu, Pb, Cd	Ita et al. (2008)
Volvariella diplasia	Cu, Fe, Cd, Pb, Ni	Lamrood and Ralegankar (2013)
V. volvacea	Cu, Zn, Fe, Cd, Pb, Ni	Lamrood and Ralegankar (2013)
Deuteromycota		
Alternaria alternate	Cd, Cr, Ni, Cu	Levinskaite (2001)
Fusarium oxysporum	Cr	Amatussalam et al. (2011)
F. solani	Cr, Zn, Ni	Sen and Dastidar (2011) and Sen (2013)

tolerant soil fungi like *Aspergillus flavus*, *Aspergillus niger*, *Rhizopus arrhizus* and *Fusarium oxysporum* has reported by many authors (Kurniati et al. 2014; Akinpelu 2014). The entophytic fungi *Lasiodiplodia theobromae* was isolated from *Boswellia ovalifoliolata* has been reported remarkable detoxification on selected heavy metals (Aishwarya et al. 2016). Mercury tolerant indigenous fungus mobilizing the mercury pollution were isolated from agricultural soil (Hindersah et al. 2018).

The source of mercury in soil comes from modern agriculture practices and industrial effluents. The filamentous fungi *Trichoderma harzianum* and *Aspergillus niger* solubilized the stable mercury and lead (Shoaib et al. 2012; Sarkar et al. 2010). High concentrated heavy metal pollutant tolerant filamentous fungi like *Aspergillus* and *Penicillium* species have been reported. Interestingly the diverse heavy metal tolerant filamentous fungi like *Trichoderma ghanese* and *Rhizopus microspores* have been reported by Zafar et al. (2007) and Oladipo et al. (2014).

Mushroom can accumulate heavy metals and deposited intracellularly, mostly sporocarps uptake maximum from the cultivation substrates. However, the molecular mechanisms of heavy metals accumulation by fungi are not yet studies. Heavy metals accumulate mostly in percentage on the fungal cell wall and also found in cytoplasm and vacuoles (Blaudez et al. 2000). The high degree of heavy metal in the soil created an adverse effect on the fungal population, toxic effects on the cell membrane and interfered with solute transport and enzyme secretion. Fungal extracellular metabolites and their chelating properties may be involved in the immobilization the heavy metals. The predominant soil fungi *Aspergillus niger* release/excretion of organic acid and oxalates compound, those compound may be solubilized/immobilized by the metals (Gadd 1999). Therefore, a different group of heavy metals tolerant fungus is being exploited in industrial wastewater and soil treatment process.

Copper metal, on the other hand, polluted the water and soil ecosystems. Copper are considered harmful to human and animal health (Rojas et al. 2017). The toxic element coppers taken up by uptake by the plants/tree are released to the environment through leaching and degradation processes. The increasing amount of copper (Cu) is witnessed in the forest ecosystem and Cu inhibit the wood rot fungi growth. However, the copper is an essential microelement and an active component in copper-containing oxidase enzymes like laccase. Wood degrading fungi are capable of breaking down toxic compounds. The brown rot fungi (*Fibroporia radiculosa*) overcome the copper pollution by secreting extracellular oxalate that can convert their metabolites to insoluble oxalate crystals (Akgul and Akgul 2018). Other brown-rot fungi such as *Serpula, Postia, Fibroporia* and *Wolfiporia* are copper tolerant fungal species (Green and Clausen 2003). *Pleurotus* species play a crucial role in biosorption/bioremediation of copper and other metals from the environment (Kapahi and Sachdeva 2017; Vaseem et al. 2017). The yeast species *Saccharomyces cerevisiae* can precipitate/bind the copper metals around their cell walls (Thippeswamy et al. 2012).

Recent studies have shown that mycorrhizal fungi are effective at removing heavy metals from soil. Introduce the ectomycorrhiza on Cu^{2+} and Cd^{2+} contaminated soil drastically reduces the metal concentration in soil (Blaudez et al. 2000). According to Jin et al. (2018), limited scientific evidence has been reported on *Actinomycetes* bioremediation. Endophytic fungi are associated with phytoremediation and them colonization rapidly on heavy metal contaminated soil. The endophytic fungi secrete siderophore molecules as iron (Fe) binding agents and indole-3-acetic acid (IAA). Siderophore is involved in the iron uptake and chelating the toxic compounds, and both these agents contribute to the plant growth (Friesen et al. 2011). However, endophytic fungi promote plant growth, trigger the mineral uptake and protect them from abiotic stress (Friesen et al. 2011; Sanchez-Fernandez et al. 2016). Another microfungi fungi *Trichoderma harzianum* effectively eliminated the nickel (Ni) from the contaminated soil, including *Aspergillus* and

9 Mycoremediation: An Elimination of Metal and Non-metal Inclusions from... 247

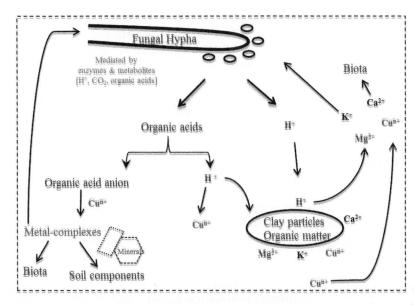

Fig. 9.4 The mechanism of organic and inorganic substance uptake and release by fungi. (Figure courtesy of Gadd 2004)

Clonostachys rosea can accumulate the silver from rock dump sites (Cecchi et al. 2017). The laboratory experiment revealed that *Aspergillus niger* had high resistance to cobalt and drastically removed the media's pollutions. Mycoremediation mechanism was revealed by Gadd (2004, 2007). Fungi interact with chemicals and organic substances through their enzymes and metabolites. However, the formation of atoms/compounds and chelating agents interact on minerals and other substances (Fig. 9.4).

9.3 Nonmetal Inclusion Elimination

9.3.1 Plastic and Oil Degradation

Plastics are synthetic or semi-synthetic polymeric material derived from petrochemicals. Most of the environmental issues in the current context occur due to the dramatic usage of plastics. Plastic wastes pollute the soil, atmosphere and water ecosystems, they recalcitrant the material and accumulate in the environment over long years. Fragmented micro/nano plastic can be easily distributed to the environment and might enter the ocean and accumulate in the forest and agricultural lands. The persistence of plastic fragments/pellets create harmful effects on aquatic animals and disturb the food web. Un-conditioning burning of plastic emits the greenhouse gasses and dioxins. In recent, global warming is a severe issue on environmental concern. The natural degradation processes are done through open sunlight and microbial interaction and other environmental processes. High intensity of sunlight and UV radiation can oxidize the plastic and this process may stimulate the microbial interaction. The filamentous and basidiomycetes fungi are dispersed/degrade the plastic pollutions from the soil. In recent years much promising research on finding potential microorganisms towards plastic degradation have been investigated.

Fungi can degrade complex polymeric materials such as lignin and may conversely depolymerize/break-down plastic. Fungal Lipases, cutinases and esterases enzymes can degrade different forms of plastic and depolymerize plastic in the form of monomers and may utilize their growth (Fig. 9.5). Bacteria and fungi can degrade the organic/inorganic toxic contaminants as sources of energy and convert them into carbon dioxide, water and biomass. The wood degrading *Pleurotus* spp., *Lentinus* and Rigidoporus species utilize the polyurethane (PU) instead of carbon. Endophytic and filamentous fungi can be capable of degrading the low-density polyethylene (LDPE) reported by Nowak et al. (2010). *Aspergillus japonicas* degrade low-density

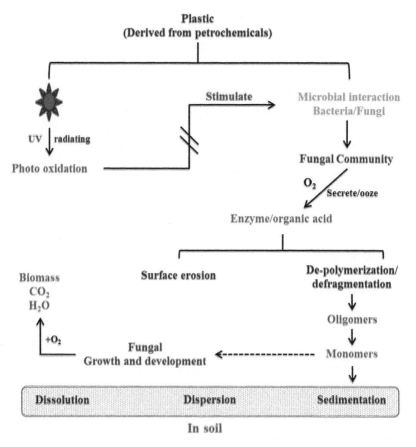

Fig. 9.5 Biodegradation of plastic in soil by the fungal community

polyethylene (industrial plastic) in laboratory-scale experiments by Raaman et al. 2012. In 2011 a student from Yale University, USA, discovered polyurethane consumed/degrade fungal species *Pestalotiopsis microspora* in Amazon rain forest and few edible mushrooms capable of breaking down and eating the plastic (Russell et al. 2011). However, fungi are unable to degrade the polyethylene (PE), whereas the number of fungi that have reportedly could degrade polyurethane and LDPE.

High consumption of petroleum hydrocarbons causes adverse effects on the ecosystem. The carcinogenic and mutagenic components from the petrochemicals can harm plant growth and soil health. The seepages and disaster can intensively affect soil physiochemical property and texture. However, the soil percolation, porosity and fertility of the soil have been affected permanently. The petroleum chemical remediation process is an elaborating process and cost-effective. However, the microbial (bacteria/fungi) degradation approach was cost and effective process. The detoxification approach is named as biostimulation. The sufficient soil nutrient enhances the microbial populations and hence, increases the rate by degradation/ mineralization processes. The petroleum products stimulate the microbial growth in soil, whereas the enzyme activity was altered. Soil enzymes were determined by the type of pollution in the soil. Photo-oxidation/UV irradiations stimulate the intake of hydrocarbons, fungal oxygenases and peroxidase are key enzymes involved in the degradation process. The enzyme activated by oxidative (incorporation of oxygen) process. The aerobic degradation of aliphatic hydrocarbons (n-alkanes) was initiated with an oxidation process and alkanes converted into fatty acid as an end product. The microbial interaction and attachment on the oil droplet surface were unclear.

However, the fungal immobilization system was actively involved in the hydrocarbon and PAHs degradation process. Hydrocarbon aging thus results in a reduced rate of degradation in the early stages. The widespread availability of hydrocarbon on soil surface may increases penetration into the pores leading to reduce microbial intake and bioavailability. Long time availability of hydrocarbon may have altered the chemical nature of the soli. Diverse filamentous and white-rot fungi have the potential to degrade and dissolution the toxic substance into no-toxic metabolites. PAHs mineralized through extracellular ligninolytic enzymes such as manganese peroxidase, laccases and lignin peroxidase. Those enzymes respond to lignin biodegradation and also participate in the different aromatic and organic complexes. Yeast genera, Yarrowia and Pichia were isolated from the oil-spilled soil and reported high potential degradation of petrochemical and diesel oil.

Fungi utilized the hydrocarbons as a source (carbon) of energy, most probably the polycyclic hydrocarbons (PAHs) in petroleum products (Sun et al. 2010). Copiotrophic fungi (molds) are anticipated to the rich organic and carbon sources and effectively utilize diesel oil. For the laboratory experiment aspect, few efficient filamentous fungi like *Aspergillus niger, A. japonicas, Penicillium glabrum* and *Cladosporium cladosporioides* were utilized in the remediation process. Fungi degrade the alkanes through the mono/di-terminal oxidization process. The corresponding alcohols are modified to aldehydes and fatty acids. Iida et al., (2000) reported that the eukaryotic P450 enzyme system was involved in the petrochemical degradation in yeast species. Besides P450 enzyme actively participated in alkane's degradation under aerobic conditions. Young (2012) reported six different white-rot fungi capable of degrading the short-chain alkane cultivated on oil containing spawn inoculum. Fungi degrade the oil inclusion in the water purifying units, and this may be applied in the Eco-Machines.

9.3.2 Polycyclic Aromatic Hydrocarbons (PAH)

Polycyclic aromatic compounds are carcinogenic/or mutagenic and they persistent in the environment. The structure is constructed only with carbon and hydrogen atoms, PHs constitute several benzene rings infused form. PAHs are naturally present in the oil and are released by burning petroleum products, coal and oil drilling (Cerniglia and Sutherland 2001). PAH are generated through industrial activities and natural combustion processes. These compounds are hydrophobic with low water solubility. Thus PAH are readily adsorbed on organic matters such as soils and sediments. The presence of light/high molecular weight aromatics is showed a low degradation coefficient. Most fungi degrade the PAH to their extracellular enzymatic reaction and synergic action with other soil microorganisms. However, the PAH degradation proportion was negligible in marsh soil compared to forest soil, due to high salinity and the enzyme action. The high salinity and slurry conditions inhibited the enzyme action most of the ligninolytic capacity of fungi. Few white-rot fungi and other anamorphic ascomycetes have halotolerant efficiency, whereas other fungi may be affect by a high salinity environment. *Phlebia* species are high salinity tolerance and modify the lignin under a saline environment (Li et al. 2003). Zygomycetes fungi were capable of degrading the PAHs by their cytochrome enzymes. Also, PAHs can be degraded by other white-rot basidiomycetes fungi due to their excellent extracellular enzymes. Fungal extracellular enzymes degradation of PAHs mirrors resamples with lignin and both water-insoluble with fused benzene rings and stereo irregular (Harvey and Thurston 2001). Ligninolytic enzymes degrade PAHs molecules and fragment in the large hydrophobic particles to pass through the cell walls.

Further, the oxidative enzyme activity on aromatic rings creates PAH-quinones that may mineralize. Litter-decomposing fungi had a high level of Mn-peroxide activity compare with wood decomposer fungi and efficiently degrade such organ pollutants (Steffen et al. 2007). The key enzyme laccase was involved in the first stage of PAH oxidize through the downstream process by fungal peroxidases. Relative numbers of scientific evidence were documented on PAHs remediation by *Agrocybe praecox, Bjerkandera adusta, Irpex lateus, Phlebia* spp. *Pleurotus ostreatus* and *Trametes versicolor* (Beaudette et al. 1998; Novotny et al. 2000; Kamei et al. 2005; Tuomela et al. 1999). Basidiomycetes white-rot fungi break down the lignin substrate, and the remaining cellulose represents the white color or yellow. White-rot lignicolous fungi produce enzymes such as laccases and peroxidases enzymes instead of polysaccharides. The Mn-Peroxide catalyzed the H_2O_2-dependent oxidation process. The phenolic components from the lignin substrate were oxidized by chelated Mn^{3+}. However, *Stropharia rugosoannulata*, the

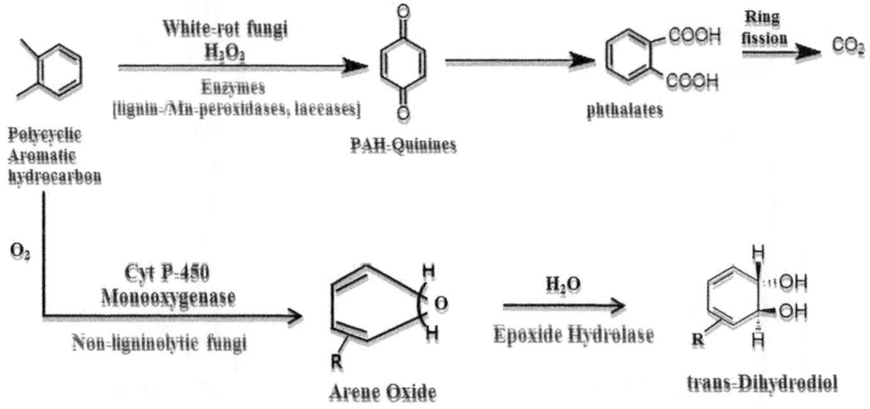

Fig. 9.6 Mechanism of polycyclic aromatic hydrocarbon (PAH) degradation by fungi. (Adapted from Cerniglia and Sutherland 2001)

decomposing litter fungi, and the fungi exhibited high degradation capacity on PAH (Steffen et al. 2007). The mechanism for PAHs degradation process has similar to lignin degradation.

Meanwhile, enzyme activity (intercellular cytochrome P-450) and other mechanisms like manganese peroxidase-mediated lipid peroxidation were involved in the initial degradation process (Fig. 9.6). Polycyclic aromatic hydrocarbons are converted into PAH-quinines and phthalates, forming of ring fission phthalates converted to carbon dioxide. The polycyclic aromatic hydrocarbons were converted to highly oxidized products such as trans-dihydrodiol by cytochrome P450 monooxygenase enzyme (Fig. 9.6). For example, *Cunninghamella elegans* a filamentous fungus was isolated from soil and capable of mineralizing PAH with varying degradation rates.

9.3.3 TNT (2-methyl-1,3,5-trinitrobenzene) Elimination

Trinitrotoluene (TNT) is a yellow solid explosive and carcinogenic toxic substance under group C. Synthesis and usage of TNT in the battlefield and military training has led to its extensive distribution. When prolongs exposed the TNT, creating adverse side effects on animals and plants (EPA 1991). The mammalian organ systems turn TNT in more harmful products and damage the genome function and metabolic processes. Low solubility TNT migrated through the surface soil and washed gradually into groundwater systems. Fungi play an essential role in clearing the environmental pollutants, instead of a non-biological process. Microorganisms like bacteria and fungi utilize TNT as a nitrogen source in aerobic conditions. Nitrogen is released from TNT and finally, nitrogen reduces as ammonium and is

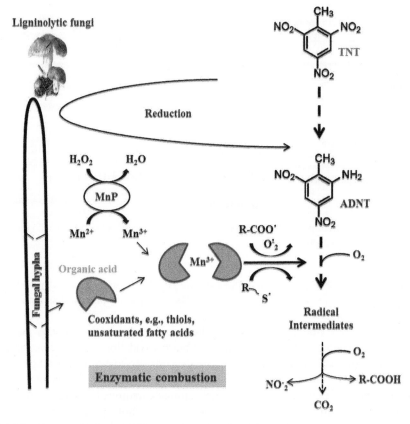

Fig. 9.7 Mycoremediation of TNT contaminated soil with ligninolytic fungi

integrated with their carbon skeletons (Fig. 9.7). Fungal bioremediation processes in TNT contaminated sites is practical and applicable in the full-scale remediation process.

In recent, strains-destructors of environment toxicant mainly belonging to white-rot fungi and bacteria have been investigated (Lee et al. 2009; Solyanikova et al. 2012). *Aspergillus niger* and Mucor species can degrade the TNT and they utilized as a source of carbon (Kutateladze et al. 2018). Mushroom ligninolytic and manganese peroxide enzyme systems cause a minimal amount of degradation. The Agaricales fungi *Agrocybe praecox* is a brown spored edible mushroom that has been involved in the TNT bioremediation process. However, both white-rot fungi, namely *Phlebia* species and *Pleurotus ostreatus*, were involved in the remediation process. White-rot fungi produced high amounts of MnP in TNT contaminated soil rather than laccase (Anasonye et al. 2015). The ligninolytic fungi provide the most powerful and potential enzymes and degrade the TNT pollutants. In resent, other white-rot fungi and litter decomposing fungi have been investigated.

Mycoremediation process was achieved in laboratory scale by liquid culture technique (Eilers et al. 1999), by their enzymes (Van Aken et al. 1999) and few studies on soil environment (Spiker et al. 1992; Fritsche and Hofrichter 2000).

9.4 Conclusions and Future Prospect

Mycoremediation is natural restoration processes that detoxify or removes contaminants in the soil and other environments. Naturally, fungi are cosmopolitans, high stress-tolerant and have high enzyme capacity to degrade the organic/inorganic pollutants in soil and water ecosystem. Several fungi have been successfully identified, which can grow in heavy metal contaminated soils and accumulate or detoxify the heavy metals. The use of the microbial approach for substantial metal tolerance and remediation is an eco-friendly and economical approach. The chapter discusses the types of fungi involved in the mycoremediation and about their metabolites. Both fungal enzymes and their metabolites are involved in the remediation process, and they utilized the pollutant as a carbon source. The fungal enzymes like lignocellulose, laccases, lignin, Mn-peroxidases and cytochrome P450 monooxygenases are involved in the mycoremediation. Most of the research findings revealed that the fungi had a high capacity to remove the heavy metals from polluted soil. Which evidence numbers of fungi were isolated, filamentous and non-filamentous fungi were isolated from contaminated soils. Fungi can uptake, incorporated and assimilated the heavy metals, and fungi tolerate the heavy metal toxicity. Mushroom other fungi were involved in plastic degradation and the assimilation process. From a future perspective, identifying fungi from the contaminated site and identifying the substrate specified enzymes and their coding genes and metabolites produced during the remediation process. To identification of the specific fungal gene, regulations may use to identify the fungal diversity. Furthermore, identification of pollutant/stress-tolerant fungi from contaminated soil and their participation towards mycoremediation and addition efficient biomarkers identification are warranted.

Acknowledgments First and second authors thank National Institute of Horticultural and Herbal Science, (Project No. PJ01419605), Rural Development Administration, Republic of Korea.

References

Acosta-Rodriguez I, Cardenas-Gonzalez JF, Martinez-Juarez VM, Rodriguez Perez A, Moctezuma-Zarate MG, Pacheco-Castillo NC (2018) Biosorption of heavy metals by *Candida albicans*. Advances in bioremediation and phytoremediation. In: Shiomi N, (ed) INTECH. doi:https://doi.org/10.5772/intechopen.72454

Aishwarya S, Venkateswarulu N, Vasudeva RN, Vijaya T (2016) Screening and identification of heavy metal-tolerant endophytic Fungi *Lasiodiplodia theobromae* from *Boswellia ovalifoliolata* an endemic plant of Tirumala hills. Asian J Pharm Clin Res 10(3):488–491. https://doi.org/10.22159/ajpcr.2017.v10i3.16697

Akgul A, Akgul A (2018) Mycoremediation of copper: exploring the meal tolerance of brown rot fungi. Bio Resour 13(3):7155–7171. https://doi.org/10.15376/biores.13.3.Akgul

Akinpelu EA (2014) Bioremediation of gold mine wastewater using *Fusarium oxysporum*. Master's Thesis, Cape Paninsula University of Technology, Cape Town, South Africa. http://hdl.handle.net/20.500.11838/918

Alothman ZA, Bahkalic AH, Khiyamic MA, Alfadulc SM, Wabaidura SM, Alamd M, Alfarhane BZ (2020) Low cost biosorbents from fungi for heavy metals removal from wastewater. Sep Sci Technol 55(10):1766–1775. https://doi.org/10.1080/01496395.2019.1608242

Alpat S, Alpat SK, Çadirci BH, Ozbayrak O, Yasa I (2010) Effects of biosorption parameter: kinetics, isotherm and thermodynamics for Ni (II) biosorption from aqueous solution by Circinella sp. Electron J Biotechnol 13(5):4–5. https://doi.org/10.2225/vol13-issue5-fulltext-20

Amatussalam A, Abubacker MN, Rajendran RB (2011) *In situ* Carica papaya stem matrix and *Fusarium oxysporum* (NCBT156) mediated bioremediation of chromium. Indian J Exp Biol 49(12):925–931. https://doi.org/10.13005/bbra/1061

Anasonye F, Winquist E, Rasanen M, Kontro J, Bjorklof K, Vasilyeva G, Jorgensen KS, Steffen KT, Tuomela M (2015) Bioremediation of TNT contaminated soil with fungi under laboratory and pilot scale conditions. Int Biodeterior Biodegradation 105:7–12. https://doi.org/10.1016/j.ibiod.2015.08.003

Arbanah M, Miradatul NMR, Halim KH (2012) Biosorption of Cr(III), Fe(II), Cu(II), Zn(II) Ions from liquid laboratory chemical waste by *Pleurotus ostreatus*. Int J Biotechnol Wellness Ind 1(3):152–162. https://doi.org/10.6000/1927-3037/2012.01.03.01

Archana A, Jaitly AK (2015) Mycoremediation: utilization of fungi for reclamation of heavy metals at their optimum remediation conditions. Biolife 3(1):77–106

Arica MY, Arpa C, Kaya B, Bektas S, Denizli A, Genc O (2003) Comparative biosorption of mercuric ions from aquatic systems by immobilized live and heatinactivated Trametes versicolor and *Pleurotus sajur-caju*. Bioresour Technol 89(2):145–154. https://doi.org/10.1016/S0960-8524(03)00042-7

Bai S, Abraham TE (2001) Biosorption of Cr (VI) from aqueous solution by *Rhizopus nigricans*. Bioresour Technol 79(1):73–81. https://doi.org/10.1016/S0960-8524(00)00107-3

Beaudette LA, Davies S, Fedorak PM, Ward OP, Pickard MA (1998) Comparison of gas chromatography and mineralization experiments for measuring loss of selected polychlorinated biphenyl congeners in cultures of white-rot fungi. Appl Environ Microbiol 64(6):2020–2025. https://doi.org/10.1128/AEM.64.6.2020-2025.1998

Bilos C, Colombo JC, Skorupka CN, Rodriguez Presa MJ (2001) Sources, distribution and variability of airborne trace metals in La Plata City area, Argentina. Environ Pollut 111(1):149–158. https://doi.org/10.1016/S0269-7491(99)00328-0

Birch LD, Bachofen R (1990) Effects of microorganisms on the environmental mobility of radionucleides. In: Bollang JM, Stozky G (eds) Soil biochemistry, vol 6. Marcel Dekker, New York, N.Y, pp 483–527. https://doi.org/10.2136/sssaj1972.03615995003600060006x

Blaudez D, Botton B, Chalot M (2000) Cadmium uptake and subcellular compartmentation in the ectomycorrhizal fungus *Paxillus involutus*. Microbiol 146(5):1109–1117. https://doi.org/10.1099/00221287-146-5-1109

Cabuk A, Ilhan S, Filik C, Çalişkan F (2005) Pb2+ biosorption by pretreated fungal biomass. Turk J Biol 29:23–28

Cecchi G, Roccotiello E, Di Piazza S, Riggi A, Mariotti MG, Zotti M (2017) Assessment of Ni accumulation capacity by fungi for a possible approach to remove metals from soil and waters. J Environ Sci Health B 52(3):166–170. https://doi.org/10.1080/03601234.2017.1261539

Cerniglia CE, Sutherland JB (2001) Bioremediation of polycyclic aromatic hydrocarbonsby ligninolytic and non-ligninolytic fungi. In: Gadd GM (ed) Fungi in Bioremediation. Cambridge University Press, Cambridge, pp 136–187

Chen G, Zeng G, Tu X, Huang G, Chen Y (2005) A novel biosorbent: characterization of the spent mushroom compost and its application for removal of heavy metals. J Environ Sci (China) 17:756–760

Chigu NL, Hirosue S, Nakamura C, Teramoto H, Ichinose H, Wariishi H (2010) Cytochrome P450 monooxygenases involved in anthracene metabolism by the white-rot basidiomycete *Phanerochaete chrysosporium*. Appl Microbiol Biotechnol 87:1907–1916. https://doi.org/10.1007/s00253-010-2616-1

Christie P, Li X, Chen B (2004) Arbuscular mycorrhiza can depress translocation of zinc to shoots of host plants in soils moderately polluted with zinc. Plant Soil 261(1–2):209–217. https://doi.org/10.1023/B:PLSO.0000035542.79345.1b

Congeevaram S, Dhanarani S, Park J, Dexilin M, Thamaraiselvi K (2007) Biosorption of chromium and nickel by heavy metal resistant fungal and bacterial isolates. J Hazard Mater 146 (2007):270–277. https://doi.org/10.1016/j.jhazmat.2006.12.017

Cunningham SD, Ow DW (1996) Promises and prospects of phytoremediation. Plant Physiol 110:715–719. https://doi.org/10.1104/pp.110.3.715

D'Annibale A, Rosetto F, Leonardi V, Federici F, Petruccioli M (2006) Role of autochthonous filamentous fungi in bioremediation of a soil historically contaminated with aromatic hydrocarbons. Appl Environ Microbiol 72(1):28–36. https://doi.org/10.1128/AEM.72.1.28-36.2006

Eilers A, Rungeling E, Stündl UM, Gottschalk G (1999) Metabolism of 2,4,6-trinitrotoluene by the white-rot fungus *Bjerkandera adusta* DSM 3375 depends on cytochrome P-450. Appl Microbiol Biotechnol 53(1):75–80. https://doi.org/10.1007/s002530051617

El-Gendy MMA, Hassanein NM, El-Hay IH, El-Baky A, Doaa H (2011) Evaluation of some fungal endophytes of plant potentiality as low-cost adsorbents for heavy metals uptake from aqueous solution. Aust J Basic Appl Sci 5:466–473

El-Morsy ESM (2004) Cunninghamella echinulata a new biosorbent of metal ions from polluted water in Egypt. Mycologia 96(6):1183–1189. https://doi.org/10.1080/15572536.2005.11832866

EPA. Health Effects Summary Tables. Annual FY-91 (1991) Prepared by thy office of health and environmental assessment, environmental criteria remedial repose. Washington, D. C. OERR 9200. a.6-303 (91-1). NTIS PB91-921199

Fan T, Liu Y, Feng B, Zeng G, Yang C, Zhou M, Zhou H, Tan Z, Wang X (2008) Biosorption of cadmium(II), zinc(II) and lead(II) by *Penicillium simplicissimum*: isotherms, kinetics and thermodynamics. J Hazard Mater 160(2–3):655–661. https://doi.org/10.1016/j.jhazmat.2008.03.038

Fourest E, Roux J-C (1992) Heavy metal biosorption by fungal mycelial by-products: mechanisms and influence of pH. Appl Microbiol Biotechnol 37:399–403. https://doi.org/10.1007/BF00211001

Friesen ML, Porter SS, Stark SC, von Wettberg EJ, Sachs JL, Martinez-Romero E (2011) Microbially mediated plant functional traits. Annu Rev Ecol Evol Syst 42:23–46. https://doi.org/10.1146/annurev-ecolsys-102710-145039

Fritsche W, Hofrichter M (2000) Aerobic degradation by microorganisms. In: Klein J (ed) Environmental processes-soil decontamination. Wiley-VCH, Weinheim, pp 146–155

Gadd GM (1999) Fungal production of citric and oxalic acid: importance in metal speciation, physiology and biogeochemical processes. Adv Microb Physiol 41:47–92. https://doi.org/10.1016/S0065-2911(08)60165-4

Gadd GM (2004) Mycotransformation of organic and inorganic substances. Mycologist 18 (2):60–70. https://doi.org/10.1017/S0269-915X(04)00202-2

Gadd GM (2007) Geomycology: biogeochemical transformations of rocks, minerals, metals and radionuclides by fungi, bio-weathering and bioremediation. Mycol Res 111:3–49. https://doi.org/10.1016/j.mycres.2006.12.001

Gadd GM (2001) Preface. In: Gadd GM (ed) Fungi in bioremedation. Cambridge University Press, Cambridge, pp xi–xiii

Gadd GM, de Rome L (1988) Biosorption of copper by fungal melanine. Appl Microbiol Biotechnol 29:610–617

Galun M, Galun E, Siegel BZ, Keller P, Lehr H, Siegel SM (1987) Removal of metal ions from aqueous solutions by Penicillium biomass: kinetic and uptake parameters. Water Air Soil Pollut 33:359–371

Green F, Clausen CA (2003) Copper tolerance of brown-rot fungi: time course of oxalic acid production. Int Biodeterior Biodegradation 51(2003):145–149. https://doi.org/10.1016/S0964-8305(02)00099-9

Hala YEK, Eman MET (2009) Optimization of batch process parameters by response surface methodology for mycoremediation of chrome-VI by a chromium resistant strain of marine *Trichoderma Viride*. Am Eurasian J Agric Environ Sci 5:676–681. http://www.idosi.org/.../14.pdf

Harvey PJ, Thurston CF (2001) The biochemistry of ligninolytic fungi. In: Gadd GM (ed) Fungi in Bioremediation. Cambridge University Press, Cambridge, pp 27–51

Hassanein NM, El-Gendy MMA, El-Hay Ibrahim HA, El Baky DHA (2012) Screening and evaluation of some fungal endophytes of plant potentiality as low-cost adsorbents for heavy metals uptake from aqueous solution. Egypt J Exp Biol (Bot) 8(1):17–23

Hepper CM, Smith GA (1976) Observation's on the germination of Endogone spore. Trans Br Mycol Soc 66(2):189–194. https://doi.org/10.1016/S0007-1536(76)80044-7

Hindersah R, Asda KR, Herdiyantoro D, Kamaluddin NN (2018) Isolation of mercury-resistant Fungi from mercury-contaminated agricultural soil. Agriculture 8(3):33–40. https://doi.org/10.3390/agriculture8030033

Iida T, Sumita T, Ohta A, Takagi M (2000) The cytochrome P450ALK multigene family of an n-alkane-assimilating yeast, *Yarrowia lipolytica*: cloning and characterization of genes coding for new CYP52 family members. Yeast 16(12):1077–1087. https://doi.org/10.1002/1097-0061(20000915)16:12<1077::AID-YEA601>3.0.CO;2-K

Ismail K, Akar T, Tunali S (2005) Biosorption of Pb(II) and Cu(II) from aqueous solution by pretreated biomass of *Neurospora crassa*. Process Biochem 40:3550–3558. https://doi.org/10.1016/j.procbio.2005.03.051

Ita BN, Ebong GA, Essien JP, Eduok SI (2008) Bioaccumulation potential of heavy metals in edible fungal Sporocarps from the Niger Delta region of Nigeria. Pak J Nutr 7:93–97. https://doi.org/10.3923/pjn.2008.93.97

Jin Y, Luan Y, Ning Y, Wang L (2018) Effects and mechanisms of microbial remediation of heavy metals in soil. A critical review. Appl Sci 8(8):1336–1343. https://doi.org/10.3390/app8081336

Johannes C, Majcherczyk A (2000) Natural mediators in the oxidation of polycyclic aromatic hydrocarbons by laccase mediator systems. Appl Environ Microbiol 66(2):524–528. https://doi.org/10.1128/AEM.66.2.524-528.2000

Kadirvelu K, Senthilkumar P, Thamaraiselvi K, Subburam V (2002) Activated carbon prepared from biomass as adsorbent: elimination of Ni(II) from aqueous solution. Bioresour Technol 81(1):87–90. https://doi.org/10.1016/S0960-8524(01)00093-1

Kamei I, Suhara H, Kondo R (2005) Phylogenetical approach to isolation of white-rot fungi capable of degrading polychlorinated dibenzo-p-dioxin. Appl Microbiol Biotechnol 69(3):358–366. https://doi.org/10.1007/s00253-005-0052-4

Kanaly R, Hur H (2006) Growth of Phanerochaete chrysosporium on diesel fuel hydrocarbons at neutral pH. Chemosphere 63(2):202–211. https://doi.org/10.1016/j.chemosphere.2005.08.022

Kapahi M, Sachdeva S (2017) Mycoremediation potential of *Pleurotus* species for heavy metals: a review. Bioresour Bioprocess 4(1):32–41. https://doi.org/10.1186/s40643-017-0162-8

Kurniati E, Arfarita N, Imai T, Higuchi T, Kanno A, Yamamoto K, Sekine M (2014) Potential bioremediation of mercury-contaminated substrate using filamentous fungi isolated from forest soil. J Environ Sci (China) 26(6):1223–1231. https://doi.org/10.1016/S1001-0742(13)60592-6

Kutateladze L, Zakariashvili N, Khokhashvili I, Jobava M, Alexidze T, Urushadze T, Kvesitadze E (2018) Fungal elimination of 2,4,6-trinitrotoluene (TNT) from the soils. Euro Biotech J 2(1):39–46. https://doi.org/10.2478/ebtj-2018-0007

Kuzhali SS, Manikandan N, Kumuthakalavalli R (2012) Biosorption of heavy metals from tannery effluent by using macrofungi. Elixir Pollut 47:9005–9006

Lamrood PY, Ralegankar SD (2013) Biosorption of Cu, Zn Fe, Cd, Pb and Ni by nontreated biomass of some edible mushrooms. Asian J Exp Biol Sci 4:190–195

Lee S, Lee SY, Shin KS (2009) Biodegradation of 2,4,6-trinitrotoluene by white-rot fungus. Mycobiol 37(1):17–20. https://doi.org/10.4489/MYCO.2009.37.1.017

Lenart-Boron A, Boron P (2014) The effect of industrial heavy metal pollution on microbial abundance and diversity in soils – a review. Environ Risk Assess Soil Contam:759–784. https://doi.org/10.5772/57406

Levinskaite L (2001) Simultaneous effect of Ni, Cd and Cr on soil micromycetes. Biologija 4:13–15

Li X, Kondo R, Sakai K (2003) Studies on hypersaline-tolerant white-rot fungi IV.effects of Mn2+ and NH4+ on manganese peroxidase and Poly R-478 decolorization by the marine isolate *Phlebia* sp. MG-60 under saline conditions. J Wood Sci 49(4):355–360. https://doi.org/10.1007/s10086-002-0492-8

Lone MI, He Z, Stoffella PJ, Yang X (2008) Phytoremediation of heavy metal polluted soils and water: progresses and perspectives. J Zhejiang Univ Sci B 9(3):210–220. https://doi.org/10.1631/jzus.B0710633

Luo JM, Xiao X, Luo SI (2010) Biosorption of cadmium (II) from aqueous solutions by industrial fungus *Rhizopus cohnii*. Trans Nonferrous Metal Soc 20(6):1104–1111. https://doi.org/10.1016/S1003-6326(09)60264-8

Maheswari S, Murugesan AG (2009) Remediation of arsenic in soil by *Aspergillus nidulans* isolated from an arsenic-contaminated site. Environ Technol 30(9):921–926. https://doi.org/10.1080/09593330902971279

Majumdar SS, Das SK, Chakravarty R, Saha T, Bandyopadhyay TS, Guha AK (2010) A study on lead adsorption by *Mucor rouxii* biomass. Desalination 251(1–3):96–102. https://doi.org/10.1016/j.desal.2009.09.137

Mamun AA, Alam Z, Nik AN, Shah SR (2011) Adsorption of heavy metal from landfill leachate by wasted biosolids. Afr J Biotechnol 10(81):18869–11888. https://doi.org/10.5897/AJB11.2768

Mihova S, Godjevargova T (2001) Biosorption of heavy metals from aqueous solutions. J Int Res. pub 1-2000/01. ISSN 1311-8978

Mishra A, Malik A (2014) Novel fungal consortium for bioremediation of metals and dyes from mixed waste stream. Bioresour Technol 171:217–226. https://doi.org/10.1016/j.biortech.2014.08.047

Muraleedharan TR, Iyengar L, Venkobachar C (1995) Screening of tropical wood-rotting mushrooms for copper biosorption. Appl Environ Microbiol 61(9):3507–3508. https://doi.org/10.1128/AEM.61.9.3507-3508.1995

Nagy B, Maicaneanu A, Indolean C, Manzatu C, Silaghi-Dumitrescu L, Majdik C (2014) Comparative study of Cd(II) biosorption on cultivated *Agaricus bisporus* and wild *Lactarius piperatus* based biocomposites: linear and nonlinear equilibrium modelling and kinetics. J Taiwan Inst Chem Eng 45(3):921–929. https://doi.org/10.1016/j.jtice.2013.08.013

Nasseri S, Mazaheri A, Noori S, Rostami K, Shariat M, Nadafi K (2002) Chromium removal from tanning effluent using biomass of *Aspergillus oryzae*. Pak J Biol Sci 5:1056–1059. https://doi.org/10.3923/pjbs.2002.1056.1059

Novotny C, Erbanova P, Cajthaml T, Rothschild N, Dosoretz C, Sasek V (2000) *Irpex lacteus*, a white-rot fungus applicable to water and soil bioremediation. Appl Microbiol Biotechnol 54:850–853. https://doi.org/10.1007/s002530000432

Nowak B, Pajak J, Labuzek S (2010) Biodegradation of compositions of poly (ethylene terephthalate) with Bionolle® or starch. Composites 10:224–228

Oladipo OG, Awotoye OO, Olayinka A, Ezeokoli OT, Maboeta MS, Bezuidenhout CC (2016) Heavy metal tolerance potential of *Aspergillus* strains isolated from mining sites. Biorem J 20:287–297. https://doi.org/10.1080/10889868.2016.1250722

Oladipo OG, Olayinka A, Awotoye OO (2014) Ecological impact of mining on soils of southwestern Nigeria. Environ Exp Biol 12:179–186

Pal TK, Bhattacharyya S, Basumajumdar A (2010) Cellular distribution of bioaccumulated toxic heavy metals in Aspergillus Niger and *Rhizopus arrhizus*. Int J Pharma Biosci 1:1–6

Pattanapipitpaisal P, Brown N, Macaskie L (2001) Chromate reduction by microbacterium liquefaciens immobilised in polyvinyl alcohol. Biotechnol Lett 23:61–65. https://doi.org/10.1023/A:1026750810580

Pouli M, Agathos SN (2011) Bioremedation of PAH-contaminated sites: from pathways to bioreactors. In: Koukkou A (ed) Microbial bioremediation of non-metals. Caister University Press, Norfolk, pp 119–147

Prasad AS, Varatharaju G, Anushri C, Dhivyasree S (2013) Biosorption of Lead by *Pleurotus florida* and *Trichoderma viride*. Brit Biotechnol J 3(1):66–78. https://doi.org/10.9734/BBJ/2013/2348

Prasenjit B, Sumathi S (2005) Uptake of chromium by Aspergillus foetidus. J Mater Cycles Waste Manage 7:88–92. https://doi.org/10.1007/s10163-005-0131-8

Raaman N, Rajitha N, Jayshree A, Jegadeesh R (2012) Biodegradation of plastic by *Aspergillus* spp. isolated from polythene polluted sites around Chennai. J Acad Ind Res 1(6):313–316

Rao KR, Rashmi K, Latha J, Mohan PM (2005) Bioremediation of toxic metal ions using biomass of *Aspergillus fumigatus* from fermentative waste. Indian J Biotechnol 4:139–143. http://hdl.handle.net/123456789/5627

Rojas CMM, Velasquez MFR, Tavolaro A, Molinari A, Fallico C (2017) Use of vegetable fibers PRB to remove heavy metals from contaminated aquifers-comparisons among Cabuya fibers, broom fibers and ZVI. Int J Environ Res Public Health 14(7):684–692. https://doi.org/10.3390/ijerph14070684

Russell JR, Huang J, Anand P, Kucera K, Sandoval AG, Dantzler KW, Hickman DS, Jee J, Kimovec FM, Koppstein D, Marks DH, Mittermiller PA, Nunez SJ, Santiago M, Townes MA, Vishnevetsky M, Williams NE, Vargas MPN, Boulanger LA, Bascom-Slack C, Strobel SA (2011) Biodegradation of polyester polyurethane by Endophytic Fungi. Appl Environ Microbiol 77(17):6076–6084. https://doi.org/10.1128/AEM.00521-11

Salt DE, Smith RD, Raskin I (1998) Phytoremediation. Annu Rev Plant Physiol Plant Mol Biol 49:643–668. https://doi.org/10.1146/annurev.arplant.49.1.643

Sanchez-Fernandez RE, Diaz D, Duarte G, Lappe-Oliveras P, Sanchez S, Macias-rubalcava ML (2016) Antifungal volatile organic compounds from the endophyte Nodulisporium sp. strain GS4d2II1a: qualitative change in the intraspecific and interspecific interactions with *Pythium aphanidermatum*. Microb Ecol 71(2):347–364. https://doi.org/10.1007/s00248-015-0679-3

Sarkar S, Satheshkumar A, Jayanthi R, Premkumar R (2010) Biosorption of nickel by live biomass of *Trichoderma harzianum*. Res J Agric Sci 1:69–74

Sasek V (2003) Why Mycoremediations have not yet come into practice. In: Sasek V, Glaser JA, Baveye P (eds) The utilization of bioremediation to reduce soil contamination: problems and solutions. NATO science series (series IV: earth and environmental sciences), vol 19. Springer, Dordrecht. https://doi.org/10.1007/978-94-010-0131-1_22

Say R, Yilmaz N, Denizli A (2003) Removal of heavy metal ions using the fungus *Penicillium canescens*. Adsorpt Sci Technol 21(7):643–650. https://doi.org/10.1260/026361703772776420

Sen M (2013) Biosorption of zinc and nickel from wastewaters using nonliving cells of *Fusarium solani*. Int J Chem Tech Appl 2:63–70

Sen M, Dastidar MG (2011) Biosorption of Cr (VI) by resting cells of *Fusarium solani*. Iran J Environ Health Sci Eng 8:153–158

Seshikala D, Charya MS (2012) Effect of pH on chromium biosorption. Int J Pharma Bio Sci 2:298–302

Shoaib A, Aslam N, Aslam N (2012) Myco and Phyto remediation of heavy metals from aqueous solution. Online J Sci Technol 2:35–40

Siddiquee S, Rovina K, S Azad SA, Naher L, Suryani S, Chaikaew P (2015) Heavy metal contaminants removal from wastewater using the potential filamentous Fungi biomass: a review. J Microbial Biochem Technol 7:384–393. https://doi.org/10.4172/1948-5948.1000243

Skowronski T, Pirszel J, PawlikSkwronska B (2001) Heavy metal removal by the waste biomass of *Penicillium chrysogenum*. Water Pollut Res J Can 36(4):793–803. https://doi.org/10.2166/wqrj. 2001.042

Solyanikova IP, Baskunov BP, Baboshin MA, Saralov AI, Golovleva LA (2012) Detoxification of high concentrations of trinitrotoluene by bacteria. Appl Biochem Microbiol 48(1):21–27. https://doi.org/10.1134/S0003683812010152

Spiker JK, Crawford DL, Crawford RL (1992) Influence of 2,4,6-trinitrotoluene (TNT) concentration on the degradation of TNT in explosive-contaminated soils by the white rot fungus *Phanerochaete chrysosporium*. Appl Environ Microbiol 58(9):3199–3202. https://doi.org/10. 1128/AEM.58.9.3199-3202.1992

Steffen KT, Schubert S, Tuomela M, Hatakka A, Hofrichter M (2007) Enhancement of bioconversion of high-molecular mass polycyclic aromatic hydrocarbons in contaminated non-sterile soil by litter-decomposing fungi. Biodegradation 18(3):359–369. https://doi.org/10.1007/s10532-006-9070-x

Sun TR, Cang L, Wang QY, Zhou DM, Cheng JM, Xu H (2010) Roles of abiotic losses, microbes, plant roots, and root exudates on phytoremediation of PAHs in a barren soil. J Hazard Mater 176 (1–3):919–925. https://doi.org/10.1016/j.jhazmat.2009.11.124

Syed K, Doddapaneni H, Subramanian V, Lam YW, Yadav JS (2010) Genome-to-function characterization of novel fungal P450 monooxygenases oxidizing polycyclic aromatic hydrocarbons (PAHs). Biochem Biophys Res Commun 399(4):492–497. https://doi.org/10.1016/j. bbrc.2010.07.094. Epub 2010 Jul 30

Tan T, Cheng P (2003) Biosorption of metal ions with *Penicillium chrysogenum*. Appl Biochem Biotechnol 104:119–128. https://doi.org/10.1385/ABAB:104:2:119

Tastan BE, Ertugrul S, Donmez G (2010) Effective bioremoval of reactive dye and heavy metals by Aspergillus versicolor. Bioresour Technol 101:870–876. https://doi.org/10.1016/j.biortech. 2009.08.099

Thippeswamy B, Shivakumar C, Krishnappa M (2012) Bioaccumulation potential of *Aspergillus niger* and *Aspergillus flavus* for removal of heavy metals from paper mill effluent. J Environ Biol 33:1063–1068

Tuomela M, Lyytikainen M, Oivanen P, Hatakka A (1999) Mineralization and conversion of pentachlorophenol (PCP) in soil inoculated with the white-rot fungus *Trametes versicolor*. Soil Biol Biochem 31(1):65–74. https://doi.org/10.1016/S0038-0717(98)00106-0

Tyler G (1974) Heavy metal pollution and soil enzymatic activity. Plant Soil 41(2):303–311. https://doi.org/10.1007/BF00017258

USEPA (1997) Recent developments for in situ treatment of metal contaminated soils. Tech. Rep. EPA-542-R-97-004, USEPA, Washington DC. http://purl.access.gpo.gov/GPO/LPS32593

Van Aken B, Hofrichter M, Scheibner K, Hatakka AI, Naveau H, Agathos SN (1999) Transformation and mineralization of 2,4,6-trinitrotoluene (TNT) by manganese peroxidase from the white-rot basidiomycete *Phlebia radiate*. Biodegradation 10(2):83–91. https://doi.org/10.1023/A:1008371209913

Vaseem H, Sing VK, Singh MP (2017) Heavy metal pollution due to coal washery effluent and its decontamination using a macrofungus, *Pleurotus ostreatus*. Ecotoxicol Environ Saf 145:42–49. https://doi.org/10.1016/j.ecoenv.2017.07.001

Young, Darcy MA (2012) Bioremediation with white-rot fungi at Fisherville Mill: analyses of gene expression and number 6 fuel oil degradation. Mosakowski Institute for Public Enterprise. 12. https://commons.clarku.edu/mosakowskiinstitute/12

Zafar S, Aqil F, Ahmad I (2007) Metal tolerance and biosorption potential of filamentous fungi isolated from metal contaminated agriculture soil. Bioresour Technol 98(13):2557–2561. https://doi.org/10.1016/j.biortech.2006.09.051

Chapter 10
Photocatalytic Degradation of Dyes in Wastewater Using Metal Organic Frameworks

Thabiso C. Maponya, Mpitloane J. Hato, Edwin Makhado, Katlego Makgopa, Manika Khanuja, and Kwena D. Modibane

Abstract The current major global crisis is in the sector of water, energy and food security. Due to the growth in civilisation, industrialisation and environmental fluctuations water scarcity and freshwater availability poses major concern to the world, especially to the developing countries. Nearly, one-third of drinkable water is obtained from surface sources such as rivers, dams, lakes, and canals, and some of this water is exposed to contamination. Therefore, appropriate treatment methodologies must be implemented in order to mitigate and prevent water contamination. Photocatalysis is an approach to curb the challenge of water contamination by introducing the use various materials for the degradation of pollutants from wastewater. Metal organic frameworks (MOFs) and their composites as materials of interests have gained popularity in addressing water pollution in photocatalysis applications by acting as absorbents of light, prominent to improve performance of photocatalytic activity. This chapter focuses on the synthesis and characterization of MOFs and their composite materials as efficient photocatalysts for degradation of dyes from wastewater.

Keywords Metal organic frameworks · Wastewater treatment · Photocatalysis · Photodegradation efficiency

---

T. C. Maponya · M. J. Hato (✉) · E. Makhado · K. D. Modibane (✉)
Nanotechnology Research Lab, Department of Chemistry, School of Physical and Mineral Sciences, University of Limpopo (Turfloop), Polokwane, South Africa
e-mail: mpitloane.hato@ul.ac.za; kwena.modibane@ul.ac.za

K. Makgopa (✉)
Department of Chemistry, Faculty of Science, Tshwane University of Technology (Acardia Campus), Pretoria, South Africa
e-mail: Makgopak@tut.ac.za

M. Khanuja
Centre for Nanoscience and Nano Technology, Jamia Millia Islamia (A Central University), New Delhi, India

© The Author(s), under exclusive license to Springer Nature Switzerland AG 2021
S. Rajendran et al. (eds.), *Metal, Metal-Oxides and Metal-Organic Frameworks for Environmental Remediation*, Environmental Chemistry for a Sustainable World 64, https://doi.org/10.1007/978-3-030-68976-6_10

10.1 Introduction

Water plays a crucial role in the society since food security and supply depends on an irrigation system or rain falling directly on a field. Apart from the food security and supply, other aspects where water is essential includes aspects of socio-economic development and for the maintenance of healthy ecosystems. With increasing human population, urbanization as well as living standards, water resources such as ground water and surface water have been contaminated with pollutants such as organics, partly because of the huge production and high usage of pharmaceutical and personal cares.

The most critical issue concerning the portable water arises due to it being scattered at various places and being unequal distributed. Also, places where this portable water are found, they are extremely polluted or contaminated (Dubreuil et al. 2013; United Nations World Water Assessment (WWAP) 2016; Gleick 1993). Due to the scarcity and contamination of the portable water globally, the trends on water demand indicates that more that 1400 million will suffer the detriments of this crisis in the next decades (Dubreuil et al. 2013; United Nations World Water Assessment (WWAP) 2016). The United Nations World Water Report (2016) projected that by 2025, 1.8 billion people will be in the likelihoods to experience water shortage as their daily normality. The faster rate at which population is increasing (having an impact on rapid industrialization that comes with environmental problems) together with parameters such as global warming (having an impact on the ecosystem and the environment) creates challenge in providing sufficient and clean drinkable water (Kulshreshtha 1998; Shannon et al. 2008; Tuzen 2009). On average, agriculture sector accounts for 70% (taken out of rivers, lakes, and aquifers) of global freshwater withdrawals, while 20% is from practical usage and 10% is from domestic and smaller industrial sectors (Tuzen 2009). Due to the continuous clean water demand by the developing countries since the year 1980, there world has experienced a yearly increment of 1% of freshwater withdrawals (Shannon et al. 2008; Tuzen 2009). The issue of freshwater withdrawal intensifies water shortages, thus negatively impacting job markets since about 78% directly depend on water utilization (Kulshreshtha 1998; Shannon et al. 2008). The main sources of water pollution arise from the constant contamination by a wide range of organic and inorganic toxins originating from developed industries (Schwarzenbach et al. 2006). The main culprit for the contamination of clean water emancipate from the improper disposal and the extensive use of organic products such as hydrocarbons, detergents, carbohydrates, plasticizers, pharmaceuticals and personal products (PPCPs), and organic dyes (Zhao et al. 2011). These organic pollutants together with other water pollutants such as heavy metal ions, bacteria, and viruses are released into the environment posing great threats to human health. All these stated factors bring such an urgent need to find alternative ways to assist in removing pollutants from wastewater, especially organic pollutants. The great danger of these pollutants is that they pose great threat to the society due to their toxicity, carcinogenic nature and non-biodegradability. These biohazards have large effect on the provision of clean and fresh water that is adequate for human consumption. There are several techniques/methods used to try remedy the impact by the removal of pollutants in water. The widely explored techniques include reverse osmosis, membrane filtration, adsorption, and disinfection (Zhao et al. 2011). Unfortunately, these techniques display some limitations during the removal process. Some of the identified limitations include high operational costs, generation of secondary pollution, and the inefficiency at low concentration (Zhao et al. 2011). Recently, advanced oxidation processes (AOPs) such as photo-Fenton, heterogenous photocatalysis, sonolysis, ozonation, as well as their combined procedures displayed great potential as solutions for the removal of organic toxins from wastewater Zhao et al. 2011; Chen and Wang 2006; Fu and Wang 2011; Crini and Lichtfouse 2019). These methods have showed improved efficiency for decomposing and mineralising organic toxins at lower concentrations without the generation of secondary pollution. AOPs proceed through the formation of reactive free radicals of OH which act as oxidising agents towards organic pollutants, and thus forming carbon dioxide, water and inorganic mineral salts. Heterogeneous photocatalysis that employs metal oxides such as TiO_2, WO_3, CdS and ZnO as semiconductor starts to gain a great deal of attention due to its unique characteristics such as the ability to operate at ambient temperature and pressure, being environmental friendly and their low energy band gap. As much as this technique holds some good promise for application in wastewater treatment, its popularity is sometimes quenched by the fact that it has limitations such as the less conversion efficiency of solar energy and difficulty in separation suitable for recyclability. These limitations greatly affect the real-life environmental implementation of this technique for applications in wastewater treatment. Hence, metal organic frameworks, MOFs, (with their tuneable structures that can attain such high surface area), are the sort after materials to enhance the catalytic activity of metal oxides. Another added advantage of MOFs is that these porous crystalline materials display features of semiconductors when exposed to light. Meaning that, MOFs possess the light harvest properties like that of a photocatalyst. In this chapter, the recent developments and advances of photocatalysis for wastewater treatment are summarized. This is achieved by doing comprehensive review focusing on the understanding of reactions mechanisms involved in wastewater treatment with the primary objective being on photodegradation of organic dyes. Then, we introduce the role of metal-organic framework (MOF) as a photocatalyst for wastewater treatment. Furthermore, we describe the challenges and approaches faced by MOF and highlights on the recent studies which are on addressing the limitation encountered by MOF. Finally, a detailed summary is provided to provide future perspective on using MOF as a photocatalyst in wastewater treatment.

10.2 Wastewater Treatment Methods

The choice of selecting wastewater treatment techniques is based on assessing the initial quality of water and parameters recognized by regulations. As shown in Fig. 10.1, the procedures involved in the remediation of wastewater treatment are classified as physical, chemical, or biological treatment. Their modes of operation are based on the mechanism of removal of pollutants (Zhao et al. 2011; Chen and Wang 2006; Qu et al. 2013; Wei et al. 2013).

Also, it has been observed that the conventional wastewater treatment approaches are not efficient for potable uses, especially against water comprising of pollutants that are found in lower concentrations (Fu and Wang 2011; Crini and Lichtfouse 2019) The only way to utilize water treated using the conventional approaches is for irrigation of crops or landscapes, refilling of aquifers and non-potable urban uses. The harvested water obtained from treatment using conventional approaches is adequate for reusability in industrial applications (i.e., cooling and boiler feed) and potable urban uses (Crini and Lichtfouse 2019). All these challenges necessitate the development of improved wastewater treatment technologies that will mitigates issues of provision of potable water (i.e., some listed in Fig. 10.1). Chemical processes, which falls amongst various technologies explored for wastewater treatment, have been employed for the removal of heavy metals and organic dyes. The mode of action for water treatment in this process apply technologies such as advanced oxidation processes (AOPs). There has been extensive investigation on the AOPs during the past decade which enabled the photo-degradation of organic compounds by producing reactive oxygen species (ROS), such as hydroxyl radicals (·OH) (Bedia et al. 2019; Deng and Zhao 2015). The AOPs utilize applications

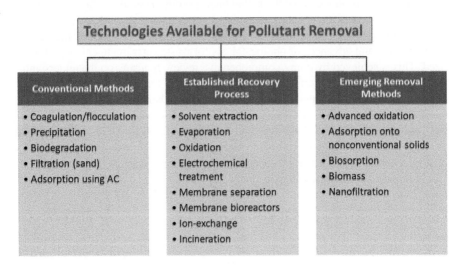

Fig. 10.1 Comparison of various technologies offered for removal of contaminants from wastewater. (Crini and Lichtfouse 2019)

such as of Fenton mechanism (iron/H_2O_2), ozonation (O_3), electrochemical oxidation, ultra-violet (UV), heterogeneous photocatalysis, or the blend of all the stated applications) (Bedia et al. 2019).

10.3 Photocatalysis

10.3.1 Background

It is well known that HP is regarded as an emerging vehicle for water detoxification in the removal of various dyes (Ribeiro et al. 2015; Singh et al. 2013; Carp et al. 2004; Ahmed 2018; Hopfield 1961; Chong et al. 2010; Babu et al. 2015). HP process involves an irradiation of photocatalyst with UV light to separate charges followed by development of the ROS (Carp et al. 2004; Ahmed 2018; Babu et al. 2015). The material which is used in HP is known as a photocatalyst. It is a semiconductor that posseses the valence band (i.e. highest occupied molecular orbital) level and conduction band (i.e. lowest unoccupied molecular orbital) level. The distance between the levels is called the energy band gap (E_g) (Hopfield 1961; Chong et al. 2010). During the irradiation of a semiconductor with the UV light, the semiconductor absorbs light with energy which is equivalent to its E_g (Bedia et al. 2019). The electrons are promoted from the valence band to the conduction band for generation of photons of charges and create an electron hole, h^+ behind (Chong et al. 2010; Babu et al. 2015). There are several pathways that photogenerated charges can take at an excited state. These include recombination, the release of the excitation energy as heat, migration to the surface of the photocatalyst or production of the reactive oxygen species (Bedia et al. 2019). Lastly, ·OH is produced through oxidation of water which is accomplished by the h^+, whereas superoxide radical anions ($O_2\cdot^-$) is generated via adsorbed oxygen reduction mechanism. On the other hand, it was seen that the protonation process may take place to oxidize this $O_2\cdot^-$ to hydroperoxyl radicals ($HO_2\cdot^-$) (Bedia et al. 2019; Babu et al. 2015). These oxidants together with a direct oxidation by h^+ are responsible for the mineralization of the organic dye to CO_2 and H_2O (Bedia et al. 2019; Deng and Zhao 2015; Ribeiro et al. 2015; Singh et al. 2013; Carp et al. 2004; Ahmed 2018; Hopfield 1961; Chong et al. 2010; Babu et al. 2015).

10.3.2 Photocatalysts for Wastewater Treatment

The fast growing interests in the area of photocatalysis for wastewater treatment has risen in the manufacture of various photocatalysts such as metal oxide and metal sulfides (Kumar and Rao 2017; Lee et al. 2016; Fagan et al. 2016; Mondal et al. 2015; Wang et al. 2015). A number of photocatalysts in water (Bedia et al. 2019; Kumar and Rao 2017; Lee et al. 2016). Nonetheless, the application of

TiO$_2$ in a powder form possesses some several drawbacks such as poor porosity, low adsorption capacity and its poor recovery from water (Bedia et al. 2019). In addition, it was seen that the photocatalytic activity of TiO$_2$ anatase, possesses a band gap energy of 3.2 eV ($\lambda \geq 387$ nm) which is high and needs to be activated by UV radiation (Kumar and Rao 2017; Lee et al. 2016). Therefore, it is necessary to modify the surface of TiO$_2$ with materials such as carbon, graphene or metal deposition (Bedia et al. 2019; Kumar and Rao 2017; Gao et al. 2014). However, it is always difficult to prepare TiO$_2$ based photocatalysts for wastewater treatment using visible and solar light. Consequently, it is imperative to explore competent, robust and cost-effective photocatalyst for replacement of the traditional ones. In the past few years, a type of crystalline materials named metal organic organic frameworks (MOFs) have received consideration in photocatalysis. These type of materials offer a wide spectrum of applications due to their structural arrangement of coordination bonds between unsaturated metal core/node and multidentate organic linkers (catalytically active) (Kojtari and Ji 2015). Furthermore, MOFs possess large surface area and well-ordered porous structures that can contribute significantly in many fields. In HP, the use of MOFs photocatalysts in HP is mainly based due to the following important factors such as encapsulation of chromophores in the internal structure of MOF, promotion of e$^-$/h$^+$ separation in the metal core and preparation of MOFs using materials with absorption bands at visible region (Bedia et al. 2019; Llabrés i Xamena et al. 2008). Moreover, some MOFs such as MOF-5 (Llabrés i Xamena et al. 2007), NTU-9 (Gao et al. 2014) and UiO-66 (Shen et al. 2013) can act as semiconductors. In these MOFs, the energy transfer takes place from the organic linker to the metal-oxo cluster (Qiu et al. 2018). Nevertheless, most MOF photocatalysts possess large band gap values caused by an insulating character of the organic linker which results in poor conductivity (Ramohlola et al. 2017a, b, c, 2018; Monama et al. 2018, 2019; Mashao et al. 2019). The large band gap limits their further application (Qiu et al. 2018). and this could be enhanced using dye sensitized materials (Yuan et al. 2015), decoration of linker or metal center (Silva et al. 2010; Fu et al. 2012) and also to combine with other semiconductors (Majedi et al. 2016). Hence, surface modification and functionalization of MOFs are required for their application as suitable photocatalytic materials.

10.4 Metal Organic Frameworks

By a careful inspection of the structure of MOFs (as shown in Fig. 10.2), one can see that the metal nodes function as both the joining points and the organic linkers whose sole purpose is to connects the ligands (Loera-Serna et al. 2012; Llabrés i Xamena et al. 2007). Due to the distinct crystallinity of MOFs, identification of the exact positions of all atoms in the framework can be observed. To properly identify a porous solid, the structure should contain permanent channels or pores which permeate through it and have dimensions huge enough to permit solvent or other molecules to migrate into the structure (Kumar et al. 2013). The other great features

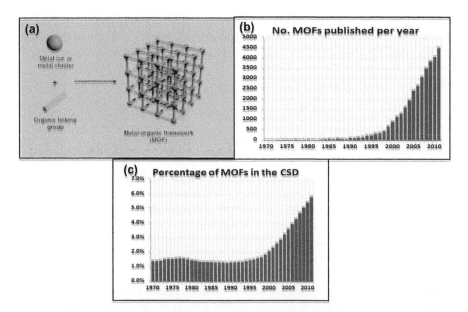

Fig. 10.2 The structure of metal organic framework. (Li and Yang 2007]

about MOF is that its structure is highly tuneable by changing the metal nodes or organic ligands. This characteristic makes it probable to acquire a suitable MOF material suitable for a specified application (Suna et al. 2017; Vlasova et al. 2016). MOFs pore permeability can range from 20 and be as high as 95 percent (Meng et al. 2017). The excellent pore permeability is associated with the high internal surface areas and due to this property, MOFs have gained much attention from the scientific community (Zhao et al. 2016; Cheetham et al. 2006).

10.4.1 Discovery of Metal Organic Frameworks

MOFs discovery occurred at the time when several scientists where studying the Zeolites (Li and Yang 2007; Tranchemontagne et al. 2008). This led to Yaghi coining the term "metal-organic frameworks, MOFs in 1995 when copper-4,4-′-bipryridal complex which displayed extended metal-organic interactions was successfully synthesised (Zhao et al. 2016; Li and Yang 2007). MOFs have its place to the general family of coordination polymers and specifically, to the 2D or 3D crystallized networks possessing porous property in comparison to the traditional coordination polymers (Qiu et al. 2018; Ramohlola et al. 2017a, b, c, 2018; Monama et al. 2018, 2019; Mashao et al. 2019; Yuan et al. 2015; Kreno et al. 2011). Currently, most researchers have turned their attention into developing and synthesis of MOFs and thus resulting in an exponential growth on scientific literature of these materials and their derivatives. Most of these researches, focusses on the experimental design and gas adsorption properties of MOFs and their derivatives for hydrogen storage applications (Campesi et al. 2010; Wang et al. 2017) and also clean energy applications due to their excellent high surface area, permanent porosity and pore size distributions (Tranchemontagne et al. 2008; Wang et al. 2017; Zhang et al. 2017).

10.4.2 Structure of Metal Organic Frameworks

The arrangement of MOFs structures can either be two-dimensional (2D) or three-dimensional (3D) networks. These networks are assembled from organic ligands and metal ion or cluster nodes (Jiang et al. 2013). Unsaturated metal sites or accessible metal sites in MOFs play a major role in influencing their adsorption performance. Transition metals such as Cu, Zn, Mn, Co are usually the ones chosen as centre connectors and they serve as Lewis acids for an activation of the coordinated organic ligands for organic transformation (Raoof et al. 2015; Kim et al. 2013; Nas et al. 2015). The adsorption capacity is due to partial positive charges of metal sides (Cheetham et al. 2006; Raoof et al. 2015). The transition metals can provide variety in geometries depend largely on their oxidation number (Tan et al. 2017; Feng et al. 2013). The linkers possess the aromatic rings in the framework, which help to maintain the structural integrity of the complex and direct the geometry of the framework. The pore volume and surface area of MOFs can be organized by Modification of organic ligands influences the surface area of the MOFs. The common used organic linkers are provided in Fig. 10.3 (Bedia et al. 2019).

Secondary Building Units (SBUs) are critical role players dictating the final geometry of MOFs. The structural and chemical properties of the SBUs and organic ligands result in the prediction of the schemes and synthesis of MOFs. It has been reported that under careful selection of reaction preconditions, multidentate linkers can form some aggregates which affect the formation of SBUs (Salunke-gawali et al. 2012). Subsequently, the SBUs will join with rigid organic links to develop MOFs with high structural stability (Yuan et al. 2016). However, the structure of the SBU is controlled by the metal-to-ligand ratio, solvent and source of anions used in balancing metal ions charges [Sun and Sun 2014; Wang et al. 2014). MOF structures with internal diameter of pores of up to 4.8 nm can extend the free space and thus results in H_2 storage (Abbasi et al. 2017). However, the empty spaces remain intact after the guest molecules have been removed. The size of the pores on the adsorbents is responsible for possible interaction between the absorbed gas molecules and the surface of the surrounding walls. Furthermore, it should be close to the kinetic diameter of H_2 molecule (0.289 nm) to endorse stronger interactions amongst H_2 molecules and the framework. The MOF (NU-100) containing <2 nm micropores possesses storage capacity of 8 wt.% [Zhao et al. 2014; Azad et al. 2016) and 7 wt.% for MOF-5 possessing 0.77 nm micropores (Langmi et al. 2013). Generally, the structure of MOFs with bigger pore sizes are more susceptible to collapse, and this affects their permanent porosity. However,

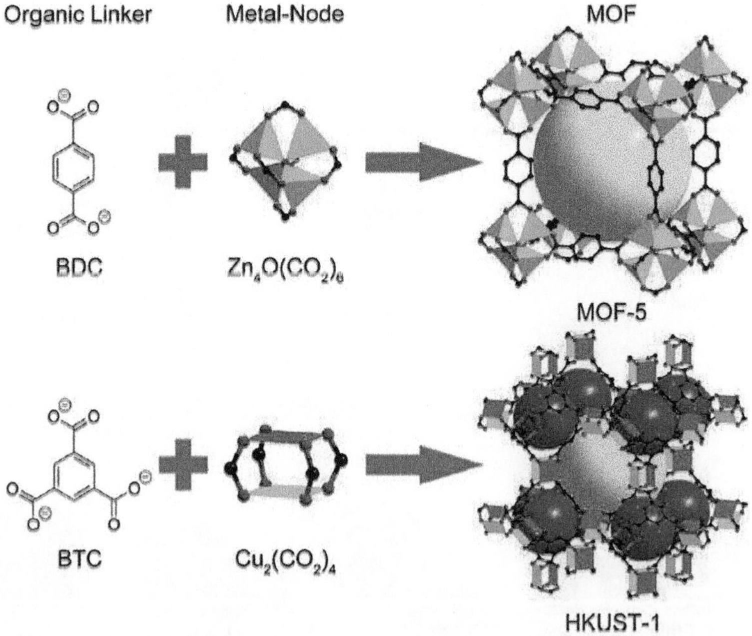

Fig. 10.3 The synthesis of MOF-5 and HKUST-1 (Cu-MOFs) using various organic linkers. Yellow and blue spheres denote the free spaces in the framework with no chemical meaning. (Bedia et al. 2019)

MOF based materials with these structures can result in enhanced performance in hydrogen gas storage (Wang et al. 2016).

10.4.3 Synthesis of Metal Organic Frameworks

Fig. 10.4a demonstrates a summary of different strategic for the synthesis of MOFs. The core aspects of MOF synthesis are to decide synthetic methodologies which can produce MOF of a good metal building units without breakdown the organic linker. Attention in this work is provided to the selected synthetic routes for preparation of MOF materials which are presented in Fig. 10.4 (Joseph et al. 2019).

10.4.3.1 Hydro/Solvothermal Preparation

The hydro/solvothermal procedure has been efficiently utilized in preparing MOFs, because it gives nanoscale structure which is difficult to obtain using conventional method (Tan et al. 2017; Loera-Serna et al. 2017,). It was seen that one of imperative factors in the synthesis of MOF is the choice of metal and organic linker followed by

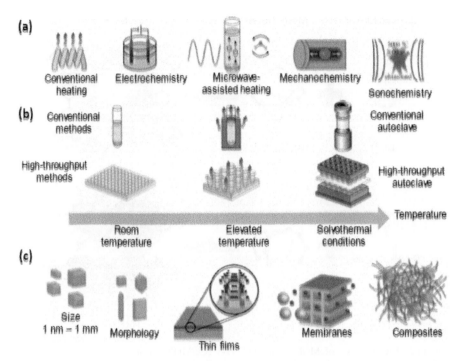

Fig. 10.4 Summary of (**a**) preparation methodologies, (**b**) potential reaction conditions (temperature), and (**c**) synthesized products of MOF structure. (Joseph et al. 2019)

temperature, concentration of metal salt, and the solubility of the reactants (Lin et al. 2012). In this case, the metal cores are attached to organic linker through functional groups to give a 3D structure of porous materials which is a paddle-wheel-like unit of cubic structure. Furthermore, the use of conventional synthesis schemes, powdered MOFs with low densities have often been synthesized (Candu et al. 2013). Nevertheless, it was observed that the low density property of MOF is not suitable for hydrogen storage application (Ramohlola et al. 2017a, b, c; Qamar et al. 2016). The hydrothermal approach is normally performed in a stainless steel autoclave as given in Fig. 10.5a. This approach allows a defined regulation on the shape and size of the material to be prepared, not like conventional approach (Ramohlola et al. 2017a, b, c). As presented in Fig. 10.5d, the epitaxial growth ZIF-67 attached to ZIF-8 benefits from harmonized unit cell parameters, and core-shell structured ZIF-8@ZIF-67 crystals were prepared (Wang and Hu 2018).

10.4.3.2 Microwave-Assisted Synthesis

Microwave-assisted preparation route has revealed to be an interesting method for fast fabrication of nanostructured porous materials (Qamar et al. 2016). Moreover, this route gives potential advantages like facile morphology control on MOF

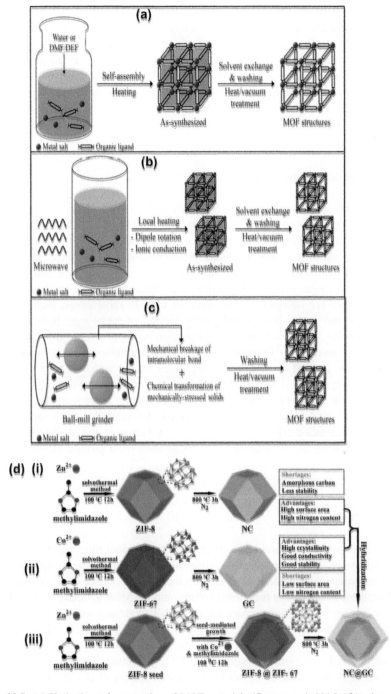

Fig. 10.5 (a) Hydrothermal preparation of MOF materials (Qamar et al. 2016), (b) microwave-assisted solvothermal synthesis of MOF materials (Lin et al. 2012; Song et al. 2006), (c) mechanochemical preparation of MOF materials (Langmi et al. 2013) and (d) schematic representation for the synthesis of (i) MOFs (ZIF-8) (ii) ZIF-67 crystals and NC@GC, and core-shell ZIF-8@ZIF-67 crystals and (iii) NC@GC via a conventional solvothermal method. (Wang and Hu 2018)

structure, and structural phase selectivity as well as narrow particle size distribution (Lee et al. 2013). The microwave-assisted in the preparation of MOFs employs heating a substrate mixture in a suitable solvent with microwave radiation for an hour in order to produce nanosized crystals (Fig. 10.5b). The quality of the crystal materials generated by microwave-assisted technique are similar to the one prepared by the ordinary solvothermal route, but microwave-assisted method achieved the preparation process very faster (Li and Yang 2007).

10.4.3.3 Mechanochemical Process

Mechanochemical method is one of the solvent-free synthetic routes employed to prepare MOF structure using only metal salt and organic linker without any solvent (Fig. 10.5c). This method uses two reaction steps, where it involves the mechanical breakdown of intramolecular bonds followed by chemical transformation (Song et al. 2006). It has several advantages over other method because chemical reactions can take place at room temperature, qualitative and quantitative yield of product are obtained in less than 20 min. In most cases, preference of the source of metal ions were metal oxides instead of metal salts because metal oxide produce in water as by-product during preparation (Wang and Hu 2018). In addition, it was shown in recent years that the mechanochemical method has been efficiently applied for the fast synthesis of MOFs with the help of liquid-assisted grinding (LAG). The LAG is used as a structure directing agent by addition of small amount of solvent into a solid reaction mixture. Moreover, the synthetic route was advanced to prepare pillared-layered structured MOFs using ion-and liquid assisted grinding (ILAG) (Tan et al. 2017). However, the method is limited to explicit MOF types which cannot yield large amount of product (Langmi et al. 2013).

10.4.3.4 Post Synthesis

MOFs has ability to combine with compound of additional functional groups into MOF framework, hence producing a different of MOF structures with varied of functional groups still maintaining the same topology (Volkova et al. 2014). Nevertheless, the introduction of functional groups onto the MOF structure remains to be a challenging problem during preparation method. This setback may be addressed by employing a post-synthesis modification (PSM) route. The PSM is one of surface modification or functionalization of MOF structure after their formation (Fig. 10.6). The addition of functional groups onto MOF structure can be obtained by noncovalent, coordinative or covalent interactions (Yuan et al. 2015; Kim et al.

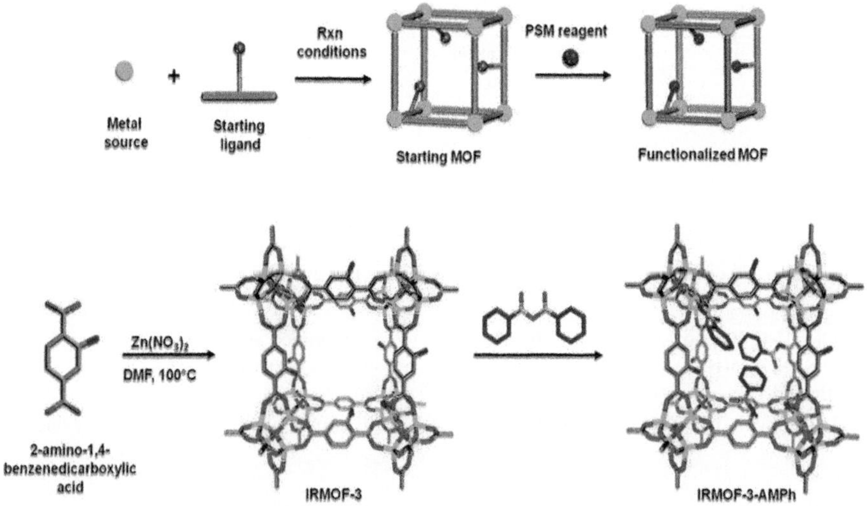

Fig. 10.6 PSM used for the synthesis of MOFs and their functionalized ligands. (Kim et al. 2013)

2013; Candu et al. 2013). It was seen that protonation and doping of the MOF structure may be used as one of the simplest approachesin carrying out PSM (Candu et al. 2013).

10.4.4 MOFs Applications

MOFs have been studied widely and found applications in many different fields such as adsorption, gas storage and separation, heterogeneous catalysis, chemical sensors, biomedicine, supercapacitors, photocatalysis, fuel cells and others (Sundriyal et al. 2018). Table 10.1 exhibits MOFs in selected various applications. For example, Sun et al. 2018 created a nanostructured Fe-Co based MOF-74 adsorbent, which is an adsorbent for the extraction of arsenic in water with maximum adsorption capacity of 292.29 and 266.52 mg/g towards As (V) and As (III), respectively. In another study (Zhang et al. 2019), the authors designed and created a new anion-pillared material (ZU-66) entrenched with molecular rotors towards the separation of CO_2/CH_4 and CO_2/N_2 gas mixtures. This behaviour improved the separation selectivity of ZU-66 for both CO_2/N_2 and CO_2/CH_4 mixtures, and obtained the high CO_2 capacity (4.56 mmol g^{-1}, 298 K, 1 bar). Yu et al. 2018, reported the use of microporous structure and multi-components as O_2, P, C, Ni, and N_2 in the MOF for producing supercapacitors with an improved performance. Their results exhibited the moderate electrochemical capacitance of 979.8 F g^{-1} at a current density of 1 A g^{-1}.

Due to a number of synthesis methods that are available for the preparation of MOFs, there is an ongoing research on the host-guest behaviour of MOFs for

Table 10.1 The use of MOFs in selected applications

MOF	Preparation method	Application	Refs.
MG@MIL-100-B	Solvothermal	Determination of the endogenous catecholamines	He et al. (2018)
ZIF-8	Ultrasonic assisted	Removal of tetracycline and oxytetracycline antibiotics	Li et al. (2019)
MIL-53(Al)-GO	Hydrothermal	Removal of arsenic	Chowdhury et al. (2018)
{[(CH$_3$)$_2$NH$_2$][Zn6(m$_3$-OH)(m$_4$-O)(NSBPDC)5(H$_2$O)$_2$].DMF.12H$_2$O}n	Hydrothermal	Adsorption and separation of organic dyes	Cui et al. (2018)
P2W18@ MIL-101(Cr)	Hydrothermal	Removal of organic dyes	Jarrah and Farhadi (2018)
ZU-66	Hydrothermal	CO_2 separation	
MIL-101@M-X-Y	Hydrothermal	CO_2 and H_2S adsorption	Alivand et al. (2019)
MG@MIL-100-B	Solvothermal	Determination of the endogenous catecholamines	He et al. (2018)
Fe$_3$O$_4$-NH$_2$@HKUST-1@PDES-MSPE	Hydrothermal	Separation of ationic dyes	Wei et al. (2019)
0.4 M-Cu@MIL-100(Fe)	Solvothermal	Adsorptive separation of C_3H_6/C_3H_8 mixtures	Yoon et al. (2019)
Fe0.05/ZIF-8	Hydrothermal	Oxygen reduction catalysts	Yang et al. (2019a)
VNU-21 (Fe$_3$(BTC)(EDB)$_2$.12.27 H$_2$O)	Solvothermal	Heterogeneous catalyst	To et al. (2019)
[(CH$_2$COOH)$_2$IM]HSO$_4$@H-UiO-66	Hydrothermal	Catalyst in biodiesel synthesis	Ye et al. (2018)
UNiMOF/g-C$_3$N$_4$	Ultrasonic assisted	Photocatalytic H_2 production	Cao et al. (2018)
Co-MB	Hydrothermal	Photocatalysis for hydrogen production	Liu et al. (2018)
[Zn3(L)$_2$(nbta)$_2$]n	Hydrothermal	Photocatalytic degradation Rhodamine B	Yang et al. (2019b)
MoS$_2$/Co-MOF	Hydrothermal	Electrocatalytic hydrogen evolution reaction	Zhua et al. (2019)
MOF-808	Hydrothermal	Biocatalyst	Baek et al. (2018)
[MIL-101(Cr)]	Hydrothermal	Biocatalyst for clean synthesis of benzoazoles	Niknam et al. (2018)
MIL-101-NH$_2$-SO$_3$H	Solvothermal	Humidity sensing	He et al. (2018)

(continued)

Table 10.1 (continued)

MOF	Preparation method	Application	Refs.
Cu/Tb@Zn-MOF	Hydrothermal	Sensor detection for aspartic acid	Ji et al. (2019)
[Co(NPDC)(bpee)].DMF.2H$_2$O	Solvothermal	Luminescent sensing for MnO$_4^-$ and Hg2þ	Li et al. (2019)
Zn-Ag$_2$O@PEDOT:PSS	Solvothermal	Battery charged by wind energy	Li et al. (2019)
UiO-66	Solvothermal	Electrolyte additive for Li-metal battery	Chu et al. (2018)
Zn-POMCF	Electrochemical	Lithium-ion batteries	Yang et al. (2019a)
Ni-MOF	Hydrothermal	Supercapacitor	Jiao et al. (2019)
CoCuNi-bdc	Hydrothermal	Supercapacitor	Mohd Zain et al. (2018)
{[(Me$_2$NH2)$_3$(SO$_4$)]$_2$[Zn2 (ox)$_3$]}n	Solvothermal	Fuel cells	Kabir et al. (2018)
NU-(1000 and NU-901)	Hydrothermal	Drug delivery	Teplensky et al. (2017)
CD-MOF	Vapor diffusion	Drug delivery	Li et al. (2017)
Cu-MOF/IBU@GM	Mechanochemical	Drug delivery	Javanbakht et al. (2019)

application in adsorption process (Vlasova et al. 2016). The focus of this chapter is on the photocatalytic degradation of wastewater pollutants by MOFs.

10.4.5 Photocatalytic Degradation of Dyes Using MOFs

The photochemical reactivity and stability of MOFs have gained them popularity for application in the degradation of dye pollutants from wastewater. Table 10.2 shows some of the MOF materials that have been studied for the degradation of organic dye pollutants such as methylene blue (MB), methyl orange (MO), and rhodamine blue (RhB). These studies revealed that MOFs have various photoactive properties based on their different structures as shown in Fig. 10.7a (Glatzmaier et al. 1990; Zeng et al. 2016; Sharma and Feng 2017). In Type I strategy, metal cores (semiconductor dots) in MOFs act as sequestered nano-semiconductors that are separated by organic linkers (Bedia et al. 2019). These types of MOFs are more effective due to their high porosity, when comparing them to conventional semiconductors and the adsorption of pollutants that are near t the semiconductor and photo-generated charges is favoured. Furthermore, the photo-response of MOFs is enhanced by the high density of photoactive dots and the organic linkers which serves as light

Table 10.2 MOF photocatalysts for the degradation of organic dyes

Material	Organic dye	Irradiation time (min)	PDE %	References
MOF, [Cu(4,4'-bipy)Cl]$_n$ + H$_2$O$_2$	MB	150	94	Zhang et al. 2018
MOF, [Co(4,4'-bipy)(HCOO)$_2$]$_n$ + H$_2$O$_2$	MB	150	55	Zhang et al. (2018)
Fe$_2$O$_3$/MIL-53(Fe)	MB	240	70	Feng et al. (2017)
Au@MIL-100(Fe)	MO	150	100	Liang et al. (2015)
Pd@MIL-100(Fe)	MO	150	100	Liang et al. (2015)
Pt@MIL-100(Fe)	MO	150	100	Liang et al. (2015)
rGO/NH$_2$-MIL-125	MB	30	100	Hong et al. (2016)
rGO/MIL-88(Fe)	RhB, MB	20	100	Wu et al. (2014)
GO/MIL-101(Cr)	MG	60	92	Fazaeli et al. (2015)
MIL-53(Fe)	Phenol	180	99	Sun et al. (2015)
NH$_2$-MIL-53(Fe)	Phenol	180	92	Sun et al. (2015)
Fe(BDC)(DMF)	Phenol	180	99	Sun et al. (2015)
BiOBr/NH$_2$-MIL-125	RhB	100	100	Zhu et al. (2016)
Bi$_2$MoO$_6$/MIL-100	RhB	90	90	Yang et al. (2017)
Ag$_3$PO$_4$/MIL-53(Fe)	RhB	90	100	Sha et al. (2015)
g-C$_3$N$_4$/MIL-125	RhB	60	100	Wang et al. (2015)
g-C$_3$N$_4$/MIL-100	RhB	240	100	Hong et al. (2016)
g-C$_3$N$_4$/MIL-53(Al)	RhB	75	100	Guo et al. (2015)
MIL-53(Fe)	RhB	50	98	Ai et al. (2014)

absorbing antennae (Bedia et al. 2019; Zeng et al. 2016). Type II MOFs involves the use of a photo-responsive dye-based organic ligand to absorb light and to transfer photogenerated charges to the metal centers (Bedia et al. 2019). In type III MOFs, the porous nature of the material serves a matrix in which the photo-active components are compressed within its structure (Zeng et al. 2016). Nevertheless, the lack of stability in water by MOFs hinders their application in photodegardation. Alvaro et al. (2007) reported on the photo-degradation of phenol in water using MOF-5. In a study conducted by Hausdorf et al. (2008) on MOF-5, the instability of the material was found to be dependent on the modification of MOF structure and the environment of the water. Laurier et al. (2013) reported on the photocatalytic degradation of rhodamine 6G under visible light (550 nm) using Fe-MOFs. Their studies showed that the photocatalytic performance of Fe-MOFs was more efficient in comparison to traditional TiO$_2$. Furthermore, they were able to maintain their structural characteristics after a photocatalytic activity. Fig. 10.7b shows photodegradation mechanism of MB using MOF where excited high-energy states of electron and hole pairs occurs upon irradiated with light and this reacts with MB resulting in their oxidation producing end products. Liu et al. (2014) reported on photocatalytic activities of Cd(II)-imidazole MOFs for degradation of the MB and methyl orange (MO) using UV light to promote the photogenerated charges that are

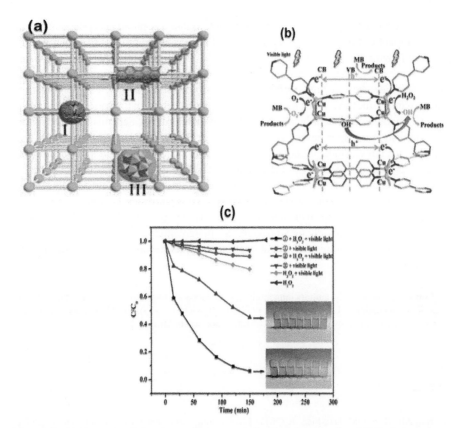

Fig. 10.7 (**a**) Example of the types of photocatalytic mechanisms using MOF photocatalyst mechanisms (Zeng et al. 2016), (**b**) photodegradation mechanism of MB and (**c**) effect of irradiation time on MB using MOF as a photocatalyst under UV radiation. (Zhang et al. 2018)

important for conversion of organic dyes. Moreover, Zhang et al. (2016) studied other types of Cd(II)-imidazole MOFs for photocatalytic retreatment of MO. They showed that the properties of the photocatalyst such as efficiency in the transference, bandgap energy, and separation of charges play important role in producing the photogeneration of charges. In addition, the high photocatalytic efficiency for the removal of MB UV radiation was observed using Zn(II)-imidazolate MOF (ZIF-8) (Jing et al. 2014).

10.5 Conclusions

This chapter encompasses the work done on MOFs for the photocatalytic degradation of organic dyes from wastewater. For example, MOFs are important materials can be employed in variety environmental applications. The significance of MOFs in wastewater treatment includes adsorption, heterogeneous photocatalysts along with enhancement of reactive oxygen generation. Photocatalytic process has been widely explored owing to its flexible design and simple operation. Thus, the use of MOFs as photocatalysts has been investigated for removal of pollutants owing to their inherent properties such as high surface area, environmental stability, easy preparation and tunable structure. The plethora of literature reported that a variety of morphological characteristics can be accomplished, which strongly depends on the method of synthesis, and thus have various effects on the removal of pollutants. In this chapter, we have demonstrated that different types of MOF have an influence on the photodegradation efficiency of organic dyes. Henceforth, the synthesis of variety of MOFs and their composites have been studied for the photocatalytic degradation of dyes such as phenol, MB, MO, MG and RhB from wastewater. Overall, these MOFs composites display improved photocatalytic efficiency towards dyes degradation. Therefore, data in this chapter provides an insight into MOF based materials for potential use as economically valuable photocatalysts for the degradation of organic dyes from wastewater.

Acknowledgements The authors immensely acknowledge the financial support from the National Research Foundation (NRF) (Grant Nos. 117727, 117984 and 118113), Sasol Foundation, Tshwane University of Technology and University of Limpopo, South Africa.

References

Abbasi AR, Karimi M, Daasbjerg K (2017) Efficient removal of crystal violet and methylene blue from wastewater by ultrasound nanoparticles Cu-MOF in comparison with mechanosynthesis method. Ultrason Sonochem 37:182–191

Ahmed N (2018) Heterogeneous photocatalysis and its potential applications in water and wastewater treatment: a review. Nanotechnology 29:342001

Ai L, Zhang C, Li L, Jiang J (2014) Iron terephthalate metal-organic framework: revealing the effective activation of hydrogen peroxide for the degradation of organic dye under visible light irradiation. Appl Catal B Environ. 148–149:191–200

Alivand MS, Shafiei-Alavijeh M, Tehrani NH, Ghasemy E, Rashidi A, Fakhraie S (2019) Facile and high-yield synthesis of improved MIL-101 (Cr) metal organic framework with exceptional CO_2 and H_2S uptake; the impact of excess ligand-cluster. Microporous Mesoporous Mater 279:153e164

Alvaro M, Carbonell E, Ferrer B, Llabrés i Xamena FX, Garcia H (2007) Semiconductor behavior of a metal-organic framework (MOF). Chem Eur J 13:5106–5112

Azad FM, Ghaedi M, Dashtian K, Hajati S, Pezeshkpour V (2016) Ultrasonically assisted hydrothermal synthesis of activated carbon-HKUST-1-MOF hybrid for efficient simultaneous ultrasound-assisted removal of ternary organic dyes and antibacterial investigation: Taguchi optimization. Ultrason Sonochem 31:383–393

Babu SG, Vinoth R, Neppolian B, Dionysiou DD, Ashokkumar M (2015) A. Diffused sunlight driven highly synergistic pathway for complete mineralization of organic contaminants using reduced graphene oxide supported photocatalyst. J Hazard Mater 291:83–92

Baek J, Rungtaweevoranit B, Pei X, Park M, Fakra SC, Liu YS, Somorjai GA (2018) Bioinspired metal-organic framework catalysts for selective methane oxidation to methanol. J Am Chem Soc 140:18208–18216

Bedia J, Muelas-Ramos V, Penas-Garzon M, Gomez-Aviles A, Rodriguez JJ, Belver C (2019) A review of photocatalytic water purification with metal organic frameworks. Catalysis 9:52

Campesi R, Cuevas F, Latroche M, Hirscher M (2010) Hydrogen spillover measurements of unbridged and bridged metal-organic frameworks-revisited. Phys Chem Chem Phys 12:10457–10459

Candu N, Tudorache M, Florea M, Ilyes E, Vasiliu F, Mercioniu I, Coman SM, Haiduc I, Andruh M, Parvulescu VI (2013) Postsynthetic modification of a metal-organic framework (MOF) structure for enantioselective catalytic epoxidation. ChemPlusChem 78:443–450

Cao A, Zhang L, Wang Y, Zhao H, Deng H, Liu X, Yue F (2018) 2D-2D heterostructured UNiMOF/g-C3N4 for enhanced photocatalytic H2 production under visible-light irradiation, ACS Sustain. Chem Eng 7:2492–2499

Carp O, Huisman CL, Reller A (2004) Photoinduced reactivity of titanium dioxide. Prog Solid State Chem 32:33–177

Cheetham AK, Rao CNR, Feller RK (2006) Structural diversity and chemical trends in hybrid inorganic-organic framework materials. Chem Commun 12:4780–4795

Chen C, Wang X (2006) Adsorption of Ni (II) from aqueous solution using oxidized multiwall carbon nanotubes. Ind Eng Chem Res 45:9144–9149

Chong MN, Jin B, Chow CWK, Saint C (2010) Recent developments in photocatalytic water treatment technology: a review. Water Res 44:299

Chowdhury T, Zhang L, Zhang J, Aggarwal S (2018) Removal of arsenic (III) from aqueous solution using metal organic framework-graphene oxide nanocomposite. Nano 8:1062e1069

Chu F, Hu J, Wu C, Yao Z, Tian J, Li Z, Li C (2018) Metal-organic frameworks as electrolyte additives to enable ultrastable plating/stripping of Li anode with dendrite inhibition. ACS Appl Mater Interfaces 11:3869–3879

Crini G, Lichtfouse E (2019) Advantages and disadvantages of techniques used for wastewater treatment. Environ Chem Lett 17:145–155

Cui YY, Zhang J, Ren LL, Cheng AL, Gao EQ (2018) A functional anionic metal organic framework for selective adsorption and separation of organic dyes. Polyhedron 161:71e77

Deng Y, Zhao R (2015) Advanced oxidation processes (AOPs) in wastewater treatment. Curr Pollut Rep 1:167–176

Dubreuil A, Assoumou E, Bouckaert S, Selosse S, Maïzi N (2013) Water modelling in an energy optimization framework – the water-scarce middle east context. Appl Energy 101:268–279

Fagan R, McCormack DE, Dionysiou DD, Pillai SC (2016) A review of solar and visible light active TiO_2 photocatalysis for treating bacteria, cyanotoxins and contaminants of emerging concern. Mater Sci Semicond Process 42:2–14

Fazaeli R, Aliyan H, Banavandi RS (2015) Sunlight assisted photodecolorization of malachite green catalyzed by MIL-101/graphene oxide composites. Russ J Appl Chem 88:169–177

Feng Y, Jiang H, Lia S, Wanga J, Jinga X, Wang Y, Chen M (2013) Metal-organic frameworks HKUST-1 for liquid-phase adsorption of uranium. Colloids Surf A Physicochem Eng Asp 431:87–92

Feng X, Chen H, Jiang F (2017) In-situ ethylenediamine-assisted synthesis of a magnetic iron-based metal-organic framework MIL-53(Fe) for visible light photocatalysis. J Colloid Interface Sci 494:32–37

Fu F, Wang Q (2011) Removal of heavy metal ions from wastewaters: a review. J Environ Manag 92:407–418

Fu Y, Sun D, Chen Y, Huang R, Ding Z, Fu X, Li Z (2012) An Amine-functionalized titanium metal–organic framework photocatalyst with visible-light-induced activity for CO_2 reduction. Angew Chem Int Ed Engl 51:3364–3367

Gao J, Miao J, Li PZ, Teng WY, Yang L, Zhao Y, Liu B, Zhang Q (2014) A p-type Ti(IV)-based metal–organic framework with visible-light photo-response. Chem Commun 50:3786–3788

Glatzmaier GC, Nix RG, Mehos MS (1990) Solar destruction of hazardous chemicals. J Environ Sci Health Part A Environ Sci Eng Toxicol 25:571–581

Gleick PH (1993) Water in crisis: a guide to the World's fresh water resources. Oxford University Press, Oxford

Guo D, Wen R, Liu M, Guo H, Chen J, Weng W (2015) Facile fabrication of g-C_3N_4/MIL-53 (Al) composite with enhanced photocatalytic activities under visible-light irradiation. Appl Organometallic Chem 29:690–697

Hausdorf S, Wagler J, Mossig R, Mertens FORL (2008) Proton and water activity-controlled structure formation in zinc carboxylate-based metal organic frameworks. J Phys Chem A 112:7567–7576

He X, Yu Y, Li Y (2018) Facile synthesis of boronic acid-functionalized magnetic metal organic frameworks for selective extraction and quantification of catecholamines in rat plasma. RSC Adv 8:41976e41985

Hong J, Chen C, Bedoya FE, Kelsall GH, O'Hare D, Petit C (2016) Carbon nitride nanosheet/metal–organic framework nanocomposites with synergistic photocatalytic activities. Cat Sci Technol 6:5042–5051

Hopfield JJ (1961) On the energy dependence of the absorption constant and photoconductivity near a direct band gap. J Phys Chem Solids 22:63–72

Jarrah A, Farhadi S (2018) Dawson-type polyoxometalate incorporated into nanoporous MIL-101 (Cr): preparation, characterization and application for ultrafast removal of organic dyes. Acta Chim Slov 66:85–102

Javanbakht S, Nezhad-Mokhtari P, Shaabani A, Arsalani N, Ghorbani M (2019) Incorporating Cu-based metal-organic framework/drug nanohybrids into gelatin microsphere for ibuprofen oral delivery. Mater Sci Eng C 96:302–309

Ji G, Zheng T, Gao X, Liu Z (2019) A highly selective turn-on luminescent logic gates probe based on post-synthetic MOF for aspartic acid detection. Sens Actuators B Chem 284:9–95

Jiang H, Makal TA, Zhou H (2013) Interpenetration control in metal – organic frameworks for functional applications. Coord Chem Rev 257:2232–2249

Jiao Y, Hong W, Li P, Wang L, Chen G (2019) Metal-organic framework derived Ni/NiO microparticles with subtle lattice distortions for high-performance electrocatalyst and supercapacitor. Appl Catal B Environ 244:732–739

Jing HP, Wang CC, Zhang YW, Wang P, Li R (2014) Photocatalytic degradation of methylene blue in ZIF-8. bRSC Adv 4:54454–54462

Joseph L, Jun B-M, Jang M, Park CM, Munoz-Senmache JC, Hernandez-Maldonado AJ, Heyden A, Yu M, Yoon Y (2019) Removal of contaminants of emerging concern by metal-organic frameworks nanoadsorbents: a review. Chem Eng J 369:928–946

Kabir MD, Kim HJ, Choi SJ (2018) Highly proton conductive Zn (II)-based metal organic framework/Nafion composite membrane for fuel cell application. Sci Adv Mater 10:1630–1635

Kim J, Kim DO, Kim DW, Sagong K (2013) Synthesis of MOF having hydroxyl functional side groups and optimization of activation process for the maximization of its BET surface area. J Solid State Chem 197:261–265

Kojtari A, Ji H-F (2015) Metal organic framework micro/nanopillars of Cu(BTC)$3H_2O$ and Zn (ADC)·DMSO. Nano 5:565–576

Kreno LE, Leong K, Farha OK, Allendorf M, Van Duyne RP, Hupp JT (2011) Metal-organic mramework materials as chemical sensors. Chem Rev 112:1105–1125

Kulshreshtha SN (1998) A global outlook for water resources to the year 2025. Water Resour Manag 12:167–184

Kumar SG, Rao KSRK (2017) Comparison of modification strategies towards enhanced charge carrier separation and photocatalytic degradation activity of metal oxide semiconductors (TiO_2, WO_3 and ZnO). Appl Surf Sci 391:124–148

Kumar RS, Kumar SS, Kulandainathan MA (2013) Microporous and mesoporous materials efficient electrosynthesis of highly active $Cu_3(BTC)_2$-MOF and its catalytic application to chemical reduction. Microporous Mesoporous Mater 168:57–64

Langmi LM, Ren J, North B, Mathe M, Bessarabov D (2013) Hydrogen storage in metal-organic frameworks: a review. Electrochim Acta 128:368–392

Laurier KGM, Vermoortele F, Ameloot R, De Vos DE, Hofkens J, Roeffaers MBJ (2013) Iron(III)-based metal-organic frameworks as visible light photocatalysts. J Am Chem Soc 135:14488–14491

Lee YR, Kim J, Ahn WS (2013) Synthesis of metal-organic frameworks: a mini review. Korean J Chem Eng 30:1667–1680

Lee KM, Lai CW, Ngai KS, Juan JC (2016) Recent developments of zinc oxide based photocatalyst in water treatment technology: a review. Water Res 88:428–448

Li Y, Yang RT (2007) Gas adsorption and storage in metal-organic framework MOF-177. Langmuir 23:12937–12944

Li H, Lv N, Li X, Liu B, Feng J, Ren X, Zhang J (2017) Composite CD-MOF nanocrystals-containing microspheres for sustained drug delivery. Nanoscale 9:7454–7463

Li N, Zhou L, Jin X, Owens G, Chen Z (2019) Simultaneous removal of tetracycline and oxytetracycline antibiotics from wastewater using a ZIF-8 metal organic framework. J Hazard Mater 366:563e572

Liang R, Jing F, Shen L, Qin N, Wu L (2015) M@MIL-100(Fe) (M = Au, Pd, Pt) nanocomposites fabricated by a facile photodeposition process: efficient visible-light photocatalysts for redox reactions in water. Nano Res 8:3237–3249

Lin K, Adhikari AK, Ku C, Chiang C, Kuo H (2012) Synthesis and characterisation of porous HKUST-1 metal organic frameworks for hydrogen storage. Int J Hydrog Energy 37:13865–13871

Liu L, Ding J, Huang C, Li M, Hou H, Fan Y (2014) Polynuclear CdII polymers: crystal structures, topologies, and the photodegradation for organic dye contaminants. Cryst Growth Des 14:3035–3043

Liu Y, Zhang F, Wu P, Deng C, Yang Q, Xue J, Wang J (2018) Cobalt (II)-based metal-organic framework as bifunctional materials for Ag (I) detection and proton reduction catalysis for hydrogen production. Inorg Chem 58:924–932

Llabrés i Xamena FX, Corma A, Garcia H (2007) Applications for Metal-Organic Frameworks (MOFs) as quantum dot semiconductors. Phys Chem C 111:80–85

Llabrés i Xamena FX, Casanova O, Galiasso R, Tailleur H, Garcia A, Corma A (2008) Metal organic framework (MOFs) as catalysts: a combination of Cu^{2+} and Co^{2+} MOFs as an efficient catalyst for tetralin oxidation. J Catal 255:220–227

Loera-Serna S, Oliver-Tolentino MA, López-Núñez ML, Santana-Cruz A, Guzmán-Vargas A, Cabrera-Sierra R, Beltrán HI, Flores J (2012) Electrochemical behavior of $[Cu_3(BTC)_2]$ metal-organic framework: the effect of the method of synthesis. J Alloys Compd 540:113–120

Loera-Serna S, Ortiz E, Beltrán HI (2017) First trial and physicochemical studies on the loading of basic fuchsin, crystal violet and black Eriochrome T on HKUST-1. New J Chem 41:3097–3105

Majedi A, Davar F, Abbasi AR (2016) Metal-organic framework materials as nano photocatalyst. Int J Nano Dimens 7:1–14

Mashao G, Modibane KD, Mdluli SB, Iwuoha EI, Hato MJ, Makgopa K, Molapo KM (2019) Polyaniline-cobalt benzimidazolate zeolitic metal-organic framework composite material for electrochemical hydrogen gas sensing. Electrocatalysis 10:406–419

Meng Z, Lu R, Rao D, Kan E (2017) Catenated metal-organic frameworks: promising hydrogen purification materials and high hydrogen storage medium with further lithium doping. Int J Hydrog Energy 38:9811–9818

Monama GR, Mdluli SB, Mashao G, Makhafola MD, Ramohlola KE, Molapo KM, Hato MJ, Makgopa K, Iwuoha EI, Modibane KD (2018) Palladium deposition on copper (II) phthalocyanine/metal organic framework composite and electrocatalytic activity of the modified electrode towards the hydrogen evolution reaction. Renew Energy 119:62–72

Mohd Zain NK, Vijayan BL, Misnon II, Das S, Karuppiah C, Yang CC, Jose R (2018) Direct growth of triple cation metal-organic framework on a metal substrate for electrochemical energy storage. Ind Eng Chem Res 58:665–674

Monama GR, Modibane KD, Ramohlola KE, Molapo KM, Hato MJ, Makhafola MD, Mashao G, Mdluli SB, Iwuoha EI (2019) Copper(II)Phthalocyanine/metal organic frameworks (CuPc/

MOF) composite with improved electrocatalytic efficiency for hydrogen production. Int J Hydrogen Energy 44:18891–18902

Mondal C, Singh A, Sahoo R, Sasmal AK, Negishi Y, Pal T (2015) Preformed ZnS nanoflower prompted evolution of CuS/ZnS p–n heterojunctions for exceptional visible-light driven photocatalytic activity. New J Chem 39:5628–5635

Nas A, Dilber G, Durmuş M, Kantekin H (2015) The influence of the various central metals on photophysical and photochemical properties of benzothiazole-substituted phthalocyanines. Spectrochim Acta A Mol Biomol Spectrosc 135:55–62

Niknam E, Panahi F, Daneshgar F, Bahrami F, Khalafi-Nezhad A (2018) Metal organic framework MIL-101 (Cr) as an efficient heterogeneous catalyst for clean synthesis of benzoazoles. ACS Omega 3:17135–17144

Qamar M, Adam A, Merzougui B, Helal A, Abdulhamid O, Siddiqui MN (2016) Metal – organic framework-guided growth of Mo_2C embedded in mesoporous carbon as a high- performance and stable electrocatalyst for the hydrogen evolution reaction. J Mater Chem A Mater Energy Sustain 4:16225–16232

Qiu J, Zhang X, Feng Y, Zhang X, Wang H, Yao J (2018) Modified metal-organic frameworks as photocatalysts. Appl Catal B Environ 231:317–342

Qu X, Alvarez PJ, Li Q (2013) Applications of nanotechnology in water and wastewater treatment. Water Res 47:3931–3946

Ramohlola KE, Masikini M, Mdluli SB, Monama GR, Hato MJ, Molapo KM, Iwuoha EI, Modibane KD (2017a) Electrocatalytic hydrogen production properties of polyaniline doped with metal organic frameworks. In: Kaneko S, Mele P, Endo T, Tsuchiya T, Tanaka K, Yoshimura M, Hui D (eds) Carbon-related materials in recognition of Nobel lectures by Prof. Akira Suzuki in ICCE. Springer Nature Publishers, pp 373–389., ISBN 978-3-319-61650-6

Ramohlola KE, Masikini M, Mdluli SB, Monama GR, Hato MJ, Molapo KM, Iwuoha EI, Modibane KD (2017b) Electrocatalytic hydrogen evolution reaction of metal organic frameworks decorated with poly (3-aminobenzoic acid). Electrochim Acta 246:1174–1182

Ramohlola KE, Masikini M, Mdluli SB, Monama GR, Hato MJ, Molapo KM, Iwuoha EI, Modibane KD (2017c) Electrocatalytic hydrogen production properties of poly(3-aminobenzoic acid) doped with metal organic frameworks. Int J Electrochem Sci 12:4392–4405

Ramohlola KE, Masikini M, Mdluli SB, Monana GR, Hato MJ, Molapo KM, Iwuoha EI, Modibane KD (2018) Polyaniline-metal organic framework nanocomposite as an efficient electrocatalyst for hydrogen evolution reaction. Compos Part B Eng 137:129–139

Raoof JB, Hosseini SR, Ojani R, Mandegarzad S (2015) MOF-derived Cu/nanoporous carbon composite and its application for electro-catalysis of hydrogen evolution reaction. Energy 90:1075–1081

Ribeiro AR, Nunes OC, Pereira MFR, Silva AMT (2015) An overview on the advanced oxidation processes applied for the treatment of water pollutants defined in the recently launched directive 2013/39/EU. Environ Int 75:33–51

Salunke-gawali S, Kathawate L, Puranik VG (2012) MOF with hydroxynaphthoquinone as organic linker : molecular structure of $[Zn(Chlorolawsone)_2(H_2O)_2]$ and thermogravimetric studies. J Mol Struct 1022:189–196

Schwarzenbach RP, Escher BI, Fenner K, Hofstetter TB, Johnson CA, Von Gunten U, Wehrli B (2006) The challenge of micropollutants in aquatic systems. Science 313:1072–1077

Sha Z, Chan HSO, Wu J (2015) Ag2CO3/UiO-66 (Zr) composite with enhanced visible-light promoted photocatalytic activity for dye degradation. J Hazard Mater 299:132–140

Shannon MA, Bohn PW, Elimelech M, Georgiadis JG, Mariñas BJ, Mayes AM (2008) Science and technology for water purification in the coming decades. Nature 452:301–310

Sharma VK, Feng M (2017) Water depollution using metal-organic frameworks-catalyzed advanced oxidation processes: a review. J Hazard Mater. https://doi.org/10.1016/j.jhazmat. 2017.09.043

Shen LJ, Liang SJ, Wu WM, Liang RW, Wu L (2013) CdS-decorated UiO–66(NH$_2$) nanocomposites fabricated by a facile photodeposition process: an efficient and stable visible-light-driven photocatalyst for selective oxidation of alcohols. J Mater Chem A 1:11473–11482

Silva CG, Luz I, Llabrés i Xamena FX, Corma A, García H (2010) Water stable Zr–benzenedicarboxylate metal–organic frameworks as photocatalysts for hydrogen generation. Chem Eur J 16:11133–11138

Singh S, Barick KC, Bahadur D (2013) Fe3O4 embedded ZnO nanocomposites for the removal of toxic metal ions, organic dyes and bacterial pathogens. J Mater Chem A 1:3325

Song Y, Yan B, Chen Z (2006) Hydrothermal synthesis, crystal structure and luminescence of four novel metal – organic frameworks. J Solid State Chem 179:4037–4046

Sun Y, Sun W (2014) Influence of temperature on metal-organic frameworks. Chin Chem Lett 25:823–828

Sun Q, Liu M, Li K, Zuo Y, Han Y, Wang J, Song C, Zhang G, Guo X (2015) Facile synthesis of Fe-containing metal-organic frameworks as highly efficient catalysts for degradation of phenol at neutral pH and ambient temperature. Cryst Eng Comm 17:7160–7168

Sun J, Zhang X, Zhang A, Liao C (2018) Preparation of Fe-Co based MOF-74 and its effective adsorption of arsenic from aqueous solution. J Environ Sci 80:197e207

Suna Y, Xua W, Dia C, Zhua D (2017) Metal-organic complexes-towards promising organic thermoelectric materials. Syn Met 225:22–30

Sundriyal S, Kaur H, Bhardwaj SK, Mishra S, Kim K-H, Deep A (2018) Metal organic frameworks and their composites as efficient electrodes for supercapacitor applications. Coord Chem Rev 369:15–38

Tan M, Chen R, Yang S, Liu Q (2017) Nonlinear optical properties of Pb-La metal-organic chelidamic acid frameworks. Opt Mater 66:197–200

Teplensky MH, Fantham M, Li P, Wang TC, Mehta JP, Young LJ, Fairen-Jimenez D (2017) Temperature treatment of highly porous zirconium-containing metal organic frameworks extends drug delivery release. J Am Chem Soc 139:7522–7532

To TA, Vo YH, Nguyen HTT, Ha PTM, Doan SH, Doan TLH, Li S, Le HV, Tu TN, Phan NTS (2019) Iron-catalyzed one-pot sequential transformations: synthesis of quinazolinones via oxidative Csp3H bond activation using a new metal-organic framework as catalyst. J Catal 370:11–20

Tranchemontagne DJ, Hunt TR, Yaghi OM (2008) Room temperature synthesis of metal-organic frameworks: MOF-5 and MOF-74. Tetrahedron 64:8553–8557

Tuzen M (2009) Toxic and essential trace elemental contents in fish species from the black sea. Turkey. Food Chem Toxicol 47:1785–1790

UNiMOF/g-C3N4 for enhanced photocatalytic H2 production under

United Nations World Water Assessment (WWAP). The United Nations World Water Development Report 2016; WWAP: 2016, Paris, France

Vlasova EA, Yakimov SA, Naidenko EV, Kudrik EV, Makarov SV (2016) Application of metal–organic frameworks for purification of vegetable oils. Food Chem 190:103–109

Volkova EL, Vakhrushev AV, Suyetin M (2014) Improved design of metal-organic frameworks for efficient hydrogen storage at ambient temperature: a multiscale theoretical investigation. Int J Hydrog Energy 39:8347–8350

Wang L, Hu X (2018) Recent advances in porous carbon materials for electrochemical energy storage. Chem Asian J. https://doi.org/10.1002/asia.201800553

Wang LJ, Deng H, Furukawa H, Gándara F, Cordova KE, Peri D, Yaghi OM (2014) Synthesis and characterisation of metal-organic framework-74 containing 2, 4, 6, 8, and 10 different metals. Inorg Chem 53:5881–5883

Wang C, Lin H, Xu Z, Cheng H, Zhang C (2015) One-step hydrothermal synthesis of flowerlike MoS$_2$/CdS heterostructures for enhanced visible-light photocatalytic activities. RSC Adv 5:15621–15626

Wang L, Han Y, Feng X, Zhou J, Qi P, Wang B (2016) Metal – organic frameworks for energy storage: batteries and supercapacitors. Coord Chem Rev 307:361–381

Wang H, Zhu LQ, Zou R, Xu Q (2017) Metal-organic frameworks for energy applications. Chem 2:52–80

Wei X, Kong X, Sun C, Chen J (2013) Characterization and application of a thin-film composite nanofiltration hollow fiber membrane for dye desalination and concentration. Chem Eng J 223:172–182

Wei X, Wang Y, Chen J, Xu P, Xu W, Ni R, Zhou Y (2019) Poly (deep eutectic solvent)-functionalized magnetic metal-organic framework composites coupled with solid-phase extraction for the selective separation of cationic dyes. Anal Chim Acta 1056:47e61

Wu Y, Luo H, Wang H (2014) Synthesis of iron(iii)-based metal-organic framework/graphene oxide composites with increased photocatalytic performance for dye degradation. RSC Adv 4:40435–40438

Yang J, Niu X, An S, Chen W, Wang J, Liu W (2017) Facile synthesis of Bi2MoO6-MIL-100 (Fe) metal-organic framework composites with enhanced photocatalytic performance. RSC Adv 7:2943–2952

Yang JJ, Mi JL, Yang XJ, Zhang P, Jin LN, Li LH, Ao Z (2019a) Metal-organic framework derived n/c supported austenite nanoparticles as efficient oxygen reduction catalysts. Chem Nano Mat 5:525–530

Yang L, Cui X, Zhang Y, Wang Q, Zhang Z, Suo X, Xing H (2019b) Anion pillared metal-organic framework embedded with molecular rotors for size-selective capture of CO from CH and N. ACS Sustain Chem Eng 7:3138–3144

Ye C, Qi Z, Cai D, Qiu T (2018) Design and synthesis of ionic liquid-supported hierarchically porous Zr metal-organic framework as a novel brønsted Lewis acidic catalyst in biodiesel synthesis. Ind Eng Chem Res 58:1123e–1132c

Yoon JW, Kim A-R, Kim MJ, Yoon T-U, Kim J-H, Bae Y-S (2019) Low-temperature Cu (I) loading on a mesoporous metal-organic framework for adsorptive separation of C3H6/C3H8 mixtures. Microporous Mesoporous Mater 279:271e277

Yu F, Li J, Liang JY, Hu SQ, Lu W, Li B, Zhou HC (2018) Hierarchical nickel/phosphorus/nitrogen/carbon composites templated by one metal-organic framework as highly-efficient supercapacitor electrode materials. *J Mater Chem A* 7:2875e–2883e

Yuan Y-P, Yin L-S, Cao S-W, Xu G-S, Li C-H, Xue C (2015) Improving photocatalytic hydrogen production of metal–organic framework UiO-66 octahedrons by dye-sensitization. Appl Catal B Environ 168:572–576

Yuan B, Yin XQ, Liu XQ, Li XY, Sun LB (2016) Enhanced hydrothermal stability and catalytic performance of HKUST-1 by incorporating carboxyl-functionalized attapulgite. ACS Appl Mater Interfaces 8:16457–16464

Zeng L, Guo X, He C, Duan C (2016) Metal-organic frameworks: versatile materials for heterogeneous Photocatalysis. ACS Catal 6:7935–7947

Zhang C, Ma D, Zhang X, Ma J, Liu L, Xu X (2016) Preparation, structure and photocatalysis of metal-organic frameworks derived from aromatic carboxylate and imidazole-based ligands. J Coord Chem 69:985–995

Zhang H, Nai J, Yu L, Lou (David) XW (2017) Metal-organic-framework-based materials as platforms for renewable energy amd environmetal application. Joule 1:77–107

Zhang M, Wang L, Zeng T, Shang Q, Zhou H, Pan Z, Cheng Q (2018) Two pure MOF-photocatalysts readily prepared for the degradation of methylene blue dye under visible light. Dalton Trans 47:4251–4258

Zhang L, Jiang K, Zhang J, Pei J, Shao K, Cui Y, Qian G (2019) A low-cost and high performance microporous metal-organic framework for separation of acetylene from carbon dioxide. ACS Sustain Chem Eng 7:1667e1672

Zhao G, Li J, Ren X, Chen C, Wang X (2011) Few-layered graphene oxide nanosheets as superior sorbents for heavy metal ion pollution management. Environ Sci Technol 45:10454–10462

Zhao M, Ou S, Wu C (2014) Porous metal-organic frameworks for heterogeneous biomimetic catalysis. Acc Chem Res 47:1199–1120

Zhao Y, Song Z, Li X, Sun Q, Cheng N, Lawes S (2016) Metal organic frameworks for energy storage and conversion. Energy Storage Mater 2:35–62

Zhu SR, Liu PF, Wu MK, Zhao WN, Li GC, Tao K, Yi FY, Han L (2016) Enhanced photocatalytic performance of BiOBr/NH$_2$-MIL-125(Ti) composite for dye degradation under visible light. Dalton Trans 45:17521–17529

Zhua M, Ma Q, Ding SY, Zhao YZ, Song WQ, Ren HP, Miao ZC (2019) A molybdenum disulfide and 2D metal-organic framework nanocomposite for improved electrocatalytic hydrogen evolution reaction. Mater Lett 239:155–158

Printed by Books on Demand, Germany